Sex and Parthenogenesis
Evolutionary Ecology of Reproductive Modes in
Non-Marine Ostracods

Sex and Parthenogenesis

Evolutionary Ecology of Reproductive Modes in Non-Marine Ostracods

Edited by

Koen MARTENS

Backhuys Publishers, Leiden, 1998

Front cover:
Front cover photo shows copula of *Eucypris virens* (left) and the ancient asexual *Darwinula stevensoni* (right), the latter with juveniles in brood space. Not to the same scale. Photos by D.J. Horne.

ISBN 90-5782-017-X

Printed in The Netherlands

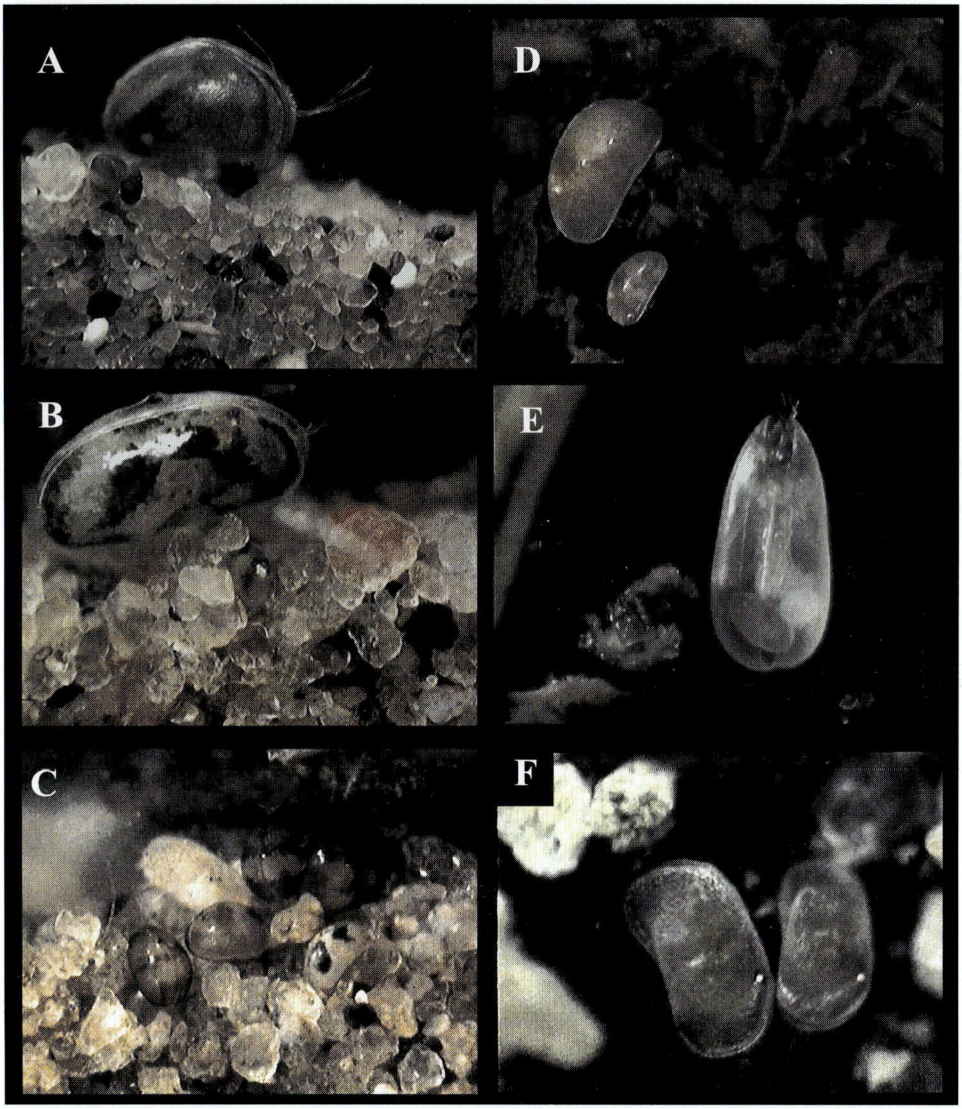

Some examples of European non-marine ostracods.

A: *Cypris pubera* (Cypridoidea, Cyprididae); adult female (approximately 2.5 mm long).

B: *Herpetocypris reptans* (Cypridoidea, Cyprididae); adult female (approximately 2.0 mm long).

C: *Cyclocypris ovum* (Cypridoidea, Candonidae); two adult females, center and lower left (approximately 0.5 mm long). + Cypridopsis vidua + Potamocypris similis?

D: *Candona candida* (Cypridoidea, Candonidae); adult female (above) and juvenile (below)(approximately 1mm long).

E: *Darwinula stevensoni* (Darwinuloidea, Darwinulidae); adult female with a large juvenile and an egg in the brood pouch (approximately 0.6 mm long).

F: *Limnocythere inopinata* (Cytheroidea, Limnocytheridae); two adult females (approximately 0.6 mm long).

All from original high-resolution video recordings in micro-aquaria by D.J. Horne.

TABLE OF CONTENTS

PREFACE

TO MATE OR NOT TO MATE!

Sex is the Queen of problems in evolutionary biology.
(Bell, 1982: 19)

Discussions on ecology and evolution invariably end up hinging around questions dealing with reproductive modes. Why do some organisms reproduce with, others without sex, while some have both types of reproduction? What are the underlying mechanisms: are there environmental factors triggering mode of reproduction, or is it entirely genetically determined? How do asexual lineages originate and how long can they persist? What are the causes for geographical parthenogenesis? Some of these questions have received seemingly satisfying answers during the past decades, others remain largely unsolved to date.

The initial steps which led to the present book were indirectly taken in 1992, at the conference of the International Society of Limnology (SIL) in Barcelona, where some of the present authors decided to start a multidisciplinary project on evolutionary ecology of reproductive modes in non-marine ostracods. Between 1994 and 1996, we were supported by a grant from the European Union, which allowed intensive exchange of post-graduate and post-doctoral researchers between the different laboratories and which also enabled us to organise several workshops, where results could be presented and discussed. Now, after six years of collaboration, during which we integrated techniques and theories from fields as different as palaeontology (spanning the entire range from Palaeozoic to Holocene), systematics, ecology and both allozyme and DNA-based molecular approaches, we believe we have achieved a novel and integrated view on reproductive ecology in general, while further confirming non-marine ostracods as an excellent model group for this field of research. Our conviction that (non-marine) ostracods have great potential to study evolutionary ecology of reproductive modes was initially based upon isolated discoveries. For example, palaeontology had shown that at least one non-marine ostracod lineage (Darwinulidae) seemed to defy the axiom that asexual lineages are doomed to early extinction; allozyme studies, one of the newest techniques applied to ostracod research at that stage, revealed an unexpectedly high clonal diversity in other asexual ostracod lineages, and morphological research showed an uneven distribution of potential effects of sexual selection on evolutionary rate in various groups. Ostracods also have other special attributes: they are one of the few living invertebrate groups with such an extensive and extended fossil record, but are also rather unique as a fossil group with such a diverse modern representation. In most cases, and in spite of certain restrictions discussed further in this book, the fossil record will not only yield organismal, but also ecological and behavioural data. Ostracod species furthermore have an exceptionally wide range of different reproductive modes, which seem linked to underlying (and detectable) genetic breeding systems, while they also display a wide range of ecological strategies and geographically segregated distribution patterns. The diversity of different patterns and processes within the non-marine ostracods is truly astonishing, and the present book aims to document this. The heterogeneity amongst the different researchers of our network (in spite of the fact that we all work on ostracods, we tended to look at this group from very different angles), allowed for important cross-fertilization of ideas, but also caused many of our discussions to be, let us say, very animated. The old belief that progress (in science) results from conflict rather than from agreement was amply corroborated. But these discussions, always conducted with a strong sense of mutual respect and with a solid dose of humour, have taught the different participants much about each other's disciplines.

Only integration of data resulting from the application of different techniques allows the development of novel and comprehensive evolutionary scenarios. However, an important advantage can also turn out to be a main weakness and this is also true for the present book. By combining different approaches, we think we are able to present a new synthesis on reproductive ecology in non-marine ostracods, but presenting data from very diverse fields makes it difficult to define the target group of readers for the book. We wish to introduce the impact of reproductive mode to ostracod workers in general, illustrate recent developments in molecular biology to morphologists and palaeontologists, stress the wealth and depth of the fossil record to molecular biologists and sell the importance of ostracods as a model group to evolutionary biologists in general. The canvas is, indeed, very wide.

We have tried to solve this problem by dividing the book into three parts: Part I with introductory chapters; Part II with chapters attempting to integrate published and new data from the various fields of ostracod research and Part III with the synthesis, combining the above data and hypotheses into a new scenario on the evolutionary ecology of breeding systems in general and of ostracods in particular.

With few exceptions (7, 13), the different chapters should always be seen as reviews of extant knowledge, in which our new data and views have been integrated. Part of the results of our work has already been published in peer-reviewed journals, other papers are still in preparation or in press, but all relevant data are summarized here and placed in a broader context. However, readers might find it useful at times to refer to the original papers, especially with regard to materials and methods used.

In the first part, chapter 1 introduces genetic aspects of reproductive modes and is especially intended for non-molecular biologists. Chapter 2 outlines sex determination mechanisms in ostracods and introduces the problems of skewed sex ratios and rare males. Chapters 3 and 4 summarise present-day ideas on history, taxonomy and morphology of non-marine ostracod lineages, especially to non-ostracod workers. All of these chapters can either be read in full, or be referred to while reading the chapters of parts II and III.

Part II starts with two chapters on geographical distribution of reproductive modes, relying mostly on analysis of information in the NODE database (Non-marine Ostracod Distribution in Europe – Ed. D.J. Horne), which presently holds nearly 10,000 records of non-marine ostracods from European inland waters. NODE can be considered as one of the main long-term results of our collaboration thus far. Chapter 5 analyses extant distribution (showing that geographical parthenogenesis is restricted to certain groups within non-marine lineages), chapter 6 discusses fossil distributions of reproductive modes at different time scales. Sex determination and the origin of asexual lineages are discussed in several chapters (introduced in chapters 1 and 2), but a special case is the suppression of male production by endocytic bacteria. The history of this very recent branch of research in reproductive biology is briefly reviewed and the potential presence of such bacteria in the ostracod *Heterocypris incongruens* are presented in chapter 7.

The four following chapters primarily deal with the relationship between morphology and breeding system. Chapters 8 and 9 summarize the impact of reproductive mode on intraspecific and, whenever possible, clonal morphological variability on valves and limbs respectively. Chapter 10 discusses morphology and reproductive behaviour: the possible effect of recognition systems and sexual selection on the diversity of male genitalia in sexual populations and the importance of brood care in ostracods. Data from these four chapters are then combined into chapter 11, where the long standing problem of clonal taxonomy is discussed, a paradox even the modern synthesis of evolution could not solve. Our improved understanding of breeding systems in non-marine ostracods contributes towards a partial solution. The next three chapters deal with ecological aspects of breeding systems. Chapter 12 reviews extant ideas and hypotheses on population dynamics, life histories and reproductive modes; chapter 13 introduces the concept of metapopulation dynamics for ostracod populations. Even a relatively simple metapopulation model already provides a possible explanation for the existence of geographical parthenogenesis in certain species and lineages. Chapter 14 discusses clonal ecology in the three main ostracod lineages, and in so doing touches the core of the new views on reproductive systems in ostracods. The final two chapters of part II deal with genetic variability in different breeding systems: chapter 15 primarily discusses clonal diversity using allozyme techniques; chapter 16 uses divergence in nuclear and mitochondrial DNA sequences to illustrate how genetic diversity can be used to estimate age of divergence of different (clonal) lineages, as well as to reconstruct molecular phylogenies. The latter chapter presents the first database of DNA sequences obtained for non-marine ostracods.

The single chapter (17) in Part III, finally, attempts to integrate all of the above into a new synthesis. The latter is a personal interpretation of the data and ideas presented in the previous chapters and must therefore be seen as a subjective summary of the book as a whole.

The present volume would never have been possible without the help, financial and other, of several people and institutions. I will attempt to cite all those who contributed directly, but beg pardon from those I might have forgotten. Apart from those figuring as authors on one or more chapters in this book, the following colleagues participated at some stage or other in the development of the project: F. Cocks, D. Connaghan, J. Sawle and I. Slipper (Chatham) assisted with the construction of the GIS database NODE; M.E. Montenegro (Paris), M. Gonzalez-Mozo (Madrid) and Y. Yin (Mondsee) conducted morphological analyses of cytherid and cypridinid limbs; P. Alcorlo (Madrid), F. Mösslacher and P. Pospisil (Mondsee), N. Johnson (Chatham) and S. Sanchez-Herrera (Madrid) assisted with ecological and morphological analyses; G. Bucci, E. Di Masso and A. Gandolphi (Parma) assisted during the preliminary stages of the development of the DNA techniques; E. Rico (Madrid), P. Pospisil and Y. Yin (Mondsee) conducted experiments on clonal ecology. T. Vandekerckhove wishes to thank his promotors: Prof. M. Gillis, Prof. J. Swings

(Laboratory of Microbiology, WE10V, University of Ghent, Belgium) and Prof. J. Mertens (Laboratory of Ecology, WE11V, ibid.), as well as Prof. J. Quatacker (N. Goormaghtigh Institute of Pathology, GE22V, ibid.) for the use of the TEM.

The main part of our financial resources originated from the European Union (DG XII) Human Capital and Mobility program, partim networks, who funded our project on "Evolutionary Ecology of Reproductive Modes in Non-Marine Ostracoda" (Contract Number: CHRXCT/93/0253) for the period 1994-1996. W. Geiger is much indebted to the Austrian Foundation F.W.F., who supported his part of the research through projects Nr. P-1060 BIO and P-10347 BIO, attributed to D.L. Danielopol; the latter also acknowledges F.W.F. project P-11149 BIO, which (partly) supported him while contributing to chapter 10. The Austrian Academic Exchange Service generously offered a scholarship within the program "North - South Dialogue" which allowed Y. Yin (Nanjing) to work in Mondsee. The Austrian Research Council (F.W.F.) supported travel costs of Y. Yin for scientific visits to China (the Tibet-Qinghai Expedition, March 1995), Brussels (1995, 1996) and Parma (Workshop 1996). D.J. Horne and K. Martens acknowledge a joint British Council – Fund for Scientific Research (Flanders, Belgium) project on "Palaeobiology of Purbeck-Wealden ostracods", which contributed to various aspects of the work. T. Vandekerckhove is a Research Assistant of the F.W.O. – Flanders, Belgium.

Several people have assisted with the production of the illustrations in this book. Mrs C. Behen (Brussels) inked most of the line drawings in this book. J. Cillis (Brussels) assisted with S.E.M. work; K. Scurr (Hull) is thanked for his help with Fig. 6.1, Mr H. Van Paesschen (Brussels) for his assistance with some of the last-minute artwork. Mrs V. Arkosi (Brussels) corrected the language wherever necessary. To all of the above, our sincere thanks.

On a personal account, I would like to thank Dr Isa Schön (Brussels) for unrelenting help, far beyond the call of duty, especially during the final stages of the preparation of this book. To Marianne, Boris and Thomas, I would like to express my gratitude for their continued and patient endurance of my mental and/or physical absence during countless weekends and holidays.

Koen Martens
Editor

AUTHORS' ADDRESSES

Angel Baltanás, Departamento de Ecologia, Universidad Autonoma de Madrid, Ciudad Universitaria de Cantoblanco, E-28049, Madrid, Spain. Email: angel.baltanas@uam.es

Roger K. Butlin, Department of Biology, University of Leeds, Woodhouse Lane, Leeds LS2 9JT, U.K. Email: gen6rb@gps.leeds.ac.uk

Dan L. Danielopol, Institut für Limnologie, Abteilung Mondsee, Gaisberg 116, A-5310 Mondsee, Austria. Email: dan.danielopol@oeaw.ac.at

Walter Geiger, c/o Angel Baltanás, Departamento de Ecologia, Universidad Autonoma de Madrid, Ciudad Universitaria de Cantoblanco, E-28049, Madrid, Spain. Email: geigerw@arrakis.es

Huw I. Griffiths, Research Institute for Environmental Science and Management, Department of Geography, University of Hull, Cottingham Road, Kingston-upon-Hull, Hull HU6 7RX, U.K. Email: h.i.griffiths@geo.hull.ac.uk

David J. Horne, The University of Greenwich, School of Earth and Environmental Sciences, Medway Campus, Grenville Building, Chatham Maritime, Kent ME4 4TB, U.K. Email: d.j.horne@greenwich.ac.uk

Koen Martens, Freshwater Biology, Royal Belgian Institute of Natural Sciences, Vautierstraat 29, B-1000 Brussels, Belgium. Email: martens@kbinirsnb.be.

Paolo Menozzi, Dipartimento di Scienze Ambientali, Università di Parma, Viale delle Scienze, I-203100 Parma, Italy. Email: menozzi@dsa.unipr.it.

Marina Otero, Departamento de Ecologia, Universidad Autonoma de Madrid, Ciudad Universitaria de Cantoblanco, E-28049, Madrid, Spain. Email: marina.otero@uam.es

Gianmarco Paris, Dipartimento di Elettronica e Informazione, Politecnico di Milano, Via Ponzio 34/5, I-20133 Milano, Italy. Email: paris@elet.polimi.it (formerly at the Universita di Parma).

Giampaolo Rossetti, Dipartimento di Scienze Ambientali, Università di Parma, Viale delle Scienze, I-43100 Parma, Italy. Email: rossetti@dsa.unipr.it.

Valeria Rossi, Dipartimento di Scienze Ambientali, Università di Parma, Viale delle Scienze, I-43100 Parma, Italy. Email: valeria@dsa.unipr.it

Isa Schön, Freshwater Biology, Royal Belgian Institute of Natural Sciences, Vautierstraat 29, B-1000 Brussels, Belgium. Email: schoen@kbinirsnb.be.

Tom Vandekerckhove, University of Ghent, Laboratories of Microbiology and Animal Ecology, K.L. Ledeganckstraat 35, B-9000 Ghent, Belgium. Email: tom.vandekerckhove@rug.ac.be.

CHAPTER 1

INTRODUCTION TO REPRODUCTIVE MODES

R.K. Butlin, I. Schön and H.I. Griffiths

Abstract

Parthenogenetic reproduction is widespread but patchy in its taxonomic distribution. This is apparently because it provides short-term benefits but long-term costs. However, while many costs and benefits for parthenogenesis, as opposed to sexual reproduction, have been put forward, there is little consensus on their relative importance or on the time scales over which they operate. Here, we review the major costs and benefits: the two-fold advantage arising from the absence of males and eight proposed disadvantages of parthenogenesis (Müller's ratchet, Mutation load, Fisher-Müller evolutionary rate, Red Queen effect, Fluctuating Selection, DNA repair, Sib-competition and Tangled-bank). Ostracod examples are introduced where appropriate but detailed discussion is delayed until later chapters. These theories suggest that parthenogenetic lineages should have short evolutionary lives. Therefore, we go on to consider ways in which lineages might persist without sex, with particular reference to the 'ancient asexuals', especially the darwinulid ostracods.

Introduction

An integral part of the accepted wisdom of evolutionary biology is that sex is an essential component of the evolutionary process. It provides the means whereby new genetic configurations can be generated, and acts as the very plastic that is moulded by natural and sexual selection (Short 1994). Despite this, many organisms reproduce without sex, they are either entirely or facultatively parthenogenetic. Not only does this represent an apparent contradiction to some basic evolutionary tenets, but parthenogenesis itself is believed to have some costs. For example, lacking the 'purging' action of recombination, deleterious mutations may accumulate within the genome of a parthenogen through the action of Müller's ratchet, leading to mutational meltdown within long-term parthenogenetic lineages (Lynch *et al.* 1993). On the other hand, the costs of sexual reproduction are also high, especially the two-fold cost associated with producing males (Maynard Smith 1989). Perhaps it is sexual reproduction that requires an explanation, rather than parthenogenesis?

Parthenogenetic reproduction has a broad, albeit sometimes clustered distribution throughout the animal and plant kingdoms (see review by Bell 1982). Thus, although a comparatively small proportion of all multicellular organisms utilise a parthenogenetic reproductive mode, there are representatives in many phyla. This broad taxonomic distribution suggests that parthenogenesis has evolved repeatedly and

Sex and parthenogenesis: evolutionary ecology of reproductive modes in non-marine ostracods, edited by K. Martens, pp. 1–24

that it confers a short-term evolutionary advantage. However, the general absence of species-rich parthenogenetic taxa suggests that the advantage is not maintained in the longer term. While the costs and benefits of sexual reproduction have been discussed extensively (see below), the time-scales are much more difficult to study (Griffiths & Butlin 1995).

At least 19 out of 34 animal phyla are known to contain at least some parthenogens (Bell 1982), although this may be an underestimate since we know little about the reproductive mode of fossil lineages (here ostracods are an important exception, see chapter 6). Parthenogenetic reproduction is particularly widespread and conspicuous amongst the arthropod orders Insecta and Arachnida, and amongst the freshwater Crustacea. In a few taxa all species are exclusively parthenogenetic: many lineages of orbatid mites, all bdelloid rotifers and all darwinulid ostracods. These are the exception, rather than the rule. Although occurring less commonly, many vertebrate groups do include parthenogenetic forms. These apparently all arise through hybridisation and most reproduce by hybridogenesis or gynogenesis (Box 1.1) (Avise 1994). This is particularly well studied in fish such as *Poeciliopsis*, in amphibians including *Ambystoma* salamanders, and in the lizard genera *Lacerta* and *Cnemidophorus*. The only vertebrate groups in which reproduction is exclusively sexual are the birds and mammals, although, in birds, the 'Beltsville turkey' provides an unusual example of an artificially produced avian semi-parthenogen in which the unfertilised eggs develop into males (Crews 1994).

Box 1.1 Modes of reproduction: terminology and genetic consequences

In this book, we deal with the evolutionary ecology of *parthenogenetic* as opposed to *sexual reproduction* in non-marine ostracods. *Sexual reproduction*, or sometimes just sex, will be used to refer to life cycles involving an alternation of *meiosis* and *syngamy*. This is more properly described as *amphimixis* (Fig. 1.1) but, since there is little opportunity for confusion in our case, we will use the more familiar terms. In ostracods, as in most animals, meiosis occurs in the production of male and female gametes and reduces the chromosome complement from diploid to haploid while syngamy, or fertilisation, restores the diploid constitution.

Parthenogenesis is the production of offspring from unfertilised eggs, that is, from female gametes without the requirement for syngamy. It should be distinguished from *asexual reproduction* which can include vegetative propagation from groups of somatic cells but, since vegetative reproduction does not occur in ostracods, we will use the terms interchangeably. Similarly, we will ignore the distinction between *arrhenotoky*, production of haploid males from unfertilised eggs and diploid males from fertilised eggs as in Hymenoptera, and *thelytoky*, production of diploid female offspring without syngamy, because arrhenotoky is unknown in ostracods.

Thelytokous parthenogenesis can happen by a number of different mechanisms (Fig. 1.1). The simplest, and probably the most important in ostracods, is *apomixis*. In apomixis, meiosis is simply suppressed so that the oocyte is produced by mitosis and retains its diploid constitution. By contrast, in *automixis* meiosis is retained and either the diploid constitution is restored after meiosis or the chromosome complement is doubled before meiosis so that the products of the reduction division are diploid. Doubling of chromosome numbers either before or after meiosis may happen by mitosis without cell division: *endomitosis*. Alternatively, two of the products of mitosis may fuse to generate the new diploid cell from which the embryo is derived. In this latter case, automixis becomes very much like *self-fertilisation* or *autogamy* where male and female gametes are produced by the same individual and fuse to generate the zygote.

Box 1.1 Continued.

The *genetic consequences* of these different modes of reproduction are central to debates about their evolutionary advantages and disadvantages. In sexual reproduction, the offspring are genetically unlike their parents for three reasons: because they receive half of their genes from each parent, because of random assortment of homologous chromosomes at meiosis in each parent, and because of *crossingover* between homologous chromosomes at meiosis. Crossing-over at chiasmata involves exchange of genetic material between homologous chromosomes (which, of course, originated in different individuals) and thus generates new combinations of alleles within each chromosome. (An *allele* is an alternative form of a gene at a particular *locus* or site on a chromosome. Allele and locus will be used wherever the catch-all term gene is ambiguous.) The term *recombination* is used variously in the literature to refer to any or all of these processes which modify combinations of alleles. Here we will use it to refer to all three, and we will use crossing-over if we intend to refer specifically to exchange between homologous chromosomes. Used in this way, sexual reproduction and recombination are almost interchangeable: an evolutionary explanation for sex is an explanation for recombination, or vice-versa. One proviso is that selection may alter recombination rates within sexual populations by modifying the number of chromosomes or the number and distribution of chiasmata.

Apomictic reproduction has opposite genetic consequences to sex. Offspring are identical to their mother, except for new mutations. Thus, any *heterozygosity* (presence of different alleles at one locus on the two homologous chromosomes in a diploid) present in the mother is retained and heterozygosity increases as a result of mutation.

The consequences of automictic reproduction are more complex because they depend on the mechanism for the restoration of chromosome numbers. Offspring may be genetically different from their mother because of re-assortment of chromosomes and crossing-over during meiosis. However, all modes of automictic reproduction involving fusion of the products of meiosis tend towards producing increasing homozygosity and loss of genetic variability within lineages. For example, in a heterozygous (Aa) mother meiosis produces a 1:1 ratio of A and a gamete. If these fuse at random, the offspring genotypes are in the ratio 1 AA : 2 Aa : 1 aa, that is 50% homozygotes and 50% heterozygotes. Since the homozygotes can produce only homozygous offspring, and the heterozygotes will again produce only 50% heterozygous offspring, it is clear that the lineage will rapidly tend to complete homozygosity. Heterozygosity may be preserved where there is a regular pattern of fusion of the products of meiosis, depending on whether fusion between sister nuclei happens between first or second division. However, even in these cases heterozygosity is only preserved in some parts of chromosomes, depending on the distribution of cross-overs (see Maynard Smith 1989 for a simple explanation). Mechanisms involving endomitosis produce homozygosity immediately if endomitosis follows meiosis, but are equivalent to apomixis if it precedes meiosis.

Finally, some special forms of parthenogenesis should be mentioned, not because they occur (so far as we know) in ostracods, but because they contribute to general discussions about the evolution of parthenogenesis. *Cyclical parthenogenesis*, or *heterogony*, is common in aphids and cladocerans. Typically, reproduction is apomictic while environmental conditions are good, but switches to sexual when conditions deteriorate, often with the sexual generation producing stress-resistant eggs. In *gynogenesis*, reproduction is parthenogenetic (apomictic or automictic), but mating occurs and the meeting of sperm and egg appears to be necessary to initiate embryonic development although fertilisation does not occur. In *hybridogenesis*, fertilisation does occur to restore the diploid constitution, but all offspring are female and they eliminate the paternally derived set of chromosomes when they produce gametes by meiosis. Thus, the maternal set of chromosomes is effectively inherited parthenogenetically.

Examples of all of these modes of reproduction, further details of the mechanisms and the genetic consequences, and a survey of the occurrence of parthenogenesis can be found in Bell (1982).

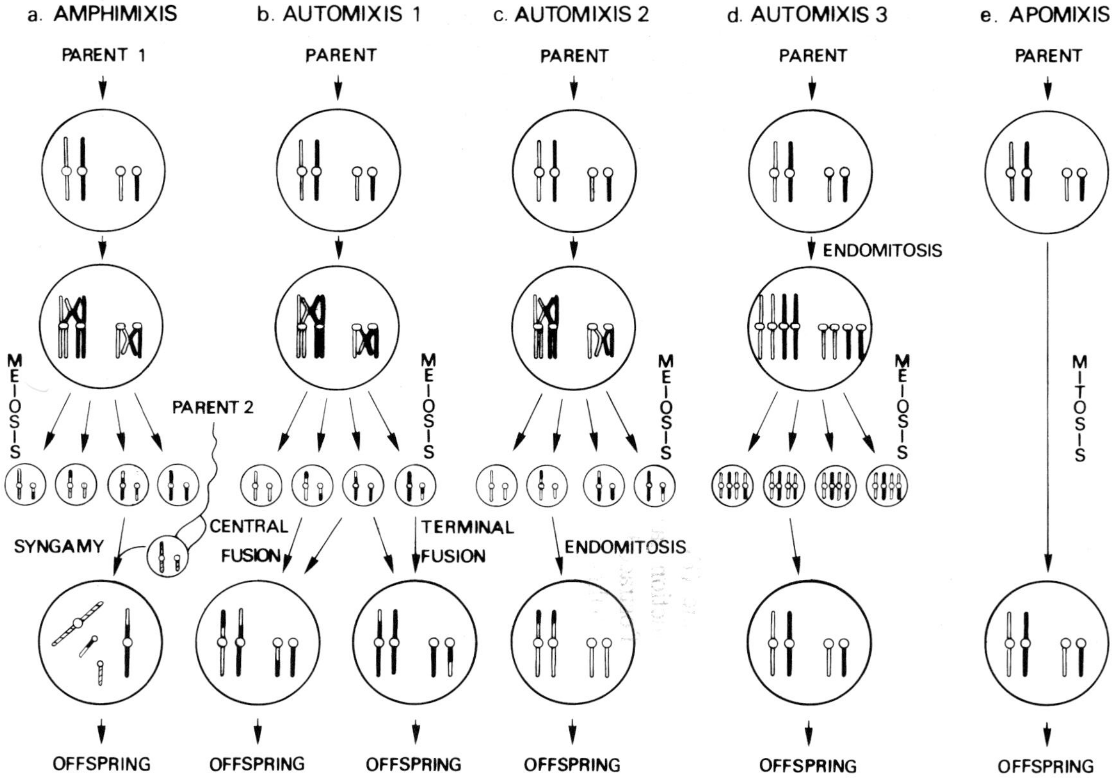

Fig. 1.1 A comparison between the mechanisms and genetic consequences of the five major modes of reproduction. a. Amphimixis or sexual reproduction. Meiosis and syngamy result in the maximum genetic variability among offspring. With self-fertilisation amphimixis becomes more like automixis. b. Automixis 1. Meiosis is followed by fusion of the products to restore the diploid chromosome complement. Homozygosity increases, except for loci between the centromere and the cross-over (central fusion) or for loci beyond the cross-over (terminal fusion). c. Automixis 2. Meiosis is followed by endomitosis (chromosome doubling without cell division) producing complete homozygosity. d. Automixis 3. Meiosis is preceded by endomitosis and involves pairing of sister chromatids. Heterozygosity is preserved. e. Apomixis. Meiosis is replaced by mitosis. Heterozygosity is preserved and accumulates over generations due to mutation.

Freshwater ostracoda have long been known to include a large number of species that are either entirely parthenogenetic, or exhibit 'geographical parthenogenesis': the occurrence of widespread parthenogenetic populations of a species that also has geographically restricted sexual populations (Vandel 1928, see chapter 5). However, the only attempt at a comprehensive review of parthenogenesis within the group has been by Bell (1982). Bell was not an ostracod specialist, but incorporated a concerted attempt to understand the patterns of parthenogenesis in ostracods in his much wider taxonomic survey. This led him to conclude (correctly) that parthenogenesis among freshwater ostracods is most marked in the Darwinulidae, which appear to lack males altogether and are candidate 'ancient asexuals' (Butlin & Griffiths 1993, Judson & Normark 1996), and most frequent in the species-rich Cyprididae in which many genera show species without males, with only rare males, or with geographical parthenogenesis. In contrast, the Candonidae are predominantly sexually reproducing, as are the Cytheroidea, while the isopod- and crayfish-commensal Entocytheridae are apparently all sexual. Most, but perhaps not all (chapter 5), marine ostracods reproduce sexually. These patterns will be discussed in detail in later chapters.

The mechanisms behind the genesis and maintenance of parthenogenetic reproduction in Ostracoda are not well understood. For many years, it was argued that some species were cyclically parthenogenetic (Box 1.1) (as are many Cladocera and aphids), although this mode of reproduction has not been confirmed in any ostracod species. Furthermore, the huge size of ostracod spermatozoa led Lowndes (1935) to argue that they could not be functional, so that even apparently sexual species were parthenogenetic. This, too, is now known to be invalid (see Wingstrand 1988). In those ostracod species investigated, apomictic parthenogenesis (Box 1.1) seems to be the norm, although robust data are available only rarely (Havel & Hebert 1989). Work by Turgeon & Hebert (1994) suggests that there is abundant spontaneous 'switching' to parthenogenesis from sexual reproduction among freshwater ostracods, and that such spontaneous transitions, coupled with hybridisation between parthenogenetic and sexual lineages, may be the 'key process that governs clonal diversity in ..... ostracods'.

In the present chapter, we aim to examine the evolutionary costs and benefits of parthenogenesis, as opposed to sexual reproduction, in order to provide a solid theoretical framework for subsequent discussions of the patterns that can be observed in freshwater ostracods.

An overview of costs and benefits

> "If we bear in mind that in sexual propagation twice as many individuals are required in order to produce any number of descendants and if we further remember the important morphological differentiations which must take place in order to render sexual propagation possible, we are led to the conviction that sexual propagation must confer immense benefits on organic life. I believe that such beneficial results will be found in the fact that sexual propagation may be regarded as a source of individual variability, furnishing material for the operation of natural selection."
> (August Weismann 1887, quoted in Kondrashov 1993)

More than a century of research effort in evolutionary biology has passed since this remarkable statement was made but the underlying question remains: what are the benefits of sexual reproduction that outweigh its undoubted costs? That there is a net benefit to sexual as opposed to parthenogenetic reproduction (see Box 1.1 for definitions) in the circumstances most commonly encountered in nature, is beyond dispute given the frequencies and taxonomic distributions of the different reproductive modes. Furthermore, there is little debate about the costs of sexual reproduction, in particular the cost mentioned by Weismann that arises from the requirement for males in sexual lineages and is nowadays usually known as the 'cost of males'. This cost can be as much as two-fold, as discussed below. To this major disadvantage may be added the 'costs of mating': the energetic costs of finding and courting mates and of maintaining structures involved in courtship and copulation, plus the predation and disease risks involved in mating.

There is much less consensus about the advantages to sexual reproduction. There is no shortage of ideas. Indeed, it has become necessary to classify the many available hypotheses in order to identify the ways in which they differ or converge (e.g. Kondrashov 1993). The lack of consensus arises partly from the very different viewpoints of those studying the problem: in particular whether the problem is essentially ecological or genetical. However, the absence of critical experimental data to distinguish among hypotheses is the most serious problem, and part of the objective of this book is to show how ostracods can be used to tackle some of the issues involved. Because our intention is to deal specifically with ostracods, some of the issues concerning sexual reproduction more generally will not be considered here. The origin of sexual reproduction is the first of these: we will work from the assumption that ostracods are derived from a common sexually reproducing ancestor and that parthenogenetic lineages have been derived from sexual descendants of this ancestor. Some of these parthenogenetic lineages may be very old, especially the 'ancient asexual' Darwinuloidea, others may be very young, perhaps originating during the Holocene. The origin of sex is considered briefly in Box 1.2.

It is also true that ostracods are generally either outcrossing sexual species or fully parthenogenetic, apomictic in those cases that have been studied (e.g. Havel & Hebert 1989) although some species contain both types of lineage (Box 1.3). They are not hermaphrodite and potentially self-fertilising, nor do they exhibit cyclic parthenogenesis so far as we know. In general, this distinction is important because it means that we are interested in evolutionary competition between sexual lineages and parthenogenetic clones, rather than degrees of selfing or levels of recombination that vary among individuals. We need to understand why parthenogenetic clones sometimes expand and persist at the expense of their sexual progenitors while in other cases sexual lineages displace parthenogens, or, indeed, how the lineages might coexist. These are processes of group selection and we will concentrate on advantages or disadvantages to groups (Maynard Smith 1978, Kondrashov 1993). This does not imply that we are only interested in long-term advantages: because the cost of males is so great, it must be countered in the short-term. Where selfing or cyclic parthenogenesis occur, selection can act on individuals as opposed to groups, altering the selfing rate or the frequency of sexual generations, respectively. This is a rather different problem, although at least some theories may apply to both situations and to changes in cross-over rates within sexual lineages.

Box 1.2 The origin of sex

There can be little doubt that the common ancestor of all extant ostracods reproduced sexually. Therefore, the problems to be considered in this book are about the origin and persistence of derived parthenogenetic lineages and the ways in which sexual reproduction persists in the face of competition with these lineages. The origin of sexual reproduction itself is a separate issue and the selective pressures that favoured the evolution of sex may well be different from those which now maintain it.

Eukaryotic sex is a complex process. At the least, it involves conjugation of cells, fusion of nuclei, meiosis and crossing over. These are likely to have evolved in sequence rather than simultaneously and it is very unlikely that advantages associated with genetic recombination could have driven all stages of the process (Hickey 1993). Two major additional selective mechanisms have been proposed: molecular symbionts or selfish DNA, and DNA repair.

Mobile DNA elements such as transposons, retrotransposons, mobile introns and 'long interspersed nuclear elements' might be genomic parasites and act as selfish DNA (Orgel & Crick 1980, Doolittle & Sapienza 1980). As they not only interrupt the host genome, but also move and spread within it by cutting and pasting mechanisms, they can damage the host's genome seriously. In Drosophila, 97% of all mutations in 10 different loci could be attributed to insertions of mobile elements (Green 1988). Furthermore, some genomic parasites move preferentially into regulatory regions (Zeyl & Bell 1997) and so may produce adaptive novelties. As mobile elements act as an additional mutation source, they can increase evolutionary flexibility in the long term. However, to ensure their spread to uninfected genomes within a population, these parasites require conjugation (Hickey 1982, 1993). It could have been selection on these molecular parasites that originally favoured conjugation.

So far, no tests have been done on the distribution of mobile elements in organisms with different reproductive modes, but data on cell organelles strongly support the prediction that selfish DNA parasites will not be found in parthenogens. Whereas mitochondria and chloroplasts from eukaryotes are generally free from genomic parasites, the nuclear genome of the same host is normally full of mobile elements. Furthermore, in algae and fungi, where cell organelles are inherited biparentally, mitochondria and chloroplasts carry numerous elements of selfish DNA (Zeyl & Bell 1997). Meselson's statement that sex might only have developed as a protective mechanism against transposons (Meselson, in Concar 1996) might be provocative, but is probably wrong, as parthenogenetic reproduction turns out to be a good strategy to avoid the invasion of mobile DNA elements, and thus more mutations, from a different lineage.

Haploid organisms that have suffered double strand DNA damage have no effective option for repair. Bacteria may take up DNA from the environment (transformation), potentially from related cells that have recently died, in the hope of finding a homologous template to use in repair. An option for a eukaryotic cell is to fuse with another cell in which damage is likely to be in other loci. Pairing of homologous stretches of DNA can then permit repair by exchange of template strands. Recombination of flanking markers might then be an incidental consequence of some of these repair events (Bernstein *et al.* 1988). Although repair does not seem to be a good explanation for the maintenance of sexual reproduction (see main text), it may well have been involved in the early stages.

Box 1.3 Major types of asexuality in ostracods

The term 'asexuals' is a conflated concept, which refers to organisms with fundamentally different origins and modes of reproduction. Following Minelli (1993) and Carvalho (1994), uniparental organisms can reproduce either by self-fertilization, by automixis, by arrhenotoky, or by amictic reproduction (see also Box 1.1). In the first three types, some degree of meiosis and recombination is still possible, and these types of reproduction are here excluded from the discussion. Within the amictic asexuals, multiplication can be agametic (somatic by fission, budding or fragmentation) or gametic (with gametes, mostly eggs or cysts). The amictic and gametic asexuals can be classified as follows:

1. Genuine *ancient asexuals* do not have an extant sexual root (neither intra- nor interspecific) and do not need interaction with other species to reproduce (e.g. bdelloid rotifers, darwinulid ostracods). As far as known, these groups reproduce apomictically only, but occasional automixis cannot entirely be excluded. Mixis with rare males in darwinulids (Little & Hebert 1996) remains a theoretical possibility, but can be refuted at least on the available evidence (Rossetti & Martens 1996, Schön *et al.* 1998). New lineages originate through mutation only.
   FP1 → FP2

2. In *species with mixed reproduction,* both reproductive types are mostly geographically segregated, although bisexuals and parthenogenetic females can occur sympatrically and in such cases, intraspecific hybridisation is possible (see below). New asexual clones can theoretically originate in four different ways:
   - *Intraspecific hybridisation:* FP1 x M1 → FP2 (mostly polyploid)
   - *Interspecific hybridisation (B/B):* FB1 x M2 → FP3
   - *Interspecific hybridisation (P/B):* FP1 x M2 → FP3 (mostly polyploid)
   - *Mutational:* FP1 → FP2

   Such lineages can produce *fully asexual spin-offs*, considered both at specific and even at the generic level. Their origin as well as subsequent proliferation can occur through interspecific hybridisation of parthenogenetic females and through mutation.The so-called 'ancient asexual' anostracan *Artemia parthenogenetica* belongs in this category (Judson & Normark 1996).

Weismann suggested that the advantages of sexual reproduction would be found in the production of genetically variable offspring which would facilitate the response to natural selection. Here, we see immediately the dichotomy in approaches to explaining sex: one can concentrate on the population genetic processes that generate variability or on the ecological processes that generate appropriate selection pressures. It is best to consider the genetics first.

While it is true that the offspring of an apomictic female are identical genetically but the offspring of a sexual female are genetically variable, the same is not true at the level of the group. Sex, or recombination, has no net effect on the amount of variation in a population at equilibrium – it simply re-shuffles the existing variation. Similarly, it has no effect on associations between alleles at different loci at the population level, if the population is in linkage-equilibrium. Linkage equilibrium means that for two loci each with two alleles, A/a and B/b, the frequency of AB combinations is simply the product of the frequencies of the A and B alleles, and so on for the other possible combinations Ab, aB and ab. In other words, the occurrence of A or a alleles in an individual is independent of the genotype at the B locus. In an equilibrium sexual population, recombination creates as many new AB combinations as it breaks down. Thus, at the level of the population, the advantage to sex or recombination cannot lie in the production of variation or the redistribution of allele combinations.

The situation is different in a population with linkage disequilibrium, that is where non-random associations exist between alleles at different loci. Here sex breaks down existing associations between alleles and can create new associations. If associations between alleles influence fitness, then sex can have advantages, or disadvantages. For example, a parthenogenetic population containing the Ab and aB genotypes cannot generate the AB genotype, even if it is advantageous, whereas it would be produced in a sexual population by recombination and so sex would provide an increase in fitness. Conversely, if the Ab and aB genotypes have high fitness but the AB and ab genotypes have low fitness, the parthenogenetic population can have an advantage by avoiding the continuous production of the unfit combinations.

Genetic models for the advantage to sexual reproduction may, therefore, be classified on the basis of the origin of linkage disequilibrium, which may be either by chance or by natural selection, and the advantage to the breakdown of associations, which may arise from elimination of 'bad alleles' or the combination of 'good alleles' (Maynard Smith 1989). The major theories falling within these categories are listed in Table 1.1. These are the theories that will be considered here: they have been selected because they represent the most plausible hypotheses in general, and with specific reference to ostracods. Kondrashov (1993) provides a more comprehensive classification. In addition to these ideas which depend on the action of natural selection on genetic variation, another class of theories suggests immediate benefits to sexual reproduction (Felsenstein 1988, Kondrashov 1993). Of these, we will consider only the DNA repair hypothesis (Bernstein & Bernstein 1991). Finally, some theories depend on population structure, that is on the non-random spatial distribution of related individuals. Only the sib-competition model is considered from this class.

Table 1.1 Models for the advantage of sexual reproduction over parthenogenesis (classification adapted from Kondrashov 1993)

Variation and Selection Models		
Selection eliminates 'bad' genes	Linkage disequilibrium is generated by chance	1. Müller's Ratchet
	Linkage disequilibrium is generated by selection	2. Mutational Load
Sex creates 'good' gene combinations	Linkage disequilibrium is generated by chance	3. Fisher-Müller Accelerated Evolution
	Linkage disequilibrium is generated by selection	4. Red Queen
		5. Fluctuating Selection
Other Models		
Immediate benefit		6. DNA repair
Population Structure		7. Sib-competition
Ecological		8. Tangled-bank

Ecologically based classifications of the proposed advantages of sexual reproduction concentrate on the modes of interaction between organisms, or between organisms and their environment, that might produce the selection pressures required by some of the above mechanisms. This applies primarily to the mechanisms that rely on the creation of favourable gene combinations by sexual reproduction and which require changing environments: directional changes in the long term for the Fisher-Müller theory or in the short term for the Red Queen mechanism, or short-term changes in optimum for the Fluctuating Selection mechanism. Negative biotic interactions are particularly likely to provide the necessary changing environmental pressures, especially parasitism as discussed below. Competition is clearly a major component of the Sib-competition model. The Tangled-bank theory (Ghiselin 1974, Bell 1982) is a further ecologically driven theory: one which falls wholly outside the scheme of argument adopted so far. It will be included because it seems to be able to explain some of the patterns of parthenogenesis seen in ostracods.

Origin and fate of clones

Before considering the hypotheses concerning the costs and benefits of sex in more detail, it will be helpful to outline the origins and fates of parthenogenetic lineages (see also Box 1.3). It seems likely that most parthenogenetic non-marine ostracod lineages reproduce apomictically and are derived, at some stage, from a sexual ancestor (Chaplin *et al.* 1994). The transition from sexual to apomictic reproduction is probably quite a straightforward evolutionary step: it involves only the abolition of meiosis without the establishment of any new mechanisms since everything required for mitosis remains in place. Potentially, a mutation in any gene essential for meiosis but not for mitosis can generate an apomictic lineage in a single step (provided it occurs in a female!). In reality, the switch is probably rarely this simple because meiosis in embedded within a complex developmental pathway leading to the mature sexual egg and disruption of meiosis may interrupt this sequence. Similarly, fertilisation or syngamy may provide the stimulus for embryogenesis, or mating may stimulate egg maturation and oviposition.

The variable requirement for genetic changes in addition to the suppression of meiosis may contribute to diversity in the rates of production of parthenogenetic lineages in different taxonomic groups. The requirement for insemination in gynogenetic fish (Vrijenhoek 1994) and planthoppers (Drosopoulos 1976), for example, clearly illustrates the need for at least two independent genetic changes to produce independent parthenogenetic lineages in some cases. The absence of parthenogenesis in mammals has been attributed to the requirement of the zygote, during its development, for genomes with different patterns of DNA methylation which are contributed by the male and female parent (Markert 1988). On the other hand, some sexual lineages, including some non-marine ostracod lineages, appear to produce parthenogenetic clones frequently on an evolutionary timescale. Based on the wide taxonomic distribution of parthenogenetic reproduction and the clonal diversity of parthenogenetic lineages, Chaplin *et al.* (1994) conclude that the rate of origin of new clonal lineages in non-marine ostracods must be exceptionally high. Bell (1982) notes the considerable variation among ostracod genera in the proportions of sexual and parthenogenetic species. Thus it appears that overall there is little barrier to the origin of parthenogenetic lineages in non-marine ostracods. These patterns will be discussed in more detail in later chapters.

Automictic mechanisms for parthenogenetic reproduction (Box 1.1) are typically more complex than apomictic mechanisms. It is more difficult to envisage single mutations that could produce the required transitions and, where single steps could be involved such as the suppression of cell division in endomitosis, they are more specific. The origin of automixis may, therefore, be relatively rare. In ostracods, automixis has yet to be conclusively demonstrated although the existence of some completely homozygous clones suggests that it might occur (Turgeon & Hebert 1994).

Whichever type of parthenogenesis is produced by mutation, it is likely that the mutation will have negative effects on other components of fitness (Lamb & Willey 1979). This will tend to reduce the fecundity of parthenogenetic females relative to sexual females and thus reduce their two-fold advantage arising from the cost of males. If it is true that parthenogenetic lineages evolve more slowly, it may be difficult for modifiers to alleviate these side effects. Another important issue is whether the mutation causes complete isolation of the parthenogenetic lineage from its sexual progenitor population. At one extreme, a mutation could occur that fixes a single genotype in a parthenogenetic lineage. Alternatively, a mutation might cause a female to produce a mix of sexual and parthenogenetic offspring, or parthenogenetic offspring that can, sometimes, mate with males from the sexual population and incorporate their genes into future parthenogenetic generations. Either of these mechanisms will rapidly generate a parthenogenetic lineage that incorporates the whole range of genetic variability of the sexual population. Intermediates are clearly possible, with only occasional genetic transfer between populations. Gene exchange between sexual and parthenogenetic ostracod lineages has been suggested (Turgeon & Hebert 1994) and will be discussed below and in chapters 15 and 16.

All parthenogenetic vertebrates (Avise *et al.* 1992), and many parthenogenetic lineages in other taxa (Bell 1982), have arisen via interspecific hybridisation and carry the nuclear genomes of two or more sexual species. In some cases, it has been possible to identify the sexual progenitors and establish which was the female parent using mitochondrial DNA. Just how hybridisation leads to parthenogenesis is not clear, but parthenogenesis can potentially maintain heterotic, highly heterozygous genotypes which would rapidly be disrupted in a sexual hybrid population. In vertebrates the parthenogenetic lineages are often, but not always, gynogenetic or hybridogenetic (Box 1.1) and therefore not independent of the sexual population(s). To date, a hybrid origin has not been established for any parthenogenetic non-marine ostracod lineage (Chaplin *et al.* 1994) but such lineages may be found as genetic techniques are applied to more and more taxa (chapters 15 and 16).

A further possible mode of origin of parthenogenetic lineages is through infection with vertically transmitted microorganisms. The best known of these is the bacterial intracellular parasite *Wolbachia* (Moran & Baumann 1994). *Wolbachia* is transmitted from mother to daughter via eggs in a wide range of arthropods. Since the bacterium is at a dead-end when it finds itself in a male, it has evolved several different strategies to favour its spread at their expense: cytoplasmic incompatibility, male-killing, feminisation, and induction of parthenogenesis. A well-studied example is the wasp genus *Trichogramma* in which strains infected with *Wolbachia* reproduce parthenogenetically, but sexual reproduction can be restored by treatment with antibiotics (Stouthamer & Werren 1993, see also Pijls *et al.* 1996). *Wolbachia* is known from isopod crustacea but not yet from ostracods (but see chapter 7).

Given the taxonomic distribution of parthenogenetic reproduction in ostracods, and generally in animals, there can be little doubt that sexual reproduction was the ancestral state. Loss of sexual reproduction can occur via any of the above routes but it seems unlikely that it is ever regained. Although a single mutation can abolish sexual reproduction, parthenogenetic lineages are likely to accumulate mutations at many other loci coding for traits that are essential for sexual reproduction, but do not contribute to fitness of the parthenogen. Thus restoration of sex becomes increasingly difficult once a lineage has been fully parthenogenetic for a substantial period of time. However, this does not preclude situations such as heterogonic species that only produce males under rare adverse conditions or in some populations, or parthenogenetic lineages that retain the ability to mate with sexual populations occasionally. It is also possible that parthenogenetic lineages will produce rare, atavistic males that are not fully functional and do not contribute to future generations (Box 1.3).

Genetically distinct clones within a parthenogenetic 'species' either arise separately from a sexual population or by accumulation of mutations since a common parthenogenetic ancestor. This process is discussed further in chapter 16. Genetic divergence can include the appearance of polyploid or aneuploid (having unbalanced chromosome numbers) clones. In the absence of meiosis, these changes in chromosome number, as well as chromosomal mutations such as translocations, are less likely to be removed by selection. Ostracod cytogenetics has been studied rather little, but polyploid clones are believed to occur (Tétart 1978, Turgeon & Hebert 1994).

For most clones, the ultimate fate must be extinction. The patchy nature of the taxonomic distribution of parthenogenesis has long been viewed as evidence that clones suffer this fate sooner than sexual populations. For ostracods there is some direct evidence for this lack of persistence, even on the timescale of the Holocene (Griffiths & Butlin 1995, chapter 6) although this analysis was at the level of appearance and disappearance of 'species' from assemblages, rather than displacement of clones or lineages. There are a few lineages that have apparently escaped extinction over long periods of time and may even have diversified despite the absence of sexual reproduction: the so-called 'ancient asexuals' (Judson & Normark 1996). The most famous example is the Bdelloidea, a group of rotifers believed to be about 35 million years old and containing over 350 parthenogenetic species. However, the non-marine ostracod superfamily, the Darwinuloidea, may be an even better example (Butlin & Griffiths 1993, Judson & Normark 1996). Although they are less diverse, 26 extant species (chapter 3), they have an excellent fossil record in which sex can be inferred from valve morphology. Parthenogenesis is apparently universal in extant darwinulids and has been in fossil species for the last 100 million years or more (Griffiths & Butlin 1995, Schoen *et al.* 1996). This fascinating group will be discussed in detail later (chapters 3, 4, 15 and 16). For the present, it is important to note that for ancient parthenogens the questions to be asked are somewhat different. Firstly, there continues to be speculation about whether they really exist: Do they have some form of 'cryptic' or infrequent sex (Judson & Normark 1996)? If they are genuinely apomictic and genuinely ancient, then the question is: How have they escaped the long-term disadvantages of parthenogenesis? This contrasts with the more usual question of how sexual populations escape the short-term disadvantages of recombination. Later in this chapter, we briefly introduce some possible solutions to the problem of ancient parthenogens which will be discussed in more detail later in this book.

The major costs and benefits of sexual reproduction

The two-fold cost of sex, the cost of males, arises because energy invested in the production of males by sexual females can be diverted to the production of female offspring by parthenogenetic females. Sexual females therefore leave less grandchildren. If 100 sexual females leave two offspring each, with the help of 100 males, the size of the sexual population remains constant with 1:1 sex ratio. If 100 parthenogenetic females leave two offspring each, both female, the parthenogenetic population doubles in size in each generation. This is undoubtedly the largest and most pervasive cost of sex for most animals.

The cost of males is not restricted to animals with separate sexes. Hermaphrodites that invest equally in male and female reproductive effort, as expected in most cases for outbreeding species, suffer the same cost. However, where, for example, male effort is reduced because of selfing or local mate competition, the cost of sex is also reduced because the sexual female can produce more than half as many female offspring as the parthenogenetic female. Similarly, in species with separate sexes where there is male parental investment, the difference in potential output between sexual and parthenogenetic females, and thus the cost of sex, is reduced and may be eliminated completely. Hermaphrodite ostracods are not known but many non-marine ostracods have very large spermatozoa (Wingstrand 1988, Cohen & Morin 1990): it is conceivable that these large spermatozoa might represent a parental investment in the zygote, increasing its survival, in which case the cost of sex would be less than two-fold. The cost of sex is also reduced if parthenogenetic females have lower fecundity than sexual females, as mentioned above.

Sex has other costs. Location and courtship of mates, and the act of mating itself, are energetically expensive, expose the animals to predation risk, and facilitate transmission of parasites between mating partners. The magnitude of these costs is likely to vary with environmental parameters, especially population density and ecological complexity. These costs may, therefore, contribute to some of the broad-scale patterns observed in the incidence of parthenogenesis such as the tendency for parthenogens to be more common at higher latitudes (Bell 1982) or the contrast between high frequency of parthenogenesis in non-marine ostracods but low frequency in marine taxa (Chaplin *et al.* 1994). In mammals, mutation rates are higher in males than in females representing an additional cost to sex, but it is uncertain whether this is a widespread phenomenon (Hurst & Peck 1996).

The main genetic consequence of sexual reproduction, the breakdown of linkage disequilibrium, may itself be a disadvantage in some circumstances. In a constant, or predictable, environment, selection builds up favourable combinations of alleles across loci. The break-up of these associations is likely to reduce offspring fitness. This 'flip-side' should be kept in mind while considering the following proposed benefits of sex, which will be treated in the order introduced above.

Müller's Ratchet

A serious problem for parthenogenetic organisms is to get rid of disadvantageous mutations. The concept of 'Müller's ratchet' focuses on the frequency of mutation-free individuals in a population. In a finite population, these individuals may be lost through genetic drift. Their mutation-free genotype can easily be restored by sexual reproduction and recombination between genotypes carrying different mutations, but in a parthenogenetic population the loss is irreversible (Müller 1964). Müller compared this process with a ratchet, which 'clicks' round another notch every time the least mutated genotype is lost: first the mutation-free genotype, then genotypes with one mutation, then those with two mutations and so on. Overall, the population accumulates mutations leading eventually to its extinction. The advantage to sex lies in the breakdown of linkage disequilibrium between disadvantageous alleles which is generated by the stochastic loss of coupling (++) combinations of favourable alleles.

Müller's ratchet will click at a rate dependent on both population size and mutation rate because together they determine the number of individuals in the least mutated class and, therefore, the chance that this class will be lost. Its effect becomes even more important if the loss of individuals with advantageous genotypes also reduces the population size leading to a 'mutational meltdown', as has been shown in simulation experiments (Lynch *et al.* 1993). In endosymbiotic bacteria, which must reproduce apomictically in small populations for a long time, an accumulation of deleterious mutations has been detected, as predicted by the ratchet (Moran 1996). However, parthenogenetically reproducing cell organelles do not obviously suffer from the ratchet indicating that it can be avoided (Gabriel *et al.* 1993). Firstly, the ratchet is most important if mutations are typically mildly deleterious: if mutations have very small effects, the fitness of the population declines slowly; if mutations have large effects, they are removed by selection and the chance of losing the mutation-free class is reduced. However, the majority of mutations do appear to be either slightly deleterious or neutral (Kondrashov 1993). Secondly, the rate of the ratchet will be influenced by the population size and the frequency of mutation-free individuals. The higher their number, the slower the ratchet operates: if 100 or more such individuals exist, the ratchet will practically stop (Haigh 1978) because chance events are unlikely to result in so many individuals failing to leave offspring in any one generation. Thirdly, the total genomic mutation rate will effect the operation of the ratchet: the lower the mutation rate, the slower the ratchet turns. Polyploidy, a common phenomenon among parthenogens, also slows the ratchet provided most deleterious mutations are recessive (Mogie & Ford 1988). Even where conditions are favourable for the ratchet to operate, many clicks will be necessary before an apomictic population is driven to extinction, taking at least 100 generations unless the population size is very small (Lynch *et al.* 1993). In conclusion, Müller's ratchet alone is not sufficient to provide a short-term advantage to sexual reproduction but it is potentially a serious problem for ancient parthenogenetic lineages.

Mutational Load

The mutational load may be greater in parthenogenetic populations than in sexual populations even without the operation of the ratchet. This alternative relies not on chance loss of favourable genotypes in parthenogenetic populations, but on a more efficient removal of disadvantageous genotypes in sexual populations, a concept called the 'mutation-load-reduction theory' (Kondrashov 1988, Crow 1994). Selection tends to reduce variation among individuals in the number of deleterious mutations they carry, generating linkage disequilibrium by removing individuals with multiple mutations. In a parthenogenetic population, the rate at which disadvantageous mutations are removed slows down and, if the input of new mutations is great enough, this leads to accumulation of mutations and reduction in fitness. In a sexual population the variation in number of mutations is restored in each generation and selection continues to remove mutations effectively. Synergistic effects of mutations on fitness could speed up this process, as the accumulation of mutations might lead to disproportionately strong selection: ultimately leading to truncation selection as the number of mutations reaches a critical value (Hurst & Peck 1996). The mutation load (the average number of mutations per individual) is independent of population size, but will be strongly affected by the genomic mutation rate. If the mutation rate is low (<1 mutation per genome per generation, Kondrashov 1988), the mutation load is not a problem for a parthenogenetic population. If it is high, this mechanism can provide a short-term advantage to sex without requiring any particular pattern of environmental change.

Fisher-Müller Accelerated Evolution

This idea is the oldest and most widely cited explanation for sexual reproduction (Fisher 1930, Müller 1932). In essence, it states that a sexual population is able to evolve more rapidly than a parthenogenetic population because recombination brings together advantageous combinations of new mutations which can only be created in a parthenogenetic population when the two mutations occur sequentially within the same lineage (Fig. 1.2). So, if the initial population has the genotype ab and both new mutations A and B are advantageous but the combination AB has the highest fitness, AB can be generated by recombination between Ab and aB individuals as soon as both mutations have occurred anywhere in the population. But, in a parthenogenetic population either the A mutation must occur within a lineage carrying the B mutation, or vice versa. Thus, the sexual population evolves more rapidly and will eventually displace the parthenogenetic population.

This argument is a stochastic one relying on recombination to modify associations between genes for the following reasons. In an infinite population, whether sexual of parthenogenetic, mutation would generate the Ab, aB and AB combinations in the expected proportions, i.e. in linkage equilibrium, and the sexual population would have no advantage. However, in a finite population, the AB double mutant occurs very rarely, the Ab and aB genotypes are in excess of their expected frequencies and there is linkage disequilibrium. Break-up of this disequilibrium by recombination generates the advantageous AB genotype. Note, though, that there may be no advantage to sex in a very small population either, because the A mutation is likely to be fixed before the B mutation appears.

This is a long-term process. It may well contribute to limiting the evolutionary potential of parthenogenetic lineages, but it cannot overcome the cost of males in the short term.

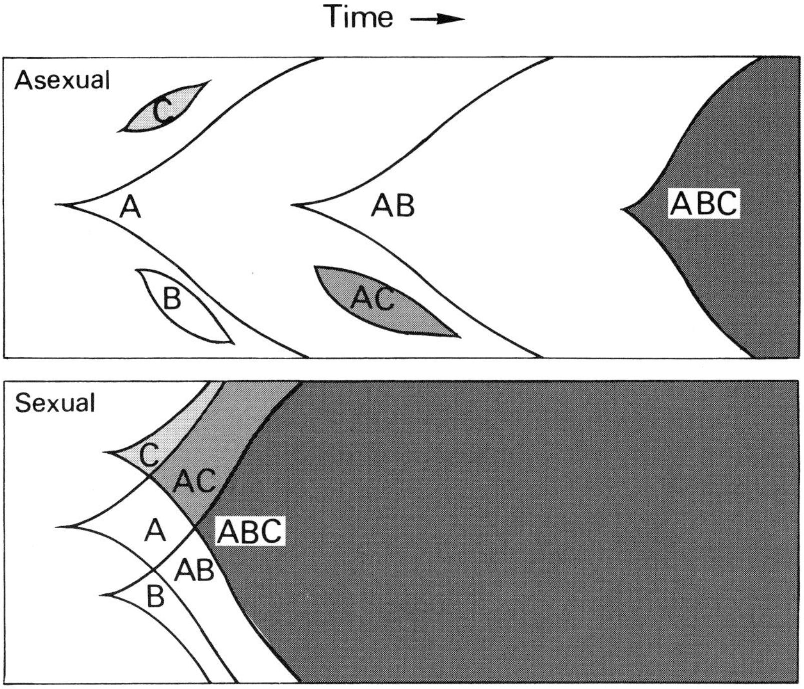

Fig. 1.2 In an asexual population, mutations A, B and C, which are all advantageous, occur in separate lineages. They cannot be brought together to form the most fit ABC genotype until all three mutations have occurred successively in the same lineage. By contrast, in a sexual population, the three mutations can increase in frequency independently and can be brought together by recombination. Thus the sexual population can evolve more rapidly (after Maynard Smith 1989).

The Red Queen

In 'Alice through the Looking Glass' (Lewis Carroll 1872), the red queen tells Alice that it takes all the running they can do just to stay in the same place. The ecological analogy is that continuous adaptive evolution is necessary to remain viable in an environment that is constantly changing in an open-ended way (Van Valen 1973). This environmental change is supposed to be driven by biotic interactions, especially by 'arms races' between hosts and parasites or predators and prey (Hamilton 1980, Eshel 1991). As in the Fisher-Müller theory, sex is proposed to increase the rate of evolution but, in this case, over a shorter term and without reliance on new mutations.

Genetically, the Red Queen process, as applied to the maintenance of sexual reproduction, works because selection in the present generation favours associations between alleles that are currently advantageous. However, if the direction of selection changes or the optimum phenotype changes between generations, these associations will hinder future response to selection. Sex breaks up associations in each generation and thereby allows the population to respond more effectively to the changing selective regime. The advantage to sex depends on the rate of change in the pattern of selection and thus on the ecological processes that constantly generate new challenges. Parasites are favoured as the explanation because they commonly have large populations and short generation times relative to their hosts. They can, therefore, rapidly evolve responses to new host resistance strategies and oblige the host to respond in turn.

Some comparative evidence supports this mechanism: parthenogenetic taxa are more common in temperate than tropical environments in line with the incidence of parasitism (Bell 1982), genes associated with resistance to parasites evolve rapidly (Hurst & Peck 1996), and recombination rates are higher in organisms with a history of strong directional selection (Burt & Bell 1987).

Fluctuating Selection

This hypothesis differs from the Red Queen in that it does not require indeterminate evolution, rather it supposes that the environment, and thus the optimum phenotype, fluctuates within a limited range. Sex may be favoured because it allows the population to track environmental fluctuations more effectively (Maynard Smith 1980).

If one considers a quantitative trait influenced by many loci and under stabilising selection, selection tends to generate 'repulsion' linkage disequilibrium, i.e. associations of increasing (+) and decreasing (-) alleles that occur in individuals of intermediate phenotype. However, if the optimum phenotype increases or decreases, ++ or - - combinations will be favoured until a new equilibrium is reached, repulsion associations impede evolutionary response and sex is favoured. A large advantage to sex requires wide fluctuations with cycle lengths that are long enough for populations to respond to selection but short enough to prevent elimination of the sexual population before sufficient change occurs.

An alternative suggestion (Roughgarden 1991) is that sex is advantageous because it *reduces* response to fluctuating selection. If, in a simple example, the environment fluctuates between two states on an appropriate time scale, a parthenogenetic population may become fixed for a single genotype in one phase of the cycle which then has low fitness in the next phase. Fluctuations in genotypic composition are damped in sexual populations which can, therefore, persist through multiple cycles.

Like the Red Queen mechanism, this hypothesis depends on particular patterns of changing selection pressures. There is really no good reason to believe that the proposed patterns are common enough in nature to explain the broad occurrence of sexual reproduction. In constant environments, or environments that change slowly or predictably, the advantages to sex are unlikely to overcome the two-fold cost and, indeed, the breakdown of linkage disequilibrium may actually be disadvantageous.

DNA Repair

These hypotheses belong to a different class of explanations for sexual reproduction: explanations that depend on direct benefits rather than the population genetic consequences of recombination (Kondrashov 1993). In essence, the suggestion is that chromosome pairing during meiosis allows the repair of mutations accumulated during the lifetime of the organism by comparison between DNA sequences of the homologues (Bernstein & Bernstein 1991, Michod 1993). Since these mutations are, at least partly, responsible for senescence, meiosis and syngamy may 'rejuvenate' the zygote at the start of the next generation. Crossing-over between distant genetic markers could be an incidental side-effect of the strand exchange needed to provide a template for repair.

This hypothesis does not seem adequate to explain sexual reproduction. It requires only diploidy and pairing of homologous chromosomes, not reduction division, syngamy or crossing-over. DNA repair may have been involved in the origin of sex (Box 1.2) and it may be an important determinant of mutation rates (see below and chapter 16).

Sib-Competition

Sib-Competition (Williams 1975, Bell 1982) is the only model that relies on population structure that will be considered here (cf. Kondrashov 1993) and it will be treated briefly. The offspring of an individual sexual female are more variable genetically than the offspring of a parthenogenetic female. If the offspring of a sexual female arrive at a patch of environment capable of supporting only a limited number of adults, there is a better chance that some of them will be well suited to the patch than if they were the offspring of a parthenogenetic female. In Maynard Smith's (1989) analogy, the sexual female buys 50 tickets in a lottery, all with different numbers, while the parthenogenetic female buys 100 tickets, all with the same number. This mechanism only works where there is significant competition between close relatives, i.e. where the population is strongly structured. Otherwise, to extend the analogy, the females are buying tickets in many different lotteries.

At first sight, it seems unlikely that this model will apply to ostracods since they are small, mobile organisms living in a relatively continuous environment. However, some ostracod populations are very patchy, on a variety of spatial scales (see chapter 13), so that its relevance cannot be excluded *a priori*.

The Tangled-Bank theory

The Tangled-bank theory (Ghiselin 1974, Bell 1982) is based on the economic analogy that it is more profitable to concentrate production on a single product in an unsaturated market, but to diversify when the market becomes saturated. In an environment in which niche space is tightly packed with interacting organisms, frequency-dependent selection favours novel or rare genotypes which utilise different resource combinations. Thus, diverse lineages have higher carrying capacities and sexual lineages outcompete parthenogenetic lineages. However, where the environment is relatively empty, the premium on diversity is reduced and selection may favour a single 'general purpose genotype' (Lynch 1984). A parthenogenetic clone may capture such a genotype and increase at the expense of sexual lineages. Selection tends to focus on the intrinsic rate of increase rather than the carrying capacity and thus the cost of males is especially severe. An alternative way of looking at this hypothesis is to argue that a single parthenogenetic clone can never eliminate a sexual population because some part of the sexual population will outcompete the clone in at least some component of the environment. The more genetically different clones exist within the locality, the more difficulty the sexual population will have. If the clones encompass all of the genotypes present within the sexual population, the sexual population must be displaced. Vrijenhoek (1994) has shown that in poeciliid fish the predicted negative relationship exists between the number of clones present in a habitat and the proportion of the total population represented by the sexually reproducing progenitor species.

This hypothesis is not readily incorporated into the classification of theories for the maintenance of sexual reproduction used here (and it is ignored by Kondrashov 1993). It can be considered a spatial analogue of the temporal variation hypothesis (the Red Queen), or a model like the sib-competition model that relies on population structure, but this misses the fact that it is driven by the diversity of local ecological opportunity rather than variation in the environment from place to place (a question of environmental grain relative to dispersal). The Tangled-bank model actually violates the initial argument made above concerning the effect of sex on genetic variation. It is based on a difference in genetic diversity between sexual and parthenogenetic lineages, whereas we have argued that sex does not generate but merely redistributes variation. Thus, it seems most powerful for explaining the persistence of sexual lineages that generate rare, genetically isolated and therefore genetically monomorphic, clones. For these species, it may help to explain the geographical distribution of sexual and parthenogenetic populations, especially the pattern of 'geographical parthenogenesis' (Lynch 1984, see chapter 5).

Persistence of parthenogenetic lineages

So far, the main focus of this chapter has been on the advantages of sexual reproduction. One or more of these mechanisms, or some additional process yet to be discovered, must counterbalance the two-fold cost of sex in the short to medium term since sexual reproduction is the most widespread reproductive mode in nature. Parthenogenetic reproduction may occur in short, sporadic episodes while the asexual lineages hold a temporary advantage. However, persistent parthenogenetic lineages, ancient parthenogens, do not fit into this picture. This section, therefore, gives an overview of current ideas concerning the ways in which these lineages might have avoided or overcome the longer-term disadvantages connected with their reproductive mode.

A good start may provide a parthenogenetic lineage with as much genetic variability as a sexual species, which might help it to persist for a considerably longer period. This might occur where parthenogenesis evolves progressively, rather than by a single step (see above). Alternatively, the parthenogenetic lineage may start off with a highly heterozygous genotype, which may confer high fitness in a range of environments (Lynch 1984), due to a multiple (polyphyletic) origin from several sexual ancestors (Parker 1979). Such hybridisation events, leading to new parthenogenetic clones or species, have been identified in a broad range of taxa by genetic analyses (Crease *et al.* 1989, Turgeon & Hebert 1995, Van Raay & Crease 1995, White *et al.* 1982, Harshmann & Futuyama 1985, Moritz *et al.* 1989, Quattro *et al.* 1991). Polyploidy is quite common in parthenogens and could be another indication of a polyphyletic origin (Havel *et al.* 1990, Little & Hebert 1994, Tétart 1978, Turgeon & Hebert 1995). Polyploidy also enables a lineage to harbour even greater fixed heterozygosity. None of these is a permanent solution, they may simply delay the inevitable.

If the parthenogenetic lineage does not disperse, but stays in close spatial proximity to the site of its origin, it might maintain high genetic variability (Crease *et al.* 1989) through occasional reproductive connection with its sexual ancestor (see next paragraph).

A little sex now and then may be sufficient to arrest Müller's ratchet and lighten the mutation load, without compromising the short-term advantages of parthenogenetic reproduction. The most prominent examples of such a strategy are cladocerans and aphids, where a cyclic change between the two different reproductive modes occurs (heterogony, Box 1.1), triggered by changes in the environment. Beside this phenomenon of cyclic parthenogenesis, other strategies that combine the advantages of both reproductive modes, have been found in crustaceans, possibly including ostracods. The diploid lineage of *Artemia parthenogenetica*, apparently a parthenogenetic species for approximately 5 million years, might combine apomictic reproduction with occasional automixis and recombination (Browne 1992). The rare recombination permitted by automixis generates new potentially advantageous genetic combinations and may enable the lineage to get rid of 'bad' genes (Table 1.1). Apparently, this is sufficient to allow its long-term persistence. The other, polyploid lineages of this species, lack the ability for recombination, however, being restricted to apomictic reproduction. Rare males are produced in all lineages, giving the impression of ongoing sexual reproduction. However, they do not mate, being the product of an accidental fusion of two haploid cells during egg synthesis, which in *Artemia* produces dysfunctional XX diploid males.

Several papers on ostracods have described exceptionally high clonal diversity in parthenogenetic lineages that co-occur with closely related, sexual species, perhaps suggesting interbreeding with the sexual relatives (Little & Hebert 1994, Turgeon & Hebert 1995). This theory is supported by the finding that some parthenogenetic clones of *Heterocypris incongruens* are apparently the result of a hybridisation with a related sexual species *H. glauca*, while others were produced through interbreeding between bisexual and parthenogenetic lineages of *H. incongruens* (Turgeon & Hebert 1994, using the name *Cyprinotus*). A similar case was mentioned for the cladoceran *Daphnia* from arctic sites (Van Raay & Crease 1995). This strategy to maintain genetic diversity can only work, however, if the species are in close geographical proximity and no genetic or behavioural mechanisms have evolved which result in reproductive isolation. In one morph of the Australian ostracod *Candonocypris novaezelandiae*, such isolation prevents females from parthenogenetic lineages at one locality from using sexual reproduction, although males of the same species are present. The same morph from another locality apparently can utilise either reproductive mode according to the availability of males (Chaplin 1992).

Most parthenogenetic ostracod species are not taxonomically isolated (Havel & Hebert 1993), so backcrossing with sexual relatives or congeners could be a common process helping them to persist. However, the strategy can only work if the necessary conditions, such as geographic, behavioural and genetic compatibility, are fulfilled. Therefore, this 'way out' is likely to be restricted to those taxa which are recent spin-offs from sexual lineages and it cannot explain the existence of ancient parthenogens like *Darwinula stevensoni*.

Although males have never been described for most parthenogenetic species, they might nevertheless be produced now and then and just escape notice due to their spatial or temporal rarity. Chaplin *et al.* (1994) created the expression 'unisexuals' for such cases and believe that this could be a significant explanation for the majority of paradoxical parthenogenetic lineages. However, so far, no species has been reported in which parthenogens indeed turned out to be unisexuals, because males were finally found (if we neglect the still unresolved controversy concerning males described for *Darwinula stevensoni,* Little and Hebert 1996). In contrast, all genetic studies on ostracods, so far, have confirmed that female-only populations reproduced exclusively parthenogenetically (Chaplin 1992, Havel *et al.* 1990, Havel & Hebert 1989, Little & Hebert 1994, Turgeon & Hebert 1995).

Brooding. By comparing the distribution of parthenogenetic reproduction and the occurrence of brooding, Lively & Johnson (1994) discovered a striking positive correlation for most fresh-water organisms, including ostracods. Brooding might allow the colonisation of more extreme habitats or, alternatively, be coupled with a restricted dispersal ability of the young. It could also enable the mother to abort damaged or deficient offspring selectively, which could reduce the costs of automictic reproduction in particular. It is not necessary, however, for parthenogenetic reproduction, as some parthenogenetic lineages (e.g. *Eucypris virens*) lack brooding. Therefore, brooding cannot act as a sufficient explanation for the evolutionary maintenance of parthenogenetic reproduction, but may facilitate it. The impact of brooding and its related behaviour will be described in more detail for the example of *Darwinula stevensoni* (chapter 10).

Metapopulations. One solution to the problem of Müller's ratchet is a sufficient population size. Even if a local (sub-)population consists of a limited number of individuals, the species might still not suffer from the ratchet if there is sufficient dispersal or extinction and recolonisation. These processes allow genotypes with low mutation loads to spread and replace those that have accumulated high loads. Dispersal within a metapopulation might also be a strategy to escape parasites since an immigrant is unlikely to encounter parasites adapted to its defences (Ladle *et al.* 1993). The metapopulation concept and its consequences for parthenogenetic reproduction will be further discussed in chapter 13.

Mutation rates and DNA repair influence both Müller' s ratchet and the mutation-load-reduction theory. The rate of lethal mutations is not important in this context, as they will be removed from the population anyway. So, we only have to consider the consequences of the rate of deleterious or slightly deleterious mutations. Müller's ratchet will be slowed down by a low genomic mutation rate which could be produced either by a generally decreased rate of deleterious and slightly deleterious mutations or by a reduced number of genes in the genome. A more efficient DNA repair could also lower the effective mutation rate per genome. In any case, the ratchet could only be slowed down, not stopped. However, its rate might become so slow that even parthenogens persisting for a long time might not suffer from it.

The effect of the mutation rate on the mutation-load-reduction theory is more direct: if the rate is low enough (e.g. through a more effective repair), there will be no need for sex. A general mutation rate (per genome per generation) of 1 or less makes sex superfluous, at least for lightening the mutation load (Kondrashov 1988). Measured rates in *Drosophila* are close to this value (see Crow 1994) but a wide range of microorganisms show a lower rate of $3x10^{-3}$ mutations per genome per DNA replication (Drake 1991, for discussion see Hurst & Peck 1996). Gabriel *et al.* (1993) suggested that another way out of the mutation load would be to stop DNA repair altogether. This can be advantageous because mutations of small effect that might result from imperfect repair are removed from the parthenogenetic population less efficiently than mutations of large effect. As already stated elsewhere (Schoen *et al.* 1996), there is no evidence, so far, that any organisms have stopped repairing DNA damage, and, as some repair systems like 'nucleotide-excision-repair' also take part in transcription, losing the ability for DNA repair could also effect other important life systems.

Possible strategies for reduction of the mutation rate in parthenogens are discussed more extensively in chapters 16 and 17.

References

Avise, J.C. 1994. *Molecular markers, natural history and evolution*, Chapman & Hall.

Avise, J.C., Quattro, J.M. & Vrijenhoek, R.C. 1992. Molecular clones within organismal clones: mitochondrial DNA phylogenies and the evolutionary histories of unisexual vertebrates. *Evol. Biol.*, **26:** 225-246.

Bell, G. 1982. *The masterpiece of nature: the evolution and genetics of sexuality,* University of California Press.

Bernstein, C. & Bernstein, H. 1991. *Ageing, sex, and DNA repair,* Academic Press.

Bernstein, H., Hopf, H.A. & Michod, R.E. 1988. Is meiotic recombination an adaptation for repairing DNA, producing genetic variation, or both? *in*: Michod, R.E. & Levin, B.R. (eds), *The evolution of sex: an examination of current ideas,* Sinauer: 139-160.

Browne, R.A. 1992. Population genetics and ecology of *Artemia*: insights into parthenogenetic reproduction. *Trends Ecol. Evol.*, **7:** 232-237.

Burt, A. & Bell, G. 1987. Mammalian chiasma frequencies as a test of two theories of recombination. *Nature*, **326:** 803-805.

Butlin, R.K. & Griffiths, H.I. 1993. Ageing without sex? *Nature*, **364:** 680.

Carvalho, G.R. 1994. Genetics of aquatic clonal organisms. *in*: Beaumont, A.R. (ed), *Genetics and evolution of aquatic organisms,* Chapman & Hall: 291-323.

Chaplin, J.A. 1992. Variation in the mode of reproduction among individuals of the ostracod *Candonacypris novaezelandiae*. *Heredity*, **68:** 411-424.

Chaplin, J.A., Havel, J.E. & Hebert, P.D.N. 1994. Sex and ostracods. *Trends Ecol. Evol.*, **9:** 435-439.

Cohen, A.C. & Morin, J.G. 1990. Patterns of reproduction in ostracodes: a review. *J. Crustacean Biol.*, **10:** 184-211.

Concar, D. 1996. Sisters are doing it for themselves. *New Scientist*, **151:** 32-36.

Crease, T.J., Stanton, D.J. & Hebert, P.D.N. 1989. Polyphyletic origins of asexuality in *Daphnia pulex*. II. Mitochondrial DNA variation. *Evolution*, **43:** 1016-1026.

Crews, D. 1994. Constraints to parthenogenesis. *in*: Short, R.V. & Balaban, E. (eds), *The differences between the sexes,* Cambridge University Press: 23-49.

Crow, J.F. 1994. Advantages of sexual reproduction. *Dev. Genet.*, **15:** 205-213.

Doolittle, W.F. & Sapienza, C. 1980. Selfish genes, the phenotype paradigm and genome evolution. *Nature*, **284:** 601-603.

Drake, J.W. 1991. Spontaneous mutation. *Annu. Rev. Genet.*, **25**: 125-146.

Drosopoulos, S. 1976. Triploid pseudogamous biotype of the leafhopper *Muellerianella farimairei*. *Nature*, **263:** 499-500.

Eshel, I. 1991. Game theory and population dynamics in complex genetical systems: the role of sex in short-term and in long-term evolution. *in*: Selten, R. (ed), *Game and equilibrium models I,* Spinger-Verlag: 6-28.

Felsenstein, J. 1988. Sex and the evolution of recombination. *in*: Michod, R.E. & Levin, B.R. (eds), *The evolution of sex: an examination of current ideas,* Sinauer: 74-86.
Fisher, R.A. 1930. *The genetical theory of natural selection,* Clarendon Press.
Gabriel, W., Lynch, M. & Burger, R. 1993. Müller's ratchet and mutational meltdowns. *Evolution*, **47:** 1744-1757.
Ghiselin, M.T. 1974. *The economy of nature and the evolution of sex,* University of California Press.
Green, M.R. 1988. Mobile DNA catalysts. *Nature*, **336:** 716-718.
Griffiths, H.I. & Butlin, R.K. 1995. A timescale for sex versus parthenogenesis: evidence from subfossil ostracods. *Proc. R. Soc. Lond., B*, **260:** 65-71.
Haigh, J. 1978. The accumulation of deleterious genes in a population – Müller's ratchet. *Theor. Pop. Biol.*, **14:** 251-267.
Hamilton, W.D. 1980. Sex versus non-sex versus parasites. *Oikos*, **35:** 282-290.
Harshmann, L.G. & Futuyama, D.J. 1985. Survivorship and growth of sexually and asexually derived larvae of *Alsophila pometaria* (Lepidoptera: Geometridae). *Am. Nat.*, **125:** 896-902.
Havel, J.E. & Hebert, P.D.N. 1989. Apomictic parthenogenesis and genotypic diversity in *Cypridopsis vidua* (Ostracoda: Cyprididae). *Heredity*, **62:** 383-392.
Havel, J.E., Hebert, P.D.N. & Delorme, L.D. 1990. Genotypic diversity of asexual Ostracoda from a low arctic site. *J. evol. Biol.,* **3:** 391-410.
Havel, J.E. & Hebert, P.D.N. 1993. Clonal diversity in parthenogenetic ostracods. *in*: McKenzie, K.G. & Jones, J.P. (eds), *Ostracoda in the Earth and Life Sciences,* Balkema: 353-368.
Hickey, D.A. 1982. Selfish DNA: a sexually transmitted nuclear parasite. *Genetics*, **101:** 519-531.
Hickey, D.A. 1993. Molecular symbionts and the evolution of sex. *J. Hered.*, **84:** 410-414.
Hurst, L.D. & Peck, J.R. 1996. Recent advances in understanding of the evolution and maintenance of sex. *Trends Ecol. Evol.*, **11:** 46-52.
Judson, P.O. & Normark, B.B. 1996. Ancient asexual scandals. *Trends Ecol. Evol.*, **11:** 41-46.
Kondrashov, A.S. 1988. Deleterious mutations and the evolution of sexual reproduction. *Nature*, **336:** 435-440.
Kondrashov, A.S. 1993. Classification of hypotheses on the advantage of amphimixis. *J. Hered.*, **84:** 372-387.
Ladle, R.J., Johnstone, R.A. & Judson, O.P. 1993. Coevolutionary dynamics of sex in a metapopulation: escaping the Red Queen. *Proc. R. Soc. Lond., B*, **253:** 155-160.
Lamb, R.Y. & Willey, R.B. 1979. Are parthenogenetic and related bisexual insects equal in fertility? *Evolution*, **313:** 774-775.
Little, T.J. & Hebert, P.D.N. 1994. Abundant asexuality in tropical freshwater ostracodes. *Heredity*, **73:** 549-555.
Little, T.J. & Hebert, P.D.N. 1996. Ancient asexuals: scandal or artefact? *Trends Ecol. Evol.*, **11:** 296.
Lively, C.M. & Johnson, S.G. 1994. Brooding and the evolution of parthenogenesis: strategy models and evidence from aquatic invertebrates. *Proc. R. Soc. Lond., B*, **256:** 89-95.
Lowndes, A.G. 1935. The sperms of freshwater ostracods. *Proc. Zool. Soc. Lond.,* **1935:** 35-48.
Lynch, M. 1984. Destabilizing hybridization, general-purpose genotypes and geographic parthenogenesis. *Quart. Rev. Biol.*, **59:** 257-290.
Lynch, M., Burger, R., Butcher, D. & Gabriel, W. 1993. The mutational meltdown in asexual populations. *J. Hered.*, **84:** 339-344.
Markert, C.L. 1988. Imprinting of genome precludes parthenogenesis, but uniparental embryos can be rescued to reproduce. *Ann. NY Acad. Sci.*, **541:** 633-638.
Maynard Smith, J. 1978. *The evolution of sex,* Cambridge University Press.
Maynard Smith, J. 1980. Selection for recombination in a polygenic model. *Genet. Res.*, **35:** 269-277.
Maynard Smith, J. 1989. *Evolutionary genetics,* Oxford University Press.
Michod, R.E. 1993. Genetic error, sex, and diploidy. *J. Hered.*, **84:** 360-371.
Minelli, A. 1993. *Biological systematics, the state of the art,* Chapman & Hall.
Mogie, M. & Ford, H. 1988. Sexual and asexual *Taraxacum* species. *Biol. J. Linn. Soc.*, **35:** 155-168.
Moran, N.A. 1996. Accelerated evolution and Müller's ratchet in endosymbiotic bacteria. *Proc. Natl. Acad. Sci. USA*, **93:** 2873-2878.
Moran, N. & Baumann, P. 1994. Phylogenetics of cytoplasmically inherited microorganisms of arthropods. *Trends Ecol. Evol.*, **9:** 15-20.
Moritz, C.C., Wright, J.W. & Brown, W.M. 1989. Mitochondrial-DNA analyses and the origin and relative age of parthenogenetic lizards (genus *Cnemidophorus*). III. *C. velox* and *C. exsanguis*. *Evolution*, **43:** 958-968.
Müller, H.J. 1932. Some genetic aspects of sex. *Am. Nat.*, **66:** 118-138.

Müller, H.J. 1964. The relation of recombination to mutational advance. *Mut. Res.*, **1:** 2-9.
Orgel, L.E. & Crick, F.H.C. 1980. Selfish DNA: the ultimate parasite. *Nature*, **284:** 604-607.
Parker, E.D., Jr. 1979. Ecological implications of clonal diversity in parthenogenetic morphospecies. *Am. Zool.*, **19:** 753-762.
Pijls, J.W.A.M., Van Steenbergen, H.J. & Van Alphen, J.J.M. 1996. Asexuality cured: the relations and differences between sexual and asexual *Apoanagyrus diversicornis*. *Heredity*, **76:** 506-513.
Quattro, J.M., Avise, J.C. & Vrijenhoek, R.C. 1991. Molecular evidence for multiple origins of hybridogenetic fish clones (Poeciliidae: *Poeciliopsis*). *Genetics*, **127:** 391-398.
Rossetti, G. & Martens, K. 1996. Redescription and morphological variability of *Darwinula stevensoni* (Brady & Robertson, 1870) (Crustacea, Ostracoda). *Bull. k. belg. Inst. Natuurw. Biol.*, **66:** 73-92.
Roughgarden, J. 1991. The evolution of sex. *Am. Nat.*, **138:** 934-953.
Schön, I., Butlin, R.K., Griffiths, H.I. & Martens, K. 1998. Slow evolution in an ancient asexual ostracod. *Proc. R. Soc. Lond. B*, **265:** 235-242.
Schoen, I., Martens, K. & Rossi, V. 1996. Ancient asexuals – scandal or artefact? *Trends Ecol. Evol.*, **11:** 296-297.
Short, R.V. 1994. Why sex? *in*: Short, R.V. & Balaban, E. (eds), *The differences between the sexes,* Cambridge University Press: 3-21.
Stouthamer, R. & Werren, J.H. 1993. Microbes associated with parthenogenesis in wasps of the genus *Trichogramma*. *J. Invertebr. Pathol.*, **61:** 6-9.
Tétart, J. 1978. Les garnitures chromosomiques des Ostracodes d' eau douce. *Trav. Lab. Hydrobiol.*, **69-70:** 113-140.
Turgeon, J. & Hebert, P.D.N. 1994. Evolutionary interactions between sexual and all-female taxa of *Cyprinotus* (Ostracoda: Cyprididae). *Evolution*, **48:** 1855-1865.
Turgeon, J. & Hebert, P.D.N. 1995. Genetic characterization of breeding systems, ploidy levels and species boundaries in *Cypricercus* (Ostracoda). *Heredity*, **75:** 561-570.
Vandel, A. 1928. La parthénogénèse géographique. *Bull. Biol. Franc. Belg.*, **62:** 164-281.
Van Raay, T.J. & Crease, T.J. 1995. Mitochondrial DNA diversity in an apomictic *Daphnia* complex from the Canadian High Arctic. *Mol. Ecol.*, **4:** 149-161.
Van Valen, L.M. 1973. A new evolutionary law. *Evol. Theory*, **1:** 1-30.
Vrijenhoek, R.C. 1994. Unisexual fish: model systems for studying ecology and evolution. *Ann. Rev. Ecol. Syst.*, **25:** 71-96.
White, M.J.D., Dennis, E.S., Honeycutt, R.L., Contreras, N. & Peacock, W.J. 1982. Cytogenetics of the parthenogenetic grasshopper *Warramaba virgo* and its bisexual relatives. 9. The ribosomal-RNA cistrons. *Chromosoma*, **85:** 181-199.
Williams, G.C. 1975. *Sex and evolution,* Princeton University Press.
Wingstrand, K.G. 1988. Comparative spermatology of the Crustacean Entomostraca. 2. Subclass Ostracoda. *Biol. Skr. Kong. Dansk. Vidensk. Selsk.*, **32:** 1-149.
Zeyl, C. & Bell, G. 1997. The advantage of sex in evolving yeast populations. *Nature*, **388:** 465-468.

CHAPTER 2

SEX DETERMINATION IN NON-MARINE OSTRACODS

I. Schön and K. Martens

Abstract

Sex in ostracods is mostly determined by genetic mechanisms, with males being the heterogamic gender. Skewed sex ratios could be the result of this sex determination system, but other factors (such as syngamy of females with both reproductive modes, sex-specific differences in life history, metapopulation structures and cytoplasmic microorganisms) might be equally important. Chromosomal changes seem to be common in non-marine ostracods, including fusion between autosomes and sex chromosomes which lead to multiple sex chromosomes and could also account for the high amount of sex-linked allozymes. Such instabilities in chromosomal structures can provoke the suppression of male production through chromosome disjunction or loss, and explain how asexuals derive from sexual populations. Chromosomal changes can also elucidate the occurrence of rare males. However, there is no evidence that functional males could be produced by parthenogenetic females, allowing reversal from asexual to sexual production; the high incidence of polyploidy, but also difficulties in restoring all genes coding for the complex structure of giant podocopid spermatozoa, might be related to this issue.

Introduction

What little information is available on sex-determination in crustaceans in general and in ostracods in particular, shows that there is no universal mechanism in these groups. Key factors triggering sex determination in the zygote can be very different, even within one and the same species, (for example in *Musca*, Bownes 1992). Generally, such mechanisms can be divided in two groups, genetic and epigenetic or environmental (Box 2.1).

For the present discussion on the impact of reproductive modes on evolutionary ecology of non-marine ostracods, sex determination mechanisms are of prime importance, as they could not only account for prominent female-biased sex-ratios observed in several natural populations of non-marine ostracods (Chaplin 1992, 1993, Chaplin *et al.* 1994), but could also be an important factor determining the complete disappearance of males from populations, thus effectively causing the origin of clonal spin-offs from sexual populations.

The following discussion deals with non-marine lineages only, and the excellent recent studies on marine myodocopid groups by Moguilevsky (1990, 1995, Moguilevsky & Whatley 1988) and the review of sex determination in marine ostra-

Sex and parthenogenesis: evolutionary ecology of reproductive modes in non-marine ostracods, edited by K. Martens, pp. 25–36

cods by Abe (1990) are used for comparative purposes only. We discuss the available data on individual sex determination, but also offer a brief review on related issues, i.e. skewed sex ratios in populations, rare male occurrences and geographical parthenogenesis. Most importantly, we wish to determine whether any of the particularities in the sex-determination systems in non-marine ostracod lineages can explain the unusually high incidence of asexual reproduction in these groups, as compared to arthropods and crustaceans in general.

Box 2.1 Sex determination in arthropods

In insects, there is no general mode valid for all taxa, but three main *genetic mechanisms* can be distinguished (Wrensch & Ebbert 1993):

1. *Male heterogamety* with chromosome combinations like XO, XY, XXO, XXY, XYY is found in several apterygote and pterygote groups.
2. *Female heterogamety* has been described for Lepidoptera and Trichoptera.
3. *Female arrhenotokous* parthenogenesis with haploid males takes places in Hymenoptera.

The following *epigenetic or environmental factors* are known to play a role in sex determination of crustaceans: temperature, (in copepods, Legrand *et al.* 1987), photoperiod, (in *Gammarus*, Bulnheim 1978), environmental conditions in general, such as availability of food, (in the cyclic parthenogens *(Daphnia)*, Fries 1964) and cytoplasmic factors, including endosymbiotic microorganisms (see chapter 7).

Meiotic drive alters the assortment of chromosomes during meiosis. Chromosomes with meiotic distortion are inherited more frequently than expected. It is most commonly observed in the heterogametic sex, where it alters the sex ratio. It mostly increases the amount of females, although in mosquitoes an excess of males is the result (Wood & Newton 1991).

Haldane's rule (Orr 1997) says that progenies of interspecific hybridisations can show an altered sex ratio, through increased sterility and even mortality of the heterogametic sex. Numerous examples are known from *Drosophila* and *Lepidoptera* (Hurst & Pomiankowski 1991).

Individual genetic sex determination in non-marine ostracods

Although most studies on karyotype morphology and sex determination systems of non-marine ostracods were undertaken in the early 1930s and 40s (e.g. Bauer 1934, 1940, Dietz 1955), they nevertheless still provide essential, and for most species the only, data available. The common pattern appears to be male heterogamy (Box 2.1), with either XO- or XY-derived systems, which are also found in marine ostracods (Abe 1990, Moguilevsky 1990, 1995, Moguilevsky & Whatley 1988). Some non-marine species possess XO systems, e.g. *Cyclocypris globosa*, *Notodromas monacha*, *Scottia pseudobrowniana* (in Dietz 1955), *Cyprideis torosa* (in Tétart 1983) and *Candona candida* (in Tétart 1978). As in *Drosophila* (Wrensch & Ebbert 1993), it is the ratio of X chromosomes to autosomes and not the absolute number of X chromosomes which seems to determine the gender. XY-systems have been reported in *Heterocypris incongruens* (Bauer 1940, Dietz 1955), *Cyclocypris laevis*, *Cyclocypris ovum*, *Cypria exsculpta* and *Cypria ophtalmica* (in Dietz 1955). The presence of one (or more) Y chromosomes does not necessarily imply that this chromosome is 'the trigger' for male development; in *Drosophila*, for example, Y is only essential for spermatogenesis, but not for sex determination *per se* (Wrensch & Ebbert 1993). This could explain why rare males (see chapters 14 and 17) have also been reported in species with XY systems (see below).

The variable number of Y chromosomes in *C. ovum* (in Dietz 1955) can be derived from secondary multiplications of one or more Y chromosomes, as the multiple Y chromosomes are heterochromatic, do not pair with other chromosomes during meiosis and show considerable size differences (Fig. 2.1: 8b, d), which are a typical result of deletions in superfluous chromosomes (Dietz 1955). However, as in a number of cases at least, the origin of additional sex-chromosomes is accompanied by a loss of autosomes, a direct reversal from autosomes to sex-chromosomes, either by losing one of the pairs of autosomes or by fusion, is also possible.

The most recent chromosomal studies on non-marine ostracods by Tétart (1967, 1978, 1983, see Fig. 2.1), focusing mainly on female karyotypes from parthenogenetic populations, revealed two widespread and exceptional results: firstly, an evolution towards multiple sex chromosomes and secondly, a high proportion of metacentric chromosomes which are the result of fusion. Both evolutionary pathways are probably linked, as multiple sex chromosomes are known to derive from fusion between sex chromosomes and autosomes (Legrand *et al.* 1987).

The resulting, multiple sex chromosome systems do not present genuine polygenic systems, as the original mechanism, such as XY or XO, remains. The fusion itself, however, could explain the unusually high amount of sex-linked allozyme loci in non-marine ostracods, whereas in the last review on sex determination in crustaceans, sex-linked genes have only been described from three other crustacean species (Legrand *et al.* 1987). Sex-linkage of specific loci differs considerably between ostracod species (see Table 2.1), further supporting this hypothesis. If the multiple sex chromosomes have partly originated from autosomes which hosted the gene loci of certain allozymes in the past, then these genes would now be found on the new sex chromosomes.

Autosomes can fuse with both sex chromosomes: Dietz (1955) observed a gradual reduction of the number of autosomes through translocations between autosomes and the X chromosome: the length of the X is much shorter in individuals of *Cyclocypris ovum* with 7 autosomes (Fig. 2.1: 8 a, b) than in those with 6 autosomes (8c, d in Fig. 2.1). In *Cypria exsculpta*, he noticed fusion between an autosome and the Y chromosome (Dietz 1955 and Fig. 2.1).

Epigenetic or environmental factors

Environmental factors known to trigger the production of sexual generations in crustaceans with epigenetically determined male production are summarised in Box 2.1, but none of these have thus far been demonstrated experimentally in non-marine ostracods.

The possible effects of endoplasmatic, symbiontic bacteria on sex-determination in non-marine ostracod populations are reviewed and discussed in chapter 7. However, some evidence might already exist, as such bacteria have been found in *Heterocypris incongruens* (chapter 7), the same species in which an irreversible switch from a sexual to an asexual reproducing population was reported within one generation (Wohlgemuth 1914).

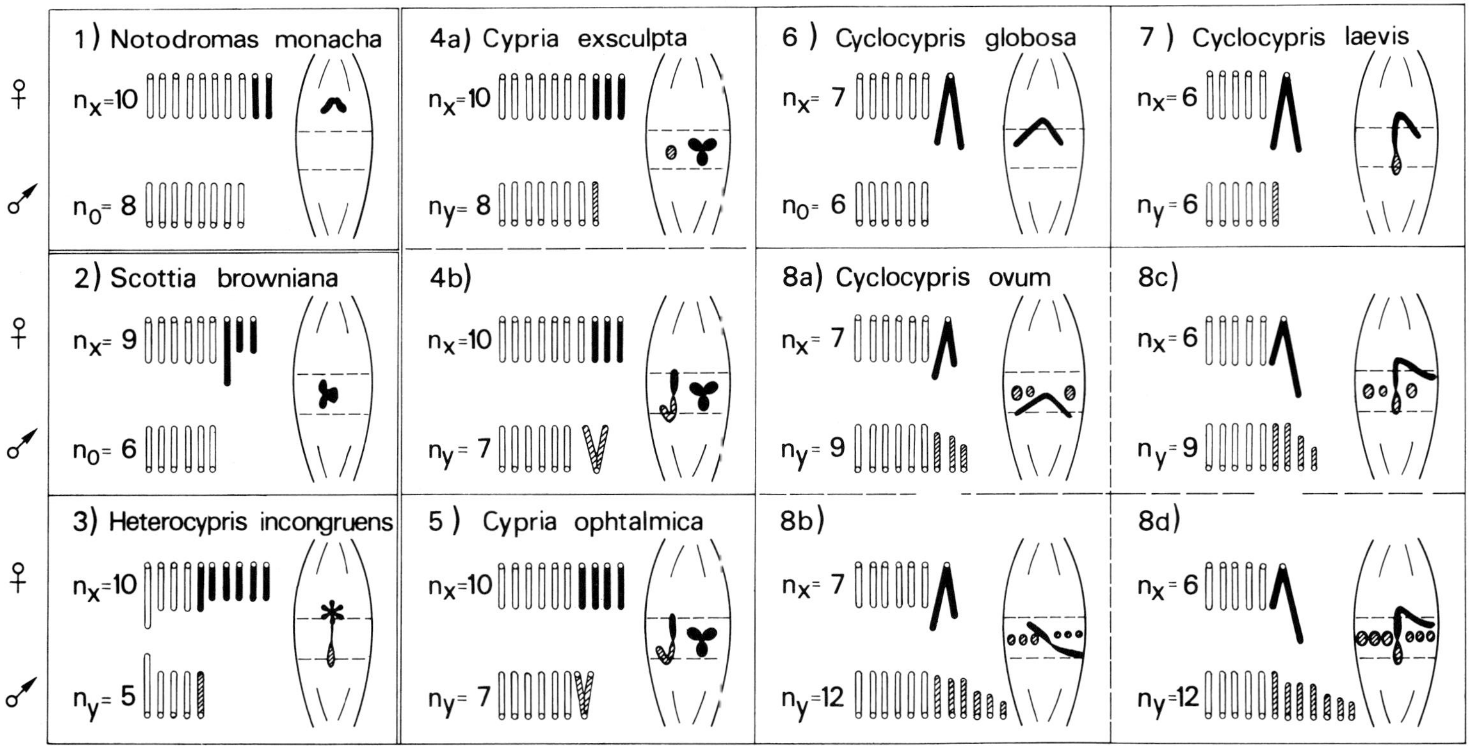

Fig. 2.1 Female and male karyotypes (left) and pairing of sex chromosomes in metaphase (right) for different lineages of Cypridoidea (after Dietz 1955). 1 = Notodromadidae, 2, 3 = Cyprididae, 4-8 = Candonidae – Cyclocypridinae. n_x = number of autosomes + X chromosomes in females, n_y = number of autosomes + Y chromosomes in males, n_o = number of autosomes in males with XO system. (Note that *Scottia browniana* is now *S. pseudobrowniana*). Autosomes are indicated with white colour, X chromosomes with black colour and Y chromosomes are shaded.

Table 2.1 Sex-linkage of allozyme loci in non-marine ostracods.
The loci MPI, PGM and GPI-1 were not sex-linked in the other screened species (Havel *et al.* 1990). In the case of MPI, this included one more (sexual) species, for PGI-1, 5 and for PGM, four other species.

species	allozyme locus	reference
Cyclocypris ampla	MPI	Havel *et al.* 1990
Cyclocypris serena	MPI	Havel *et al.* 1990
Candonocypris novaezelandiae	PGM	Chaplin 1992
Cypretta brevisaepta	MPI	Little & Hebert 1994
Bradleystrandesia splendida	MPI	Turgeon & Hebert 1995
		Havel *et al.* 1990
	PGM	Turgeon & Hebert 1995
		Havel *et al.* 1990
Bradleystrandesia deltoidea	MPI	Turgeon & Hebert 1995
		Havel *et al.* 1990
	PGM	Turgeon & Hebert 1995
		Havel *et al.* 1990
Bradleystrandesia tincta	MPI	Turgeon & Hebert 1995
	PGM	Turgeon & Hebert 1995
Heterocypris glauca	MPI	Turgeon & Hebert 1994,
	GPI-1	Havel *et al.* 1990
	GPI-1	Havel *et al.* 1990

Löffler (1990) links the presence of occasional (rare) males in *Limnocythere inopinata* to high alkalinity levels, but experimental evidence for this is lacking. In chapter 14, the occurrence of such rare males in lab-cultures of clonal females is provisionally linked to crowding effects, as such males were only produced in multi-female experiments, never in single female cultures. The main problem to study such phenomena is the low incidence of their occurrences. Thus far, male production in lab cultures was always incidental, never the result of a set-up aiming to test this particular factor. Epigenetic sex determination in non-marine ostracods remains as yet to be demonstrated.

Skewed gender ratios and rare males

Skewed gender ratios

Chaplin *et al.* (1994) compiled data on sex ratios in extant ostracods and found, on average, a 2:1 female bias. Although the analysis did not show how these ratios differ between sexual populations and populations with mixed reproductive strategies, such a deviation on average is likely to be significant. There are several possible explanations for these skewed sex-ratios.

(1) The most obvious one, but only valid for those taxa with mixed reproduction, is the sympatric occurrence of both sexual and parthenogenetic females together with males in the mixed populations. In that case, the sex ratio is bound to be female-biased.

(2) In circumstances where the presence of parthenogenetic females is very unlikely (for example, Park *et al.* (1997) reported sex ratios of 2:1 (female: male) in species of *Gomphocythere* in the ancient Lake Tanganyika, where sexual reproduction is nearly general amongst the endemic species (Martens 1997)), female excess has been explained by the presence of multiple sex-chromosomes and segregation distorters on the X chromosomes (Chaplin *et al.* 1994, Butlin *et al.* in press), but the relevance of such processes in non-marine ostracods has thus far not been demonstrated.

(3) There are also cases of male excess in a number of investigated populations. This is quite common in species of *Candonopsis* (in Klie 1938), and one species of *Gomphocythere* from Tanganyika had 2:1 male: female ratio (Park *et al.* 1997). Both this, as well as some of the situations with female excess, can be caused by a different life history of both genders, with the majority of individuals of the two sexes reaching maturity at different times of the year, of course with their peaks at least partially overlapping.

(4) The metapopulation structure of ostracod populations (either as conglomerates of populations in separate pools – chapter 13 – or with patchy distribution in seemingly continuous populations, such as ancient lakes – Martens 1997) can cause artificial, skewed sex ratios in single patches, when compared with the ratio in the total metapopulation, but they would nevertheless constitute real situations.

(5) Cytoplasmic symbionts can turn genetic males into phenotypic females (chapter 7), and if not all females of a population are infected, this will cause skewed sex ratios rather than all-female populations.

Rare males

An extreme case of distorted sex ratios is the occurrence of single (rare) males, which have thus far been reported from all three lineages of non-marine ostracods, even from the ancient asexual darwinulids. In the Anostracan *Artemia parthenogenetica*, for example, such rare males generally constitute less than 1% of the population (Browne 1992); but in ostracod populations these ratios are even lower, e.g. in chapter 14, one male in 10,000 females was reported from laboratory cultures. The males in *Artemia* are asexually produced (as in cyclic parthenogens), because they have genetic markers identical to those of their mothers (Browne 1992). However, whereas males are the heterogametic sex in ostracods, they are the homogametic sex (XX) in *Artemia*. Therefore, production of rare males must be the result of different mechanisms in both groups. In *Artemia*, they appear to be the result of the fusion of two haploid X bearing cells after the second meiotic division in automixis (Browne 1992). In non-marine ostracods there are two possibilities. In species with the XO system, the reversal from obligate parthenogenesis to the renewed production of males seems fairly simply and certainly plausible, as it requires only the loss of one (set) of the X chromosomes. If the Y chromosome is required for sex determination (see above), then it seems hard to imagine any mechanism by which Y can be regained once lost. Functionality of such rare males is discussed in chapter 17.

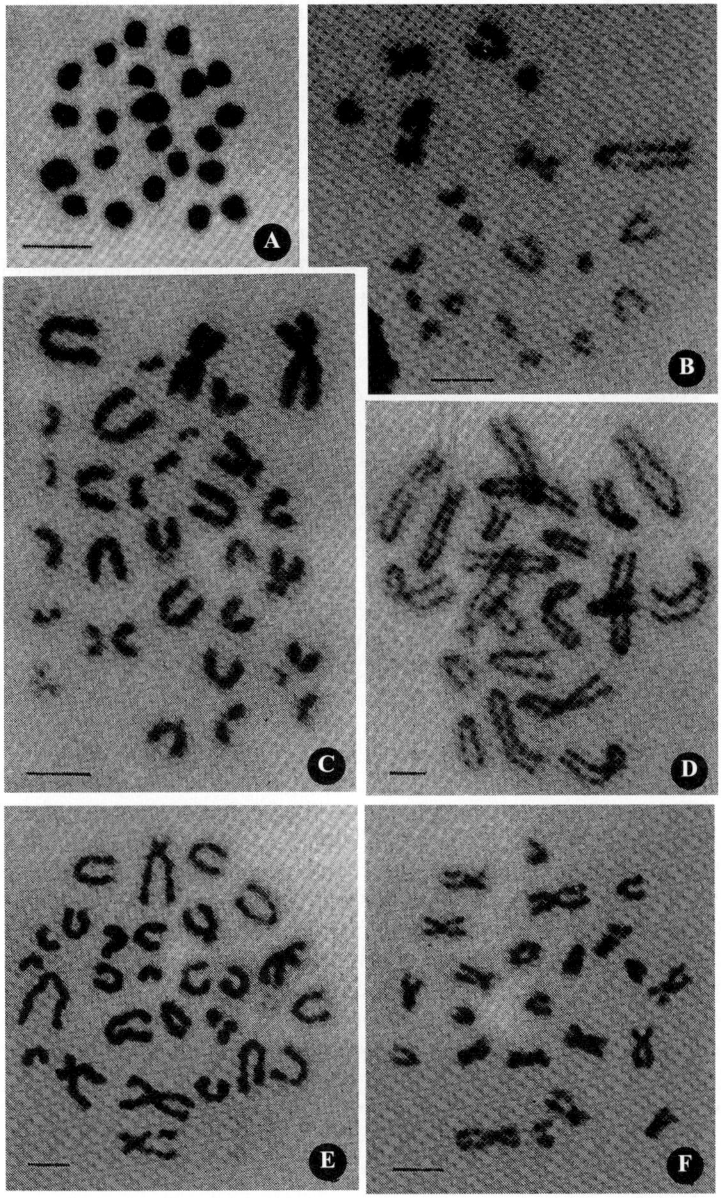

Fig. 2.2 Female karyotypes from different ostracod species, photographed during the metaphase. From Tétart (1978).
A. *Darwinula stevensoni.*
B. *Heterocypris incongruens.*
C. *Eucypris virens.*
D. *Herpetocypris reptans.*
E. *Bradleystrandesia fuscata.*
F. *Prionocypris zenkeri.*

Polyploidy, Haldane's rule and hybridisation

Polyploid females can be generated through multiplication of chromosomes in parthenogenetic lineages, e.g. by non-disjunction during mitosis. A tetraploid clone of *Heterocypris incongruens* could have originated that way (Turgeon & Hebert 1994), but could as well be the result of hybridisation, the second mechanism which can cause polyploidy. Indeed, most polyploid clones known so far seem to be the result of hybridisation events, either intra- or interspecifically, of males with parthenogenetic females: the haploid spermatozoon would fuse with an (at least) diploid apomictically produced egg cell, giving rise to triploidy or higher levels of polyploidy. Both inter- and intraspecific hybridisation have been reported in non-marine ostracods, the former e.g. for *Bradleystrandesia tincta*, *B. fuscata* and *B. reticulata* (in Turgeon & Hebert 1994), the latter for *Heterocypris incongruens* (in Turgeon & Hebert 1995) and *Eucypris virens* (chapter 15, 16). Thus, Haldane's rule (see Box 2.1) could present an additional explanation for the bias of male ratios, as in interspecific hybridisation it is always the heterogametic sex which has increased sterility or mortality or even caused all-female populations, if male mortality is absolute.

High polyploidy levels could act as barriers against reversal to sexual reproduction, as the required mechanisms would increase in complexity and chromosome disjunction during meiosis would be distorted. Polyploidy frequencies can be very high in Cyprididae, up to 100% in some taxa (chapter 15), so that this mechanism could contribute significantly to skewed sex ratios.

Giant spermatozoa

Since the impressive review by Wingstrand (1988), the morphology of spermatozoa of the extant ostracod lineages is relatively well known: eight very different types are reported from the 3 lineages with Recent representatives (Myodocopida, Platycopida, Podocopida) and can actually be used as diagnostic features at ordinal and family levels (Fig. 2.3). Wingstrand (loc.cit.) called this variation 'spectacular', when compared to extant variation in other Crustacean groups. The spermatozoa in Cypridoidea were known as the longest in the Animal Kingdom, they are between 0.2 and 1.5 cm long and can thus be 10 times longer than the animal. Recently, however, some heteropteran insects have been found to have longer spermatozoa, namely the waterbugs *Notonecta undulata* (16 mm) and *N. glauca* (12 mm) (references in Wingstrand 1988), but this still means that cypridinid ostracods have the longest spermatozoa in the Animal Kingdom *relative* to their body size. Butlin *et al.* (in press) postulate that these larger spermatozoa could be due to skewed sex ratios. Where there is an excess of females, sperm competition will be rare and males may rather invest in a few large spermatozoa than in many small spermatozoa (Butlin *et al.* in press). This hypothesis remains to be tested.

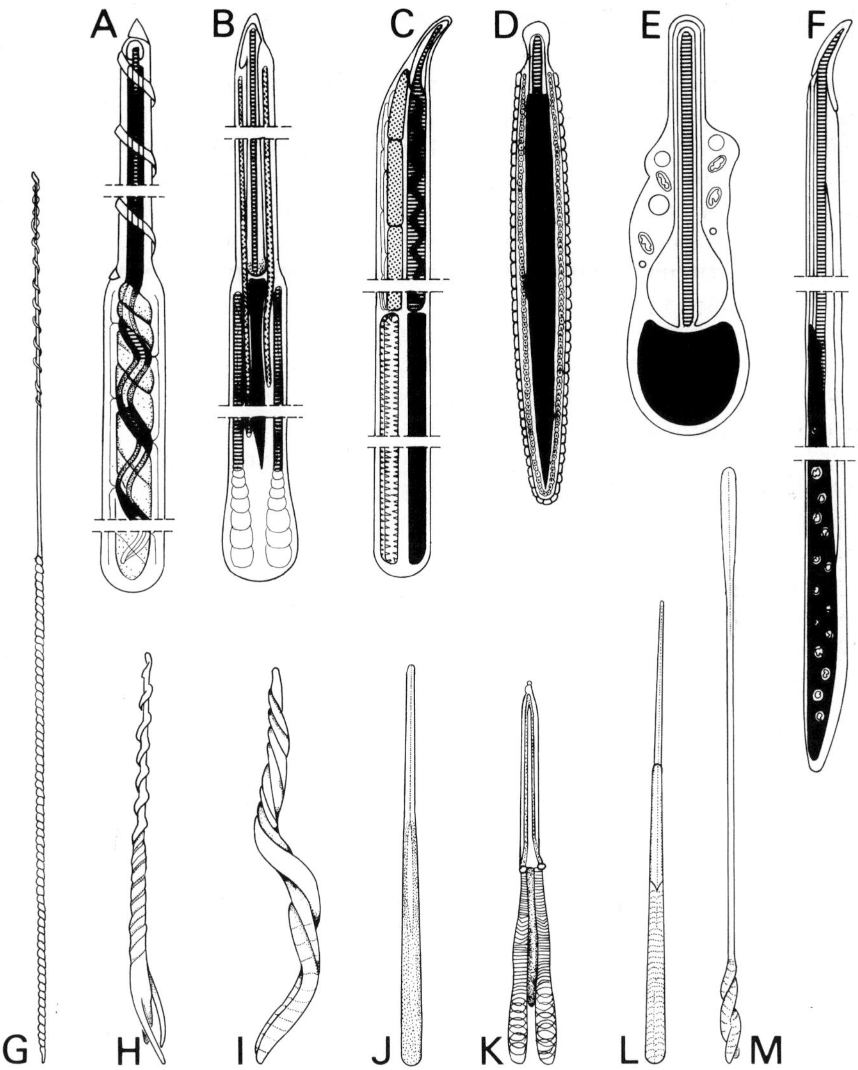

Fig. 2.3 Main types of spermatozoa in non-marine ostracods.
Diagrams (A-F) and illustrations redrawn from SEM images (G-M) of ostracod spermatozoa (all redrawn after Wingstrand 1988).

A. Cypridoidea (Podocopida).
B. Cytheroidea (Podocopida).
C. *Cytherella* (Platycopida).
D. *Polycope* (Myodocopida).
E. *Sarsiella* (Myodocopida).
F. *Conchoecia* (Myodocopida).
G. *Candona* (Cypridoidea, Podocopida).
H. *Elofsonia* (Cytheridoidea, Podocopida).
I. *Xestoleberis* (Cytheroidea, Podocopida).
J. *Paralimnocythere* (Cytheroidea, Podocopida).
K. *Xestoleberis* (Cytheroidea, Podocopida).
L. *Loxoconcha* (Cytheroidea, Podocopida).
M. *Hemicythere* (Cytheroidea, Podocopida).

Conclusions – high incidence of asexuality in non-marine ostracods

Why parthenogenesis is so exceptionally common in (non-marine) ostracods, as compared to other arthropod groups, is still unclear. The fact that sex seems to be primarily genetically determined certainly facilitates the suppression of male production. Furthermore, as the common pattern of sex determination in ostracods is male heterogamy, a non-disjunction of the two or more X chromosomes during meiosis in XO systems (or polyploidy) will be sufficient to cause female sex in the zygote. This could happen quite regularly: ostracod karyotypes generally vary to a remarkable degree, even within one species or between closely related species (Fig. 2.1), and although we cannot link this observation to any obvious genetically-based feature, it could nevertheless indicate that chromosome systems in ostracods are less stable than those of other crustaceans. The high frequency of chromosome fusion (Tétart 1978) strongly supports this view.

Furthermore, cyclic parthenogenesis does not occur in non-marine ostracods; all-female populations generally remain in their reproductive mode, with the exception of the occurrence of rare males, which are furthermore generally deemed atavistic and non-functional (see discussion in chapter 17). The genetic system simply does not allow this flexibility, as a reversal from female to male karyotype would require the change from X to a Y chromosome (in XY systems) or a loss of all X chromosomes (in XO systems) which leads to the death of that zygote. A reversal from all-female to mixed reproduction will be even more difficult. Functional males would have to be produced, which means that also the genes required for spermatogenesis have to be 're-established'. Taking the complex and special structure of ostracod spermatozoa into account (Fig. 2.3), they are almost certainly coded by several, and not one, genes. Such reversals to sexual reproduction could thus be more difficult in ostracods than in other crustacean groups.

Finally, why mechanisms of environmental sex determination are so limited in ostracods, remains an open question.

The apparent inability (or at least, the very low probability) of asexual females to revert to the production of both females and males, enables at least the Cyprididae as a group and some isolated species of Limnocytheridae, to have distribution patterns reminiscent of geographical parthenogenesis (Vandel 1928, chapter 5). Unlike what has been claimed (McKenzie 1971), the patterns in ostracods do not entirely conform to geographical parthenogenesis as originally defined, because sexuality is indeed confined to circum-Mediterranean areas, but parthenogenesis occurs in the entire area, sometimes even in the same populations with sexuals (chapters 5, 15). Species showing this distribution pattern thus have two types of females, sexual and parthenogenetic ones, and therefore, again, reversals from asexual to sexual reproduction in one individual must be very rare. It is possible that in groups which do not show geographical parthenogenesis (which is most notably the case in the European Candonidae), reversals between sexual and asexual reproduction in females are more common, making competitive exclusion of either reproductive mode in certain areas impossible (Horne & Martens in press, chapter 17). However, there are no molecular or karyological peculiarities in Candonidae thus far known, which could offer support to this hypothesis.

References

Abe, K. 1990. What the sex ratio tells us: a case from marine ostracods. *in*: Whatley, R. & Maybury, C. (eds), *Ostracoda and Global events*, Chapman and Hall: 175-185.

Bauer, H. 1934. Geschlechtschromosomen bei Ostracoden. *Naturwissenschaften*, **22:** 543.

Bauer, H. 1940. Über die Chromosomen der bisexuellen und der parthenogenetischen Rasse des Ostracoden *Heterocypris incongruens* RAMD. *Chromosoma*, **1:** 620-637.

Browne, R.A. 1992. Population genetics and ecology of *Artemia*: insights into parthenogenetic reproduction. *Trends Ecol. Evol.*, **7:** 232-237.

Bownes, M. 1992. Molecular aspects of sex determination in insects. *in*: Crampton, J.M. & Eggleston, P. (eds), *Insect Molecular Science*, Academic Press: 76-100.

Bulnheim, H.P. 1978. Variability of the modes of sex determination in littoral Amphipoda. *in*: Battaglia, B. & Beardmore, J.A. (eds), *Marine Organisms*, Plenum Press: 529-548.

Butlin, R.K., Schön, I. & Martens, K. (in press). Asexual reproduction in non-marine ostracods. *Heredity.*

Chaplin, J.A. 1992. Variation in the mode of reproduction among individuals of the ostracod *Candonocypris novaezelandiae. Heredity*, **68:** 411-424.

Chaplin, J.A. 1993. The local displacement of a sexually reproducing ostracod by a conspecific parthenogen. *Heredity*, **71:** 259-268.

Chaplin, J.A., Havel, J.E. & Hebert, P.D.N. 1994. Sex and ostracods. *Trends Ecol. Evol.*, **9:** 435-439.

Dietz, R. 1955. Zahl und Verhalten der Chromosomen einiger Ostracoden. *Z. Naturforsch., Teil b*, **10:** 92-95.

Fries, G. 1964. Über die Einwirkung der Tagesperiodik und der Temperatur auf die Generationswechsel, die Weibchengröße und die Eier von *Daphnia magna. Z. Morphol. Ökol. Tiere*, **53:** 475-516.

Havel, J.E., Hebert, P.D.N. & Delorme, L.D. 1990. Genetics of sexual Ostracoda from a low arctic site. *J. evol. Biol.*, **3:** 65-84.

Horne, D.J. & Martens, K. (in press). Geographical parthenogenesis in European non-marine ostracods (Crustacea, Ostracoda); post-glacial invasion or Holocene stability? *Hydrobiologia.*

Hurst, L.D. & Pomiankowski, A. 1991. Causes of sex ratio bias may account for unisexual sterility in hybrids: a new explanation of Haladane's rule and related phenomena. *Genetics*, **128:** 841-858.

Klie, W. 1938. Krebstiere oder Crustacea. III: Ostracoda, Muschelkrebse. *in*: Dahl, F. (ed), *Die Tierwelt Deutschlands und der angrenzenden Meeresteile,* Gustav Fisher: 230 pp.

Legrand, J.J., Legrand-Hamelin, E. & Juchault, P. 1987. Sex determination in Crustacea. *Biol. Rev.*, **62:** 439-470.

Little, T.J. & Hebert, P.D.N. 1994. Abundant asexuality in tropical freshwater ostracodes. *Heredity*, **73:** 549-555.

Löffler, H. 1990. Palaeolimnology of Neusiedlersee, Austria. I. The succession of ostracods. *Hydrobiologia*, **214:** 229-238.

McKenzie, K.G. 1971. The palaeozoogeography of freshwater Ostracoda. *in*: Oertli, H.J. (ed), *Paléoécologie des Ostracodes, Bull. Centr. Res. Pau-SNPA Suppl.,* **5:** 207-238.

Martens, K. 1997. Speciation in ancient lakes. *Trends Ecol. Evol.*, **12:** 177-182.

Moguilevsky, A. 1990. Cytogenetic studies on marine myodocopid Ostracoda: the karyotypes of some species of *Vargula* Skogsberg, 1920. *in*: Whatley, R. & Maybury, C. (eds), *Ostracoda and Global events*, Chapman and Hall: 187-208.

Moguilevksy, A. 1995. Cytogenetic studies on marine mydocopid Ostracoda: the karyotypes of some species of halocyprids. *J. micropalaentology*, **14:** 81-95.

Moguilevksy, A. & Whatley, R. 1988. Cytogenetic studies on marine myodocopid Ostracoda: the karyotypes of *Gigantocypris dracontovalis* Cannon, 1940 and *Macrocyprididina castanea* (Brady, 1897). *in*: Hanai, T., Ikeya, N. & Ishizaki, K. (eds), *Evolutionary biology of ostracoda, its fundamentals and applications*, Elsevier: 293-300.

Orr, H.A. 1997. Haldane's rule. *Ann. Rev. Ecol. System.*, **28:** 195-218.

Park, L.E., Cohen, A.S., Alin, S. & Martens, K. 1997. Sex ratio bias in freshwater ostracods: an example from Lake Tanganyika, East Africa. *Abstracts I.A.G.L.R.*

Tétart, J. 1967. Etude des garnitures chromosomiques de quelques ostracodes d'eau douce. *Bull. Soc. Zool. Franc.*, **92:** 167-177.

Tétart, J. 1978. Les garnitures chromosomiques des Ostracodes d'eau douce. *Trav. Lab. Hydrobiol. Univ. Grenoble*, **69-70:** 113-140.

Tétart, J. 1983. Les caractéristiques chromosomiques de *Cyprideis littoralis* Brady. *Sci. Geol. Bull.*, **36:** 221-222.
Turgeon, J. & Hebert, P.D.N. 1994. Evolutionary interactions between sexual and all-female taxa of *Cyprinotus* (Ostracoda: Cyprididae). *Evolution*, **75:** 561-570.
Turgeon, J. & Hebert, P.D.N. 1995. Genetic characterization of breeding systems, ploidy levels and species boundaries in *Cypricercus* (Ostracoda). *Heredity*, **75:** 561-570.
Vandel, A. 1928. La parthénogénèse géographique. *Bull. Biol. Franc. Belg.*, **62:** 164-281.
Wingstrand, K. G. 1988. Comparative spermatology of the Crustacean Entomostraca. 2. Subclass Ostracoda. *Biol. Skr. Kong. Dansk. Vidensk. Selsk.*, **32:** 1-149.
Wohlgemuth, R. 1914. Beobachtungen und Untersuchungen über die Biologie der Süßwasserostracoden; ihr Vorkommen in Sachsen und Böhmen, ihre Lebensweise und ihre Fortpflanzung. *Intern. Rev. Hydrobiol. Hydrograph. Biol. Suppl.*, **6:** 1-72.
Wood, R.J. & Newton, M.E. 1991. Sex-ratio distortion caused by meiotic drive in mosquitoes. *Amer. Natur.*, **137:** 379-391.
Wrensch, D.L. & Ebbert, M.A. 1993. *Evolution and diversity of sex ratios in insects and mites,* Chapman & Hall: 630 pp.

CHAPTER 3

AGE AND DIVERSITY OF NON-MARINE OSTRACODS

K. Martens, D.J. Horne and H. I. Griffiths

Abstract

Non-marine Ostracoda belong to three main lineages within the Podocopina which have been separate from each other for anything between 360 and 450 million years, and it can thus be expected that these lineages have developed significantly different biological strategies. Even within the Cytheroidea, multiple and independent invasions of non-marine habitats have given rise to unrelated and convergent cytheroid radiations in these habitats, although most taxa in Cytheroidea are marine. Darwinuloidea are exclusively non-marine to date, but this ancient asexual lineage has low extant diversities. Cypridoidea have radiated widely in non-marine habitats; the Cyprididae are by far the most diverse ostracod family in continental waters. Most supra-generic lineages already existed in the Mesozoic, and maybe even originated in the Palaeozoic.

Introduction

The present chapter gives an introductory overview of the classification of the Ostracoda in general and of non-marine ostracods in particular. It is clear that this subject can only be treated in the broadest of terms, as a detailed treatise would require several volumes of its own. Special attention will nevertheless be devoted to age and taxonomic position of the species and genera that are used as test-organisms in subsequent chapters.

History and phylogeny of ostracods

The Ostracoda have the most complete fossil record of any extant group of arthropods, largely thanks to two factors: the presence, in most, of calcified, bivalved shells and their small size (Moore 1961). Calcite shells have a high potential for preservation in sedimentary deposits, while their small size (length c 1 mm or less) means that the resulting fossils can be recovered from sedimentary rocks in large quantities with relative ease, using standard micropalaeontological processing techniques. Analysing several hundreds of specimens per sample is the rule, rather than the exception. Furthermore, ostracod valves tend to be preserved in exquisite detail, especially in clays, shales and mudstones.

Sex and parthenogenesis: evolutionary ecology of reproductive modes in non-marine ostracods, edited by K. Martens, pp. 37–55

Most sources put the origin of the Ostracoda in the Cambrian (c 540 million years ago – Box 3.1), but there is a growing consensus that the Cambrian order Bradoriida, which have phosphatic rather than calcite shells, are not ostracods. Recent work on fossils with phosphatised appendages has shown that the Bradoriida are not ostracods, and might not even belong to the Crustacea s.s. (Xianguang *et al.* 1996). Earlier work already suggested the same for the second group of Cambrian 'Ostracoda', the Phosphatocopina (Müller 1982). The controversy is not solved and the position of the Leperditicopida, and maybe even the Palaeocopida, within the Ostracoda is still debated. Related to this, a new discussion on the position of the Ostracoda within the Crustacea was provoked by controversial molecular data (Spears & Abele 1997), but this is outside of the scope of the present book.

Box 3.1 Simplified Phanerozoic time scale (after Odin 1994)

Era	Period	Epoch	Age (My)
			Recent
Cainozoic	Quaternary	Holocene	0.01
		Pleistocene	1.75
	Tertiary: Neogene	Pliocene	5.3
		Miocene	23.0
	Tertiary: Palaeogene	Oligocene	33.7
		Eocene	53
		Palaeocene	65
Mesozoic	Cretaceous	Upper	96
		Lower	135
	Jurassic	Upper	154
		Middle	175
		Lower	203
	Triassic	Upper	230
		Middle	240
		Lower	250
Palaeozoic	Permian		295
	Carboniferous		355
	Devonian		408
	Silurian		435
	Ordovician		500
	Cambrian		540

Our best information at present thus states that true ostracods first appeared in the Ordovician, about 500 million years ago. It is important to note that the five presently acknowledged orders of Ostracoda all originated during this geological period, i.e. about half a billion years ago and in a time span of about 50 million years (Fig. 3.1, Maddocks 1982). All evolution after that gave rise to lower level taxonomies only.

Two of these five orders are extinct: the Leperditicopida disappeared at the end of the Devonian, c 360 million years ago; the Palaeocopida became extinct at the end of the Triassic (c 200 million years ago). The three remaining orders all have extant representatives.

The Myodocopida are exclusively marine, many are pelagic and live in the open ocean; some species can attain lengths of 3 cm. The Myodocopida generally have weakly or non-mineralised valves and consequently this group has a poor fossil record.

The Platycopida and the Podocopida, on the other hand, both have shells impregnated with calcium carbonate and these hard parts are easily fossilised. The Platycopida are also exclusively marine and have relatively few extant representatives.

The Podocopida have undergone the most extensive post-Palaeozoic evolutionary radiation and have maintained their position as the most diverse group of Ostracoda since the early Mesozoic (Box 3.1). The living taxa of this group also inhabit a much wider range of environments than the other, exclusively marine, orders: they are found in deep oceans, shelf seas, estuaries, lagoons, salt marshes, athalassic freshwater and saline lakes, temporary ponds, streams, springs, groundwater, bromeliad water pouches, (semi-) terrestrial habitats such as damp mosses

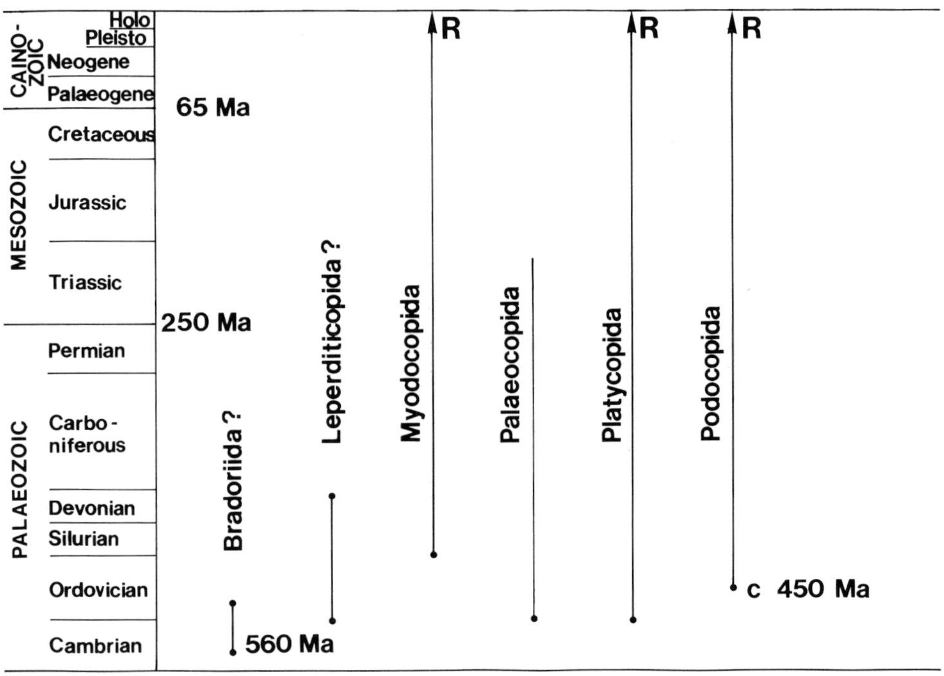

Fig. 3.1 Origin of the different orders of the Ostracoda. Bradoriida and Leperditicopida are uncertain (Based on Maddocks 1982).

and leaf litter in tropical forests; some are parasitic or commensal on echinoderms, molluscs and even on gills of other crustaceans.

The Podocopida are the only order with non-marine representatives; most of the remainder of this book will thus deal with lineages in this group only.

The taxonomy of the Podocopida has been subject to extensive discussion during the past decades. Most of the controversy focuses on the existence and the actual definition of the suborder Metacopina Sylvester-Bradley, 1961, which comprises three Palaeozoic superfamilies, of which only the Healdioidea reached into the Mesozoic, to disappear in the Cretaceous. One school maintains that the Metacopina died out with the Healdioidea, and maintain five closely allied lineages (ranked as infraorders in Podocopina by Martens (1992), but cited here as superfamilies) within the Podocopina: Sigillioidea, Darwinuloidea, Cypridoidea, Cytheroidea and Bairdioidea (Danielopol 1972, Maddocks 1982) (Fig. 3.2). Another school argues that *Saipanetta* is a living Healdioidean, and therefore that only the latter two groups are genuine Podocopina, while the former three must be regarded as extant representatives of the Metacopina (McKenzie 1982 and others). Meanwhile, even Healdioidea seem to be removed from the Metacopina (Lethiers, pers.com.). Maddocks (1976), however, stressed the podocopine aspects of *Saipanetta* and, although she maintained all five lineages in the Podocopina, Maddocks (1982) agreed to a closer relationship between the first two lineages. The Bairdioidea are then considered the oldest Podocopina, which gave rise to the Cytheroidea, whilst the Sigillioidea are an independent branch, which in turn gave rise, again independently, to the Cypridoidea and the Darwinuloidea. Other authors, however, consider the evidence for such grouping

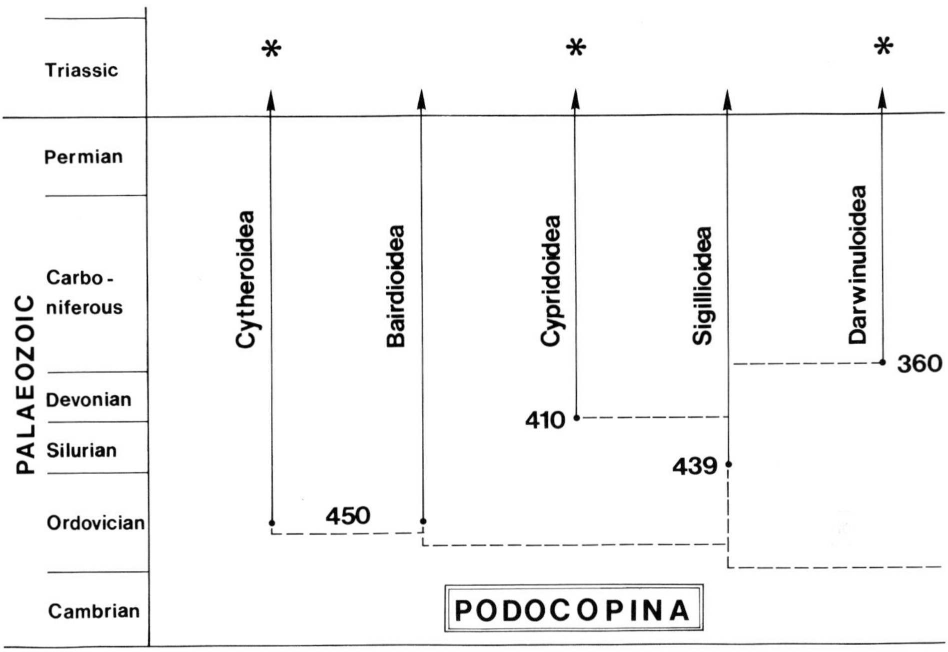

Fig. 3.2 Origin of lineages in the Podocopina. All five indicated as superfamilies, but considered infraorders by Martens (1992). (Based on Maddocks 1982)

insufficient, and retain five parallel lineages, without any preference with regard to mutually closer affinities (Danielopol 1976).

Of these five lineages, Sigillioidea and Bairdioidea are exclusively marine; Cytheroidea are largely marine, but have some important non-marine radiations; Cypridoidea are mostly non-marine with some (ancestral or reversed, this is not certain) radiations in marine environments; the Darwinuloidea, finally, are presently exclusively non-marine.

It is important to stress that the three non-marine lineages are spread over different infraorders, maybe even suborders, in order to illustrate the potential phylogenetic distances between them. Non-marine ostracods cannot be regarded as one group; they consist of three different lineages, which have been apart for 450 to 360 million years, and which invaded non-marine habitats independently of each other, most likely repeatedly in different circumstances. It should thus not be surprising that they have developed different biological strategies to cope with life in continental water bodies.

As all non-marine representatives in these lineages are always in the nominate superfamilies (Cytheroidea ...) of the three infraorders, we use these levels to indicate the three lineages in the present book (Box 3.2).

Box 3.2 Taxonomic levels

The following table illustrates hierarchical ranking of the main ostracod lineages discussed in the present chapter. * indicates a group with non-marine representatives.

Class	Ostracoda
Order 1	Leperditicopida?
Order 2	Myodocopida
Order 3	Palaeocopida
Order 4	Platycopida
Order 5	Podocopida
Suborder 1	Metacopina
Suborder 2	Podocopina
Infraorder 1	Cytherocopina*
Infraorder 2	Bairdiocopina
Infraorder 3	'Sigillioidea'
Infraorder 4	Darwinulocopina*
Infraorder 5	Cypridocopina*
Superfamily 1	Macrocypridoidea
Superfamily 2	Pontocypridoidea
Superfamily 3	Cypridoidea*
Family 1	Ilyocyprididae*
Family 2	Notodromadidae*
Family 3	Candonidae*
Family 4	Cyprididae*

Note: Super-familial ending -oidea, as in Cytheroidea, is used in accordance with the recommendations of the International Code for Zoological Nomenclature. In palaeontological literature, the name-endings -acea are commonly used, e.g. as in Cytheracea, but as ostracods are animals, we will follow the ICZN throughout this book.

Age and phylogeny of non-marine ostracods

Cytheroidea

The Cytheroidea originated c 450 million years ago, but show a significant invasion of non-marine habitats only after the Permian-Triassic extinction. Most non-marine Cytheroidea are in the family Limnocytheridae, of which Whatley & Moguilevsky (in press) situate the origin, through the subfamily Tomiellinae, in the Carboniferous (c 300-350 Myr), but the extant subfamilies, Timiriaseviinae and Limnocytherinae, only originated around or directly after the Permian-Triassic division and show moderate, but significant, radiation in non-marine habitats from the mid-Mesozoic onwards. The Timiriaseviinae was one of the most successful groups of non-marine ostracods during the Mesozoic and still has several extant genera (Colin & Danielopol 1980). The Limnocytherinae are the most speciose cytherid non-marine ostracod group at present, with at least three successful lineages (Danielopol *et al.* 1990).

The extant genus *Limnocythere sensu stricto* (in its strictest sense) is at least 150 million years old, since the Bathonian (Middle Jurassic) species *Limnocythere incerniculum* in all probability belongs to this genus (Wakefield 1994). The type species itself, *L. inopinata*, is most likely of Pleistocene age (c 2 Myr – Diebel & Pietrzeniuk 1975). The age and history of the other cytheroid non-marine lineages (see below) are far less known.

Cypridoidea

Maddocks (1982) placed the origin of the Cypridoidea in the Devonian (c 400 Myr). For a long time, it was accepted that the first non-marine cypridoideans were of Bathonian (Middle Jurassic) age (Whatley 1990a, b, 1992), but widely dispersed freshwater faunas were present in the late Palaeozoic (Late Carboniferous-Early Permian), and some of these might have been genuine Cypridoidea (Lethiers & Damotte 1993). In fact, some of the valves of non-marine forms described from that era strikingly resemble the present-day Cypridoidea (and are thus reported as *Candona*, *Cypridopsis*, etc.). However, most likely these cypridoid forms in the Palaeozoic were more closely related to the speciose Carbonitoidea (Sohn 1985), or even the Geisinidae (Moore 1961). Whether or not genuine Cypridocopina were present in the Palaeozoic thus depends on whether or not Carbonitoidea belong in this infraorder. Lethiers & Damotte (loc. cit.) also provided evidence for the fact that certain biological strategies (drought-resistant eggs and brooding, although the latter in the Darwinulocopina only) were already present in Permian non-marine faunas.

Nevertheless, the explosive radiation of the Cypridoidea in continental habitats occurs only from the Bathonian onwards (Middle Jurassic – c 165 million years ago – Fig. 3.3). Most of the newly-formed specific diversity in that period must be attributed to the now extinct, enigmatic, *Cypridea*-lineage, but several representatives of extant cypridid lineages were already present at that stage, lurking in the background of *Cypridea*-dominated communities, much as the small, early mammals were biding their time while large contemporary dinosaurs were still ruling the Earth (Horne & Martens in press and chapter 6).

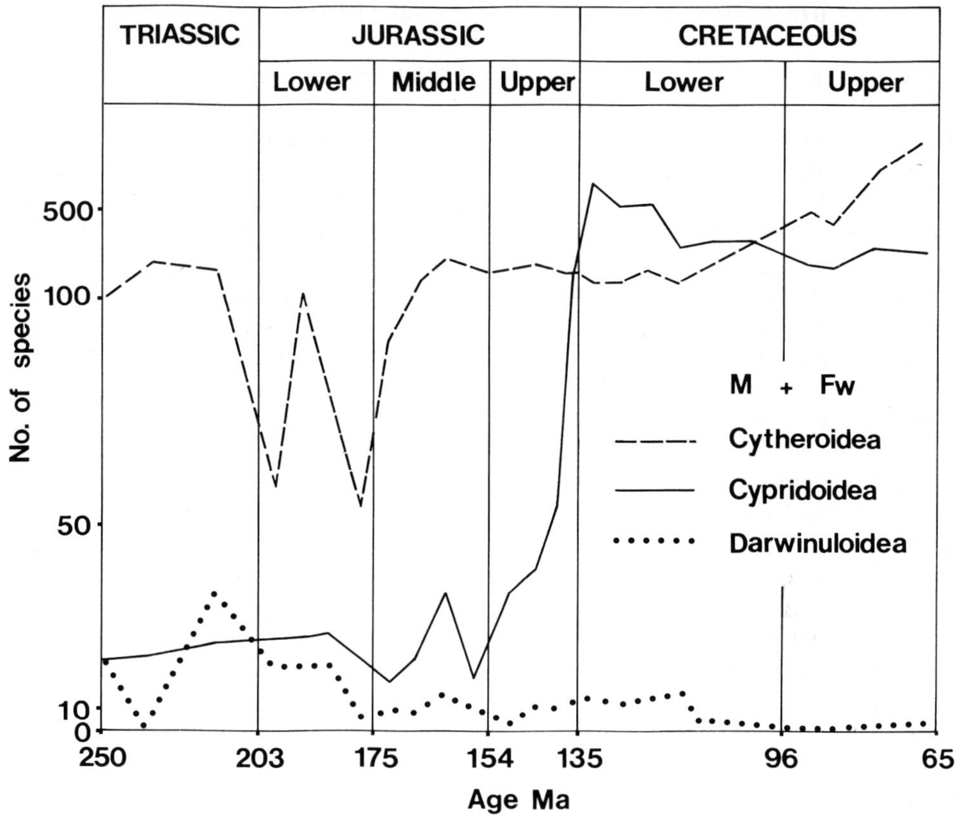

Fig. 3.3 Explosive radiation of Cypridoidea in the Upper Jurassic/Lower Cretaceous (after Whatley 1992). M = marine, FW = non-marine.

After the Cretaceous-Tertiary mass extinction (which wiped out the dinosaurs) had also caused the disappearance of most of the *Cypridea*-lineage, the direct ancestors of present-day faunas radiated widely, giving rise to diverse faunas of Cyprididae primarily in the southern, and of Candonidae primarily in the northern hemisphere (see below). Other families of the Cypridoidea, such as Notodromadidae and Ilyocyprididae, remained far less speciose.

The genus *Eucypris* is first nominally cited in the Upper Cretaceous (c 80 million years ago – Moore 1961), but it is doubtful that this Cretaceous taxon actually belongs to *Eucypris sensu stricto*. The oldest record of *E. virens* is of Pleistocene age (max. 1 Myr, Griffiths 1995).

Darwinuloidea

The Darwinuloidea originated c 360 million years ago, but like the other lineages, also the early darwinulids were exclusively marine. This group had its highest supra-specific (Sohn 1987, 1988) and specific diversity in the Permian (100-200 species in

Darwinula sensu lato (in its broadest sense) Kempf (1980), compared to about 28 living taxa, Rossetti & Martens 1998). Incidentally, this period of higher diversity coincided with the presence of sexuality in at least a number of groups in this lineage (Abushik 1990), although to relate dimorphisms in often poorly preserved assemblages to sexual differences in morphology is not always straightforward (see chapter 6).

The Darwinuloidea were significantly decimated by the Permian-Triassic extinction and have never re-established similar diversities since. The genus *Darwinula* could be as old as 350 million years, although Molostovskaja (in Abushik 1990) places most Palaeozoic *Darwinula sensu lato* in the genus *Palaeodarwinula*. If *Darwinula sensu stricto* is post-Palaeozoic, then it is at most 250 million years old, which is still exceptionally old, especially for a fully asexual genus. The type species, *D. stevensoni*, and *Microdarwinula zimmeri* are at least 20 million years old (Straub 1952).

Conclusion: invasion of continental habitats

The three extant lineages with non-marine representatives all originated in marine environments. Only at later stages in their evolution were non-marine habitats invaded and did exclusively non-marine lineages come into existence. The first massive invasion of non-marine waters occurred in the Silurian (408-435 Myr), but ostracods seem to have missed out on this first wave of continental invasion (Eldredge 1994). The first invasion of non-marine habitats by ostracods occurred in the Lower Carboniferous, which led to a large radiation of non-marine ostracods in the Upper Carboniferous, mostly comprising Carbonitoidea and Darwinulocopina. Whether or not genuine Palaeozoic Cypridocopina existed, depends on the position of the Carbonitoidea (Sohn 1985); the existence of Palaeozoic Cypridoidea depends on the taxonomic position of the Permian taxa presently referred to as *Candona*, *Cypridopsis*, etc. This revision is still in progress.

Pre-Tertiary origin of extant subfamilies

That most present-day, supra-generic lineages (families, subfamilies, tribes) already existed in the Mesozoic is first of all indicated by the superficial resemblance between present day faunas and Mesozoic or even Palaeozoic taxa (see above). It is further exemplified by a comparison of African and South American non-marine ostracod faunas (Martens 1998). Both continents started to drift apart during the Cretaceous and were certainly separated by the beginning of the Tertiary (c 65 million years ago) and although several lineages of non-marine ostracods are potentially capable of intercontinental dispersal (McKenzie & Hussainy 1968), successful intercontinental colonisation must have remained a rare event. Yet, only few supra-generic endemics occur on both continents. The Entocytheridae occur in North and South America, but not in Africa. Its nominal subfamily is therefore also absent from the Ethiopian Realm, as is the Neotropical endemic subfamily Rudjakoviellinae. Africa has only one subfamily absent from South America: the Megalocypridinae have only few representatives on the Indian subcontinent, and could therefore be regarded as Ethiopian endemics. Another 20 subfamilies have representatives on both continents, and these must already have existed before Africa and South America drifted apart.

Further evidence for Mesozoic (or earlier) origin of (sub-) families is provided by recent research on fossils with phosphatised appendages of the Cretaceous genus *Pattersoncypris* (Bate 1972, Smith unpubl.). Its striking resemblance in detailed soft part anatomy to the present-day Eucypridini demonstrates a long-term (100 million years) morphological stability and shows that certainly the subfamily Eucypridinae, if not also the nominate tribe, is at least of Mesozoic age.

Extant diversity

Figure 3.4 summarises numbers of species of the main non-marine lineages on three continents, namely Africa, South America and Europe. For North America, Asia and Australia, no recent comprehensive checklists are available. It appears that Cypridoidea have an overwhelming taxonomic success and that Darwinuloidea are at all times a minor component of the extant biodiversity in non-marine habitats. The family Cyprididae is more successful in the southern than in the northern hemisphere, while the Candonidae constitute the largest fraction of the 'other Cypridoidea' in Europe. Taxonomic diversity of representatives of the three lineages will now briefly be discussed, mainly to illustrate the taxonomic position of the organisms discussed in the remainder of this book.

Cytheroidea-Limnocytheridae

The largest radiation of Cytheroidea in non-marine habitats, and the one that concerns us most here, is in the family Limnocytheridae, which consists of two subfamilies: the Timiriaseviinae, with brooding species (and mostly with inflated brooding pouch in females) and the Limnocytherinae, which lack both brooding pouch and brooding behaviour. There are about 150-200 extant species presently known in this family.

The Timiriaseviinae were a very diverse group in Mesozoic non-marine waterbodies, but extant representatives constitute only a small part of the present-day ostracod diversity. There are six extant genera: *Metacypris* occurs in epigean waters of the Holarctic; *Kovalevskiella* is a hypogean species in the northern hemisphere; *Cytheridella*, *Gomphocythere* (syn. *Gomphodella*) and *Afrocythere* occur in lakes in the southern hemisphere, with the highest diversity in Africa; *Elpidium* has radiated in bromeliad water pouches in Central and South America.

The Limnocytherinae consist of three tribes (Danielopol *et al.* 1990), of which one (Dinarocytherini) is completely extinct and which are mainly distinguished on the basis of the hinge. Limnocytherini have a smooth cardinal bar in the left valve (LV) and simple cardinal teeth in the right valve (RV). The Leucocytherini have a crenulated cardinal bar in the LV and cardinal teeth in the RV that are generally well developed and where especially the posterior tooth often consists of 2 or 3 lobes. Dinarocytherini also have a well-developed hinge, but with the crenulated bar in the LV showing an additional anterior, and sometimes also an additional posterior tooth (amphidont) (Krstic 1987). This shows an evolution within the Limnocytherinae from well-developed amphidont hinge, over modified antimerodont to lophodont and nearly adont in *Korannacythere* (Fig. 3.5, 3.6). Note that the direction of this evolutionary scenario is entirely hypothetical.

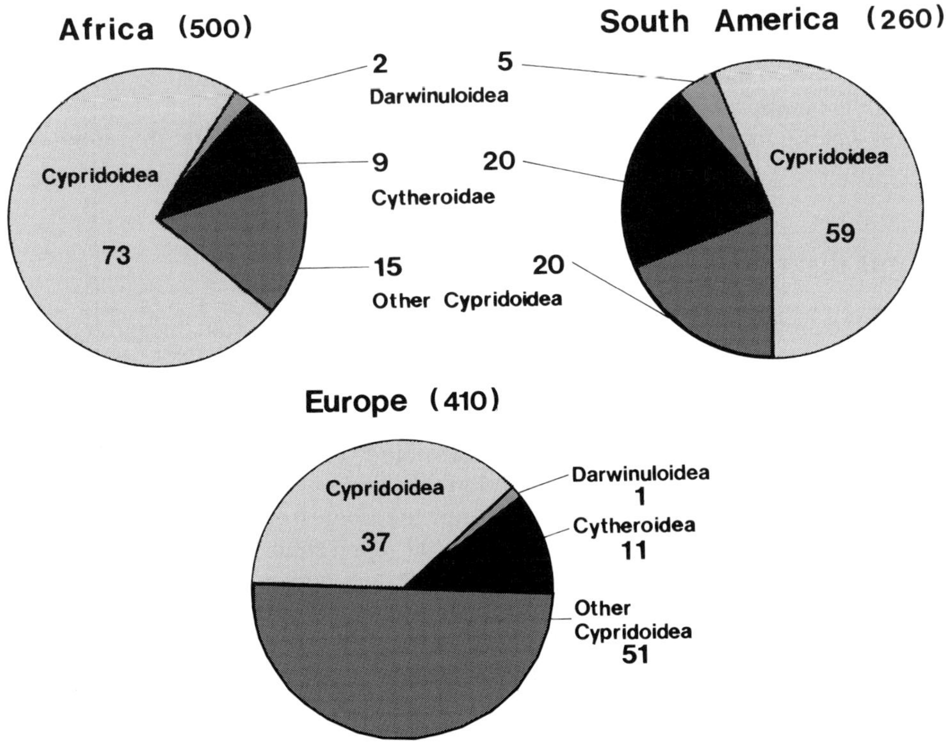

Fig. 3.4 Specific diversities in Africa, S. America, Europe (after Martens 1998 and Löffler & Danielopol 1978).

Leucocytherini have four extant and two fossil genera, but the phylogeny of this group remains at present rather obscure (Martens 1991). The nominate genus, *Leucocythere* occurs in the Palaearctic and Africa, *Ovambocythere* is endemic to South Africa. The South African species of *Leucocythere* are the most plesiomorphic in the genus. *Potamocythere* and *Athalocythere* are thus far known from the Palaearctic only (Danielopol *et al.* 1990, Martens 1991).

In the nominate tribe, the Limnocytherinae, a clear evolutionary pattern emerges, as more data on this group become available (Fig. 3.6). There are at present three main lineages within this tribe, and most characteristics are in the morphology of the hemipenis (see next chapter for anatomical descriptions) (Martens 1996, in press). The first is the actual *Limnocythere*-lineage, to which also *Limnocytherina*, *Galolimnocythere* and *Korannacythere* belong. Most North American species actually belong in *Limnocytherina*. The second group in the Limnocytherini is characterised by branched pore canals and a gradual increase in size of the caudal ramus in the hemipenis. The Palaearctic *Paralimnocythere*, especially speciose in the Balkan Lakes, still has a relatively small caudal ramus in the hemipenis, but the Pacific *Kiwicythere* already shows hyperdevelopment of this structure. The tendency towards hyperdevelopment of the ramus and setae in the hemipenis is furthered in the third lineage, which consists of the South American *Neolimnocythere* and *Paracythereis*.

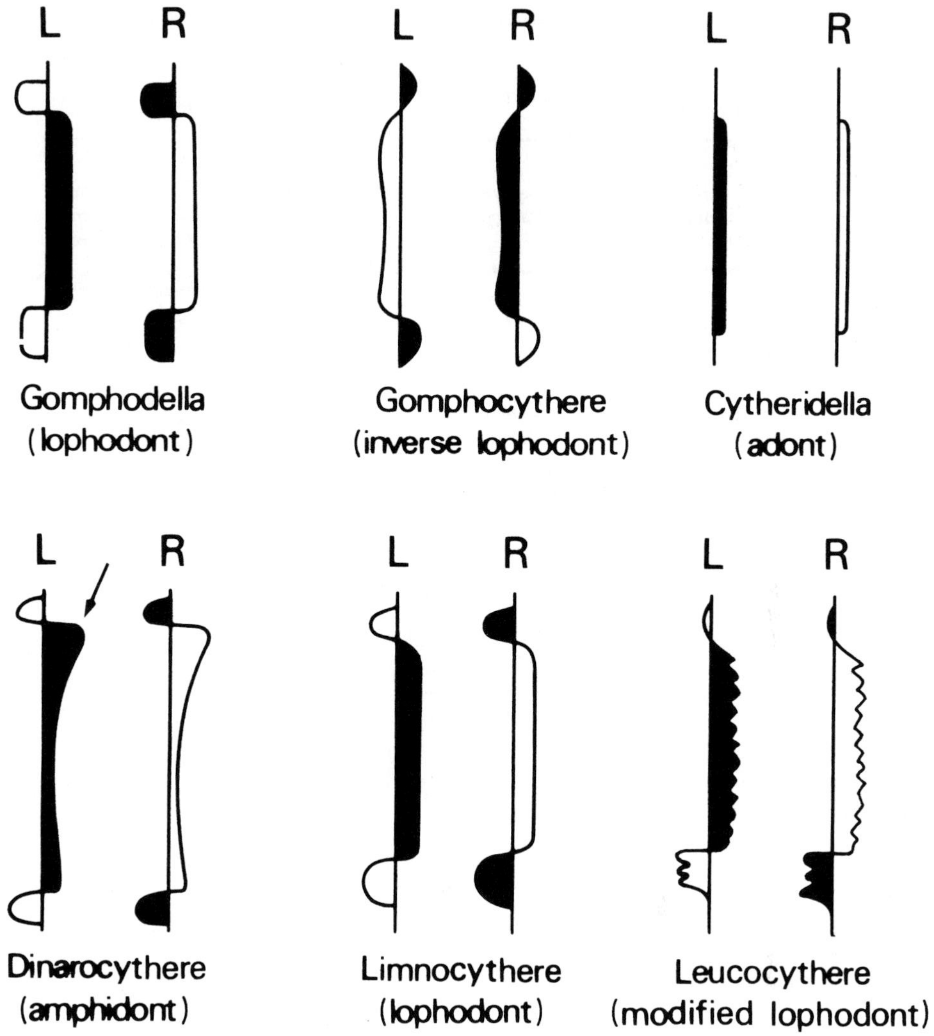

Fig. 3.5 Illustration of main hinge types in Limnocytheridae.

The radiation of Neotropical limnocytherids is ill-understood and needs urgent elaboration; especially the wide radiations on the Altiplano, in Lake Titicaca and other lagunas, will prove highly informative for the evolution of the Limnocytherinae as a whole. If the neolimnocytherids are indeed derived from a paralimnocytherid ancestor, then at least some remnants of branched pore canals must be visible in the most plesiomorphic taxa of the South American group. The alternative scenario, that neolimnocytherids and paralimnocytherids have evolved from a common ancestor, is equally plausible.

This lineage is, in the present book, represented by *Limnocythere inopinata*, the type species of the genus and consequently also of the supra-generic lineages.

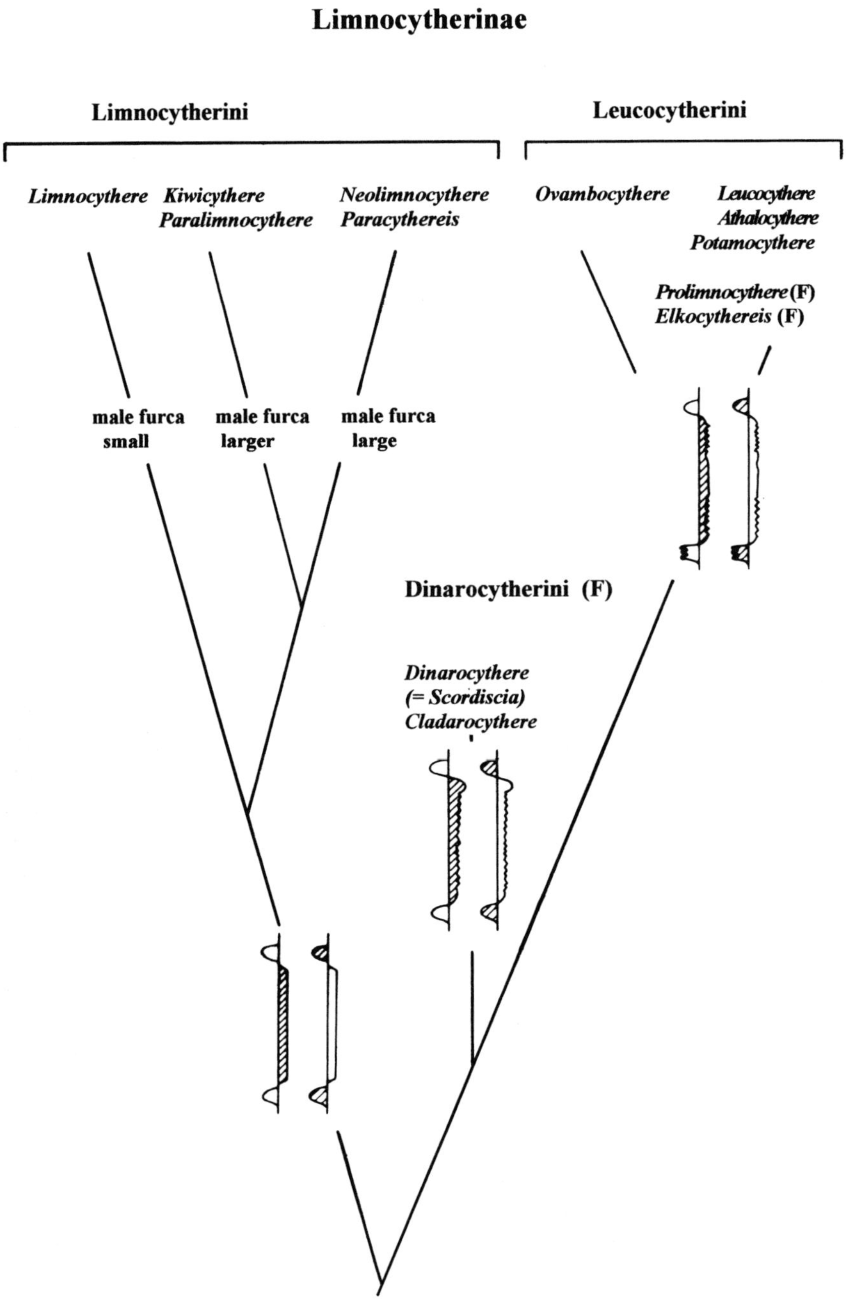

Fig. 3.6 Evolution of main groups in Limnocytherinae.

Other Cytheroidea

The Cytherideidae have large radiations in the two oldest of all extant ancient lakes: about 60 (sub-) species of the genus *Cytherissa* are described from Lake Baikal (Mazepova 1990), while an extant radiation of the *Cyprideis*-lineage consists of at least 20 species in 6 genera in Lake Tanganyika, most likely more (Wouters & Martens 1994, Martens 1994). Apart from these localities, this family is rare in non-marine habitats. The Entocytheridae, a third distinct group, constitute a special lineage of parasites or commensals on amphipod gills (Danielopol & Hart 1985). Other non-marine cytheroidean lineages are in significant radiations of *Leptocythere* in the Balkan lakes (Martens 1994), the (semi-) terrestrial Terrestricytheridae (Shornikov 1980) and at least a few representatives of the predominantly marine Xestoleberididae (McKenzie 1971, 1977). The non-marine Cytheroidea thus do not constitute one monophyletic lineage, but rather represent at least 7 separate invasions of continental habitats.

Cypridoidea-Cyprididae

This is the most widely radiated group, but at the same time also the group in which phylogenetic affinities are most obscure; Box 3.3 summarises its taxonomy. More than 1,000 extant species of Cyprididae have been described thus far, but the actual number of taxa is at least twice as high.

Ostracod carapaces in this group show an almost unlimited plasticity, they can be minute (0.4 mm or less) to giant (c 8 mm), globular or laterally flattened, smooth, spiny, striate, reticulate, bald or hirsute and so on. Furthermore, similarity in valve shape does not necessarily signify phyletic proximity, as many instances of convergence are known. Convergence in evolutionary processes within subfamilies also exists in this group: for example, size and inward displacement of anterior selvages has comparable evolutions amongst genera within the Eucypridinae, the Cypridinae and the Megalocypridinae (Martens 1986, 1989, 1990, 1992).

The morphology of the limbs, on the contrary, shows a high level of morphological stability and uniformity and small differences in soft part morphology are often sufficient to provoke the creation of new (sub-) families. Certain subterranean or terrestrial habitats, however, cause drastic adaptations, even in soft part morphology, and this can lead ostracod taxonomists astray. For example, Shornikov (1980) created a new family for the genus *Terrestricypris*, but Danielopol & Betsch (1980) have shown that these aberrant animals were in fact hyper-modified Candonidae. This combination of convergence, plasticity, uniformity and occasional hyper-modification of morphological features makes that the phylogenetic relationships amongst the 20 or so subfamilies in the Cyprididae remain unresolved to date.

As a representative species of the Cyprididae, *Eucypris virens* was selected. But as generic delimitations in the Eucypridini are at present unstable, intrageneric variability was assessed in European *Herpetocypris*, another common genus in the Cyprididae. Note that the extrapolation of the results obtained from these taxa to the Cypridoidea as a whole remains uncertain for the reasons cited above.

Box 3.3 Taxonomy of Cypridoidea, with approximate number of extant genera.

Class	Ostracoda LATREILLE, 1806	
Subclass	Podocopa G.W. MÜLLER, 1894	
Order	Podocopida SARS, 1866	
Suborder	Podocopina SARS, 1866	
Infraorder	Cypridocopina JONES, 1901	
Superfamily	Cypridoidea BAIRD, 1845	
Family	Ilyocyprididae KAUFMANN, 1900	
Subfamily	Ilyocypridinae KAUFMANN, 1900	(1 genus)
Subfamily	Pelocypridinae TRIEBEL, 1962	(1 genus)
Family	Candonidae KAUFMANN, 1900	
Subfamily	Candoninae KAUFMANN, 1900	(18 genera)
Subfamily	Paracypridinae SARS, 1923	(MARINE)
Subfamily	Cyclocypridinae KAUFMANN, 1900	(7 genera)
Family	Notodromadidae KAUFMANN, 1900	
Subfamily	Notodromadinae KAUFMANN, 1900	(3 genera)
Subfamily	Centrocypridinae HARTMANN & PURI, 1974	(1 genus)
Subfamily	Oncocypridinae DE DECKKER, 1979	(2 genera)
Subfamily	Cyproidinae HARTMANN, 1963	(1 genus)
Subfamily	Indiacypridinae HARTMANN & PURI, 1974	(1 genus)
Family	Cyprididae BAIRD, 1845	
Subfamily	Cyprinotinae BRONSTEIN, 1947	(c 7 genera)
Subfamily	Ngarawinae DE DECKKER, 1979	(1 genus)
Subfamily	Limanocypridinae HARTMANN & PURI, 1974	(1 genus)
Subfamily	Eucypridinae BRONSTEIN, 1947	(c 10 genera)
Subfamily	Cypricercinae McKenzie, 1971	(11 genera)
Subfamily	Rudjakoviellinae TRIEBEL, 1973 in MALZ	(1 genus)
Subfamily	Isocypridinae HARTMANN & PURI, 1974	(3 genera)
Subfamily	Herpetocyprellinae	(1 genus)
Subfamily	Herpetocypridinae KAUFMANN, 1900	(12 genera)
Subfamily	Cypridinae BAIRD, 1845	(7 genera)
Subfamily	Megalocypridinae ROME, 1965	(7 genera)

Box 3.3. Continued.

Subfamily	Dolerocypridinae TRIEBEL, 1961 (1 genus)
Subfamily	Hungarocypridinae BRONSTEIN, 1947 (1 genus)
Subfamily	Scottiinae BRONSTEIN, 1947 (2 genera)
Subfamily	Bradycypridinae HARTMANN & PURI, 1974 (3 genera)
Subfamily	Cyprettinae HARTMANN, 1963 (1 genus)
Subfamily	Batucyprettinae VICTOR & FERNANDO, 1981 (1 genus)
Subfamily	Cypridopsinae BRONSTEIN, 1947 (17 genera)
Subfamily	Callistocypridinae SCHORNIKOV, 1980 (1 genus)

Other Cypridoidea

There are three other families presently recognised in the Cypridoidea.

The Candonidae have the second largest non-marine ostracod radiations, with almost 200 species in Europe alone (Löffler & Danielopol 1978), probably at least 500 worldwide. Of the three subfamilies, the Candoninae have extensive radiations in ancient lakes and groundwater-associated habitats, although they also occur in epigean water bodies. The Cyclocypridinae are most speciose and comprise a number of fully sexual lineages. The Paracypridinae mostly occur in marine environments; whether this is a readaptation or constitutes an ancestral condition is impossible to say, as this group unfortunately has left a depauperate fossil record, due to their weakly calcified valves.

The Ilyocypridinae consist of only one (*Ilyocypris*), maybe two extant genera, if the South American *Pelocypris* also belongs in this group (Martens & Behen 1994). There are several dozens of species, all with well-calcified valves, which accounts for their extensive fossil record. However, most species cannot be identified from the valves, as these have small interspecific differences. Most species are known from both all-female and sexual populations.

The Notodromadidae have high supra-generic, but a low generic and specific diversity. At most 50 species in 8 genera have been described, but their high morphological divergence means that they have been divided over 5 subfamilies. Horne & Martens (in press) suspect that at least the genus *Cyprois* was already present in the Purbeck-Wealden (c 100 Myr).

Darwinuloidea-Darwinulidae

Extant taxonomic diversity in this group is low. Only the Darwinulidae have survived to Recent times, all other (sub-)families have become extinct. Within this family, only five genera are presently described: the monospecific *Microdarwinula* (the Tertiary species *M. brevis* is most likely a synonym of the Recent *M. zimmeri*) and

Darwinula s.s., as well as *Alicenula, Vestalenula* and *Penthesilenula* (Rossetti & Martens 1998), together with 28 extant species worldwide. These genera are characterised by both valve and soft part characters, but the revision of both the Palaeozoic and the Cainozoic darwinulid fauna is still in progress, so that the fossil age of these genera with extant representatives is as yet unknown.

The most common species, *Darwinula stevensoni*, the type species of the genus and consequently again of the entire lineage, is extensively investigated as a representative of this group and has turned out to be the most remarkable species of all. Several other darwinulids have been used as outgroups for molecular analyses.

Phylogenetic distribution of reproductive modes

Table 3.1 illustrates the phylogenetic distribution of the different reproductive modes. This table is preliminary at best, as most records of species do not describe the reproductive mode of the population found (see also discussion in chapter 5). The present picture shows that only Darwinuloidea can be considered ancient asexuals, while all but one subfamily of the Cyprididae clearly have mixed reproduction, almost to the exclusion of genuine sexuality in any of the subfamilies. Most of the other groups are either sexual or have mixed reproduction, but a systematic assessment of the reproductive mode of all species is necessary before any generalisation can be made.

Brooding occurs in Darwinulidae and in certain cytheroid lineages: Timiriaseviinae, *Cyprideis* and a few other, non-marine lineages. None of the Cypridoidea are brooders and this is, to say the least, remarkable. A more complete review of brooding in Ostracoda, past and present is given in chapter 10.

Table 3.1 Phylogenetic distribution of reproductive modes within the non-marine ostracods lineages. AA= ancient asexual, S= fully sexual, M= mixed reproduction, *B*= brooders, *NB*= non-brooders

Darwinuloidea	Darwinulidae *B*		AA
Non-marine Cytheroidea	Limnocytheridae	Limnocytherinae *NB*	S/M
		Timiriaseviinae *B*	S/M
	Cytherideidae	*Cytherissa* *NB*	S/M
		Cyprideis *B*	S/M
	Entocytheridae *NB*		S
Non-marine Cypridoidea *NB*	Notodromadidae		S/M
	Ilyocyprididae		M
	Candonidae	Candoninae	S/M
		Cyclocypridinae	S/M
	Cyprididae	Scottinae	S
		c 20 subfamilies	M

References

Abushik, A.F (ed) 1990. Manual of the microfauna of the USSR. Vol. 4. *Palaeozoic Ostracoda. VSE-GEI, NEDRA*, Leningrad, 356 pp (in Russian).

Bate, R.H. 1972. Phosphatised ostracods with appendages from the lower Cretaceous of Brasil. *Paleontology,* **15:** 379-393.

Colin, J.P. & Danielopol, D.L. 1980. Sur la morphologie, la systématique, la biogéographie et l'évolution des ostracodes Timiriaseviinae (Limnocytheridae). *Paléobiol. continent.,* **9:** 1-51 + 16 plts.

Danielopol, D.L. 1972. Supplementary data on the morphology of *Neonesidea* and remarks on the systematic position of the Bairdiidae (Ostracoda, Podocopida). *Proc. Biol. Soc. Wash.,* **85:** 39-48.

Danielopol, D.L. 1976. The superfamily Cyprididacea: some remarks on the phylogenetical affinities between the main ostracod Cypridacean groups. *Abhandl. Verh. naturwiss. Ver. Hamburg* (N.F.), **18/19** (Suppl.): 77-85.

Danielopol, D.L. & Betsch, J.M. 1980. Ostracodes terrestres de Madagascar: systématique, origine, adaptations. *Rev. Ecol. Biol. Sol.,* **17:** 87-123.

Danielopol, D.L. & Hart, C.W. 1985. Notes on the Center of Origin and on the antiquity of the Sphaeromicolinae, with description of *Hobbsiella,* new genus (Ostracoda, Entocytheridae). *Stygologia,* **1:** 54-70.

Danielopol, D.L., Martens, K. & Casale, L.M. 1990. Revision of the genus *Leucocythere* Kaufmann, 1892 (Crustacea, Ostracoda, Limnocytheridae), with the description of a new species and two new tribes. *Bull. k. belg. Inst. Natuurw. Biol.,* **59** (1989)**:** 63-94.

Diebel, K. & Pietrzeniuk, E. 1975. Mittel- und jungpleistozäne Ostracodenfaunen des Raums Potsdam-Brandenburg-Rathenow in stratigraphischer und ökologischer Sicht. *Z. geol. Wiss.,* **3:** 1197-1233.

Eldredge, N. 1994, *The Miner's Canary*, Virgin Books: 250 pp.

Griffiths, H.I. 1995. European Quaternary Freshwater Ostracoda: a biostratigraphic and palaeobiogeographic primer. *Scopolia,* **34:** 1-168.

Horne, D.J. & Martens, K. (in press). The importance of resting eggs for the evolutionary success of non-marine Ostracoda (Crustacea). *in*: Brendonck, L., De Meester, L. & Hairston, N. (eds), Diapause in the Crustacea. *Adv. Limnol.* **52:** 549-561.

Kempf, E. 1980. Index and Bibliography of non-marine Ostracoda. 1, Index A. *Geol. Univ. Köln, Sonderveröffentl.,* **35:** 188 pp.

Krstic, N. 1987. Three new ostracode genera from lacustrine Miocene. *Bull. Acad. Serbe Sci. Arts, Cl. Sci. nat. math., Sci. nat.,* **27**: 129-139 + 12 plates.

Lethiers, F. & Damotte, R. 1993. La grande dispersion des espèces d'ostracodes (Crustacea) d'eau douce à la fin de l'ère primaire. *C.R. Acad. Sci. Paris, Sér. II,* **316:** 427-433.

Löffler, H. & Danielopol, D.L. 1978. Ostracoda (Cyprididae). *in*: Illies, J. (ed), *Limnofauna Europaea*, Gustav Fisher: 196-208.

McKenzie, K.G. 1971. The palaeozoogeography of freshwater Ostracoda. *in*: Oertli, H.J. (ed), *Paléoécologie des Ostracodes, Bull. Centr. Res. Pau-SNPA,* **5**, Supp.: 207-238.

McKenzie, K.G. 1977. Ostracoda. *in*: *La Faune terrestre de l'île de Sainte-Hélène, 4e partie, Ann. Mus. r. Afr. centr., Sci. zool.,* **220:** 444-451.

McKenzie, K.G. 1982. Homeomorphy, persistent joker in the taxonomic pack, with the description of *Bradleycypris* gen.nov. *in*: Bate, R.H., Robinson, E. & Sheppard, L.M. (eds), *Fossil and Recent Ostracoda,* Ellis Horwood: 407-438.

McKenzie, K.G. & Hussainy, S.U. 1968. Relevance of a freshwater cytherid (Crustacea, Ostracoda) to the continental drift hypothesis. *Nature*, **220:** 806-808.

Maddocks, R. 1976. Quest for the ancestral podocopid: numerical cladistic analysis of ostracode appendages, a preliminary report. *Abhandl. Verh. Naturwiss. Ver. Hamburg. (N.S.),* **18/19** Suppl.**:** 39-53.

Maddocks, R. 1982. Part 4. Ostracoda. *in*: Abele, L.G. (ed), *The Biology of Crustacea. Vol. 1. Systematics, the Fossil Record and Biogeography,* Acad. Press: 221-239.

Martens, K. 1986. Taxonomic revision of the subfamily Megalocypridinae Rome, 1965 (Crustacea, Ostracoda). *Verhand. k. Acad. Wetensch., Lett. Schone Kunst., Kl. Wetensch.,* **48:** 81 pp. + 64 figs.

Martens, K. 1987. Homology and functional morphology of the sexual dimorphism in the antenna of *Sclerocypris* Sars, 1924 (Crustacea, Ostracoda, Megalocypridinae). *Bijdr. Dierk.,* **57:** 183-190.

Martens, K. 1989. On the systematic position of the *Eucypris clavata* – group, with a description of *Trajancypris* gen. nov. (Crustacea, Ostracoda). *Arch. Hydrobiol. Suppl.,* **83:** 227-251.

Martens, K. 1990. Revision of African Cypridini. I. The genera *Cypris* O.F. Müller, *Pseudocypris* Daday and *Globocypris* Klie. *Bull. k. belg. Inst. Natuurw. Biol.*, **60:** 127-172.
Martens, K. 1991. A new and enigmatic representative of *Leucocythere* Kaufmann, 1892 from the Eastern Cape Province (South Africa) (Crustacea, Ostracoda). *J. nat. Hist.*, **25:** 583-596.
Martens, K. 1992. On *Namibcypris costata* n.gen. n.sp. (Crustacea, Ostracoda, Candoninae) from a spring in northern Namibia, with the description of a new tribe and a discussion on the classification of the Podocopina. *Stygologia,* **7:** 27-42.
Martens, K. 1994. Ostracod speciation in ancient lakes: a review. *in*: Martens K., Goddeeris, B. & Coulter, G. (eds), *Speciation in ancient lakes, Adv. Limnol.,* **44:** 203-222.
Martens, K. 1996. On *Korannacythere* nov.gen. (Crustacea, Ostracoda) a new genus of temporary pool limnocytherids from southern Africa, with the description of three new species and a generic reassessment of the Limnocytherinae. *Bull. k. belg. Inst. Natuurw. Biol.,* **66:** 51-72.
Martens, K. 1998. Diversity and endemicity of recent non-marine ostracods (Crustacea, Ostracoda) from Africa and South America: a faunal comparison. *Verhandl. int. Verein. Limnol.,* **26:** 2093-2097.
Martens, K. (in press). On the origin and development of different recognition systems in the Limnocytherinae (Crustacea, Ostracoda, Cytherocopina): chance or selection? *in*: Horne, D.J. & Martens, K. (eds), Proceedings of the XIII international symposium for ostracodologists. *Hydrobiologia.*
Martens, K. & Behen, F. 1994. A checklist of the non-marine ostracods (Crustacea, Ostracoda) from South-American inland waters and adjacent islands. *Trav. sci. Mus. Hist. nat. Lux.*, **22:** 81 pp.
Mazepova, G. 1990. Rakuschkovye ratschki (Ostracoda) Baikala. *Nauk. Sib. Otdel. Akad. Nauk. SSSR*, Novosibirsk, 470 pp.
Moore, R.C. (ed) 1961, *Treatise on Invertebrate Palaeontology. Part Q, Arthropoda 3, Crustacea, Ostracoda.* Geol. Soc. Am. Univ. Kansas Press,.
Müller, K.J. 1982. *Hesslandona unisulcata* sp. nov. with phosphatised appendages from Upper Cambrium 'Orsten' of Sweden. *in*: Bate, R.H., Robinson, E. & Sheppard, L.M. (eds), *Fossil and Recent Ostracoda,* Ellis Horwood: 276-304.
Odin, G.S. 1994. Geological time scale. *C.R. Acad. Sci. Paris, Sér. II,* **318:** 59-71.
Rossetti, G. & Martens, K. 1998. Taxonomic Revision of the Recent and Holocene representatives of the Family Darwinulidae (Crustacea, Ostracoda), with a description of three new genera. *Bull. k. belg. Inst. Natuurw. Biol.,* **68.**
Shornikov, E.I. 1980. Ostracodes in terrestrial biotopes. *Zool. J.,* **59:** 1306-1319. (in Russian)
Sohn, I.G. 1985. Latest Mississippian (Namurian A) nonmarine ostracodes from West Virginia and Virginia. *J. Paleontol.,* **59:** 446-460.
Sohn, I.G., 1987. The ubiquitous ostracode *Darwinula stevensoni* (Brady & Robertson, 1870), redescription of the species and lectotype designation. *Micropaleontology*, **33:** 150-163.
Sohn, I.G., 1988. Darwinulocopina (Crustacea: Podocopa), a new suborder proposed for nonmarine Paleozoic to Holocene Ostracoda. *Proc. biol. Soc. Wash.,* **101:** 817-824.
Spears, T. & Abele, L.G. 1997. Crustacean phylogeny inferred from 18S rDNA. *in*: Fortey, R.A. & Thomas, R.H. (eds), *Arthropod Relationships*, Chapmann & Hall: 169-187.
Straub, E.B. 1952. Mikropaläontologische Untersuchungen im Tertiär zwischen Ehingen und Ulm a.d. Donau. *Geol. Jb.*, **66:** 433-523.
Wakefield, M.I., 1994. Middle Jurassic (Bathonian) Ostracoda from the Inner Hebrides, Scotland. *Monogr. Paleontogr. Soc., Lond.*, **148 (596):** 1-89, 11pls.
Whatley, R.C. 1990a. The relationship between extrinsic and intrinsic events in the evolution of Mesozoic non-marine Ostracoda. *in*: Kaufmann, E.G. & Walliser, O.H. (eds), Extinction events in Earth history. *Lecture Notes in Earth Sciences* **30:** 253-263.
Whatley, R.C. 1990b. Ostracoda and global events. *in*: Whatley, R. & Maybury, C. (eds), *Ostracoda and Global Events,* Chapman & Hall: 3-24.
Whatley, R.C. 1992. The reproductive and dispersal strategies of Cretaceous non-marine Ostracoda: the key to pandemism. *in*: Mateer, N.J. & Chen, P.J. (eds), *Aspects of non-marine Cretaceous geology,* China Ocean Press: 177-192.
Whatley, R.C. & Moguilevsky, A. (in press). The origins and early evolution of the Limnocytheridae. *in*: Crasquin, S., Braccini, E. & Lethiers, F. (eds), Proceedings of the 3rd European Ostracodologists Meeting. *Bull. Centr. Rech. Explo.-Prod. Elf-Aquit.*
Wouters, K. & Martens, K. 1994. Contribution to the knowledge of the *Cyprideis* species flock (Crustacea, Ostracoda) of Lake Tanganyika, with the description of three new species. *Bull. k. belg. Inst. Natuurw. Biol.,* **64:** 111-128.

Xianguang, H., Siveter, D.J., Williams, M., Walossek, D. & Bergström, J. 1996. Appendages of the arthropod *Kunmingella* from the early Cambrian of China: its bearing on the systematic position of the Bradoriida and the fossil record of the Ostracoda. *Phil. Trans. R. Soc. Lond. B*, **351:** 1131-1145.

CHAPTER 4

GENERAL MORPHOLOGY OF NON-MARINE OSTRACODS

K. Martens

Abstract

Ostracoda are bivalved Crustacea, which have their reduced bodies completely enveloped by a carapace. Reductions have occurred through oligomerisation and only four pairs of head appendages, three pairs of thoracic appendages and a pair of caudal rami (formerly called furca) are present. The morphology is remarkably uniform, also amongst the three lineages of non-marine ostracods. Most morphological evolution has occurred in the valves and in the male reproductive organs. One historic record of a putative male of *D. stevensoni* is considered to deal with an atavistic specimen.

Introduction

As a number of subsequent chapters will extensively describe and discuss morphological variability within non-marine ostracods, both inter-specific, related to sexual selection, and intra-specific and inter-clonal, a brief introduction to the general morphology of non-marine ostracods is here presented. Its aims are to illustrate the differences in morphology between the three main lineages of non-marine ostracods. It is of course impossible to represent all extant variability within this group, and only general patterns will be described.

General morphology

Ostracods are small to medium-sized bivalved crustaceans. Due to a partial reduction of the thorax and a nearly complete reduction of the abdomen (oligomerisation), the body is completely enveloped by the carapace, which consists of two lateral valves. Most limbs and organs are paired. There is at present intense discussion as to what ostracods really are, and several major taxonomic works on Ostracoda indeed do not provide a diagnosis of the subclass. The discovery of several other bivalved crustacean and non-crustacean arthropod groups (see chapter 3) has acutely shown that diagnostic features of Ostracoda must be found in the limbs (Xianguang *et al.* 1996). A diagnosis of the Ostracoda, summarised from the literature (Müller 1912, Klie 1938, Moore 1961, Van Morkhoven 1962, Maddocks 1982), is presented in Box 4.1.

Sex and parthenogenesis: evolutionary ecology of reproductive modes in non-marine ostracods, edited by K. Martens, pp. 57–75

Box 4.1 Diagnosis of the Ostracods

Small (generally 0.5-1.0 mm, up to 30 mm in some marine planktonic Myodocopida), mostly laterally compressed Crustacea, with bivalved, mostly fully calcified, carapace, completely enveloping a largely unsegmented body (exceptions exist). Most organs and appendages paired. Valves hinging dorsally, closed by central adductor muscles and without growth lines (valves are shed at each moult); hinge adont or absent, or consisting of complex sequences of teeth, bars and sockets; valve margins either simple or with complex closing mechanisms. Head with 4 pairs of appendages: two pairs of antennae, a pair of mandibulae (pre-oral) and (post-orally) a pair of maxillulae, maxillae missing. Thorax generally with 1-3 pairs of appendages, the first and the third pair possibly modified into maxillipeds, clasping organs or cleaning limbs. Abdomen generally without genuine appendages, a pair of caudal rami mostly present. Males with one or two large hemipenes, in the latter case fused at the base. Sperm pumps present either incorporated within the hemipenes or as separate Zenker's organs. Cainozoic Darwinuloidea fully parthenogenetic; other lineages are either sexual or with mixed reproduction, sexes always separate. Brooding in the carapace possible.

Valve structure (Fig. 4.1)

Most of the present description is based on Hartmann (1966).
Ostracod valves consist of an outer and an inner lamella. Along the valve margins, where these two lamellae are fused, a number of structures occur and most of these are important with regard to taxonomy and identification. The margin of the (peripheral) *calcified part of the inner lamella* is called the *inner margin*. The space between the outer lamella and the calcified part of the inner lamella is called the *vestibulum* and *radial septa*, connecting inner and outer lamellae in the vestibulum, can occur here. Part of the outer and inner lamellae can be fused (*fused zone*) and *marginal pore canals* mostly cut through this area. The inner margin of the fused zone is the *line of concrescence.* On the calcified part of the inner lamella, crest-like *inner lists* and *selvages* can occur. When valves are examined in internal view with transmitted light, valve margin, inner margin, line of concrescence, inner lists and selvage are distinguishable.

The selvage is originally a valve margin which has been inwardly displaced. This means that part of the outer lamella follows the selvage to the inside of the valve (where one would normally expect the inner lamella) and that a new valve margin is created. The selvage will thus remain associated with the primary fused zone, i.e. with marginal pore canals on which hairs occur. Inner lists, on the other hand, never bear a row of hairs. A new (secondary) fused zone with pore canals and hairs will be formed near the new valve margin. Selvages and inner lists form part of the closing mechanism of the valves, for example with the selvage of the right valve fitting between a selvage and an inner list in the left valve when valves are closed.

Valves are closed by powerful central *adductor muscles*. The attachment sites of these muscles (both central and dorsal) give typically arranged *muscle scars.* Especially the shape and number of the *central muscle scars*, situated near the centre of each valve, are of taxonomic importance.

Both valves are dorsally attached to each other by a *hinge*. In Cypridoidea, the hinge has no calcified structures and consists primarily (only?) of ligaments. In Darwinuloidea, mostly the left valve has a solid (always smooth) ridge, which fits

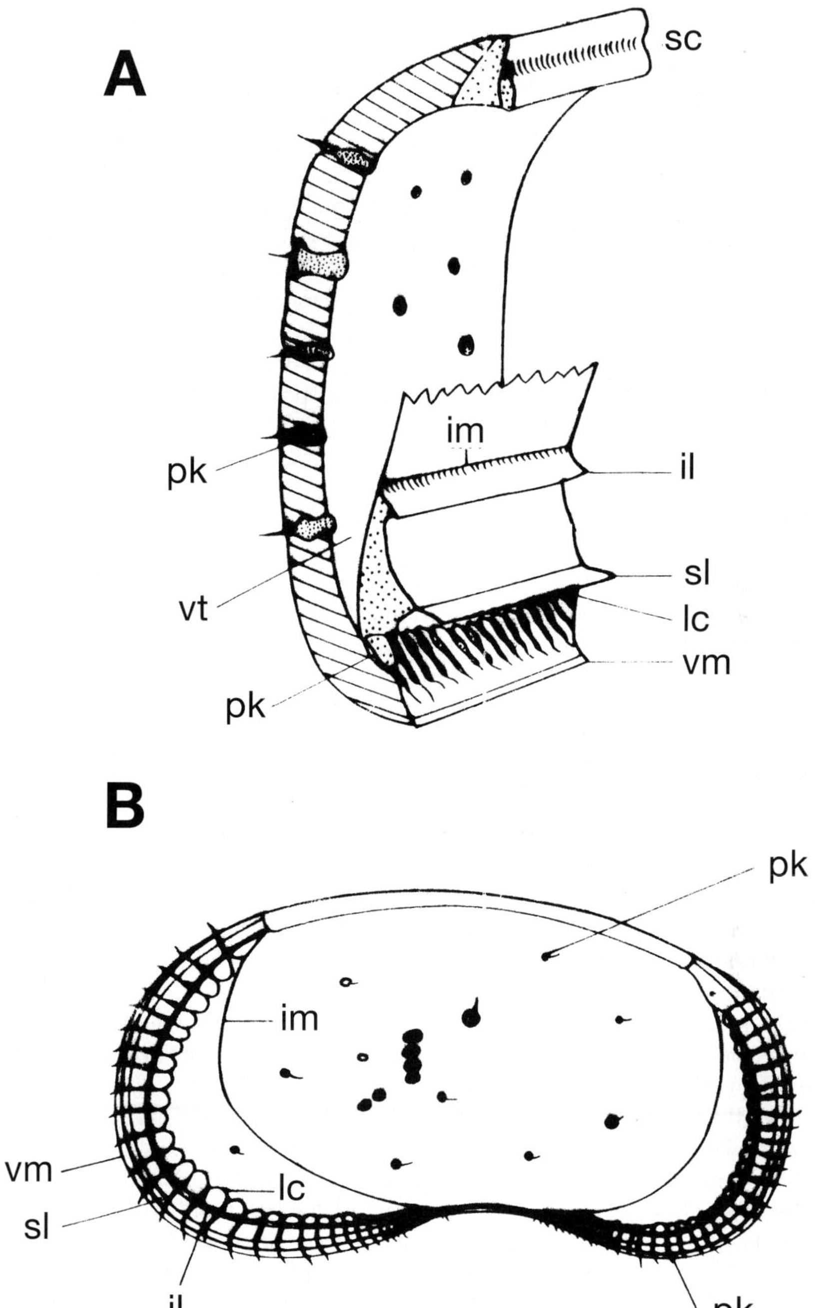

Fig. 4.1 General morphology of valves (redrawn after Hartmann 1966). A. Section of hypothetical ostracod valve. B. Hypothetical ostracod valve in lateral view.
Abbreviations: il= inner list. im= inner margin. lc= line of concrescence. sc= hinge. sl= selvage. vm= valve margin. vt= vestibulum. pk= pore-canal.

into a groove in the right valve. In *Microdarwinula*, two large tooth-like structures (one on the anterior, one on the posterior side of the dorsal margin) occur on the right valve. In Cytheroidea, the hinge shows much morphological diversity, although the most explicit shapes occur in marine taxa. In its simplest form, the hinge consists of a cardinal ridge in the left valve, a cardinal groove in the right valve, anterior and posterior cardinal teeth on the right valve, fitting into cardinal sockets in the left valve. Exceptionally, one or both teeth are missing; the ridge can be smooth or crenulate. Occasionally, hinge reversal occurs, where the ridge and sockets switch from left to right valve, groove and teeth from right to left.

The entire outer lamella has normal *pores*, bearing hairs on the external side. All sorts of carapace ornamentations (spines, nodes, tubercles, ridges, pits, etc.) can occur.

Valves of selected species are illustrated in Figs. 4.2-4.4: of all five extant genera of Darwinulidae (Fig. 4.2); showing sexual dimorphism in three genera of Limnocytheridae (Fig. 4.3) and selected species in the 4 families of the Cypridoidea (Fig. 4.4)

Limbs

Non-marine ostracod groups all belong to the Podocopina, and thus generally have the following (paired) appendages: Antennula (A1), Antenna (A2), mandible (Md), Maxillula (Mx1, and first, second and third thoracopods (T1-T3), while caudal rami (not furcae) are also present. There are generally no abdominal appendages. The limbs of one species of each of the different lineages are here illustrated, of *Cypris pubera* (for Cypridoidea – Fig. 4.5), of *Limnocythere thomasi* (for Cytheroidea – Fig. 4.6) and of *Darwinula stevensoni* (for Darwinuloidea – Fig. 4.7). Homology follows Matzke-Karatsz (1995) and Smith & Martens (in press).

The Antennae are generally the largest appendages; chaetotaxy (pattern of claws and setae on the different segments) can be complex (Broodbakker & Danielopol 1982). Both of these limbs are used for locomotion, either swimming (if natatory setae are present) or crawling (with claws and claw-like setae). The A2 is also used for feeding and in the male also during mating (see below).

Mandibles consist of a heavily sclerotised coxal part, on which a more slender palp is inserted; the latter bears a small respiratory plate. The Md is used almost exclusively for gathering, handling and chewing food particles.

The Maxillula consists of a palp (mostly two-segmented), three endites and a large respiratory plate. Endites and palp are also used during feeding. Although Md and the fifth limb also have small respiratory plates (mostly exopodites), the respiratory plate of the Mx1 is responsible for most of the ostracod's respiration.

There has been much debate as to whether the fifth limb is either of thoracic (Kesling 1951, Athersuch *et al.* 1989) or of cephalic origin (Hartmann 1977, Maddocks 1982). The latter seems plausible in Cypridoidea, where this limb in females indeed looks like a maxilliped with mostly undivided palp. In Darwinuloidea, however, this limb has a segmented palp, while in Cytheroidea, the fifth limb is a walking leg. Martens (1990b) offered morphological support for the fact that at least in cytherids, this limb is a thoracic appendage, arguing that the so-called sternum is actually the division between head and thorax and that the fifth

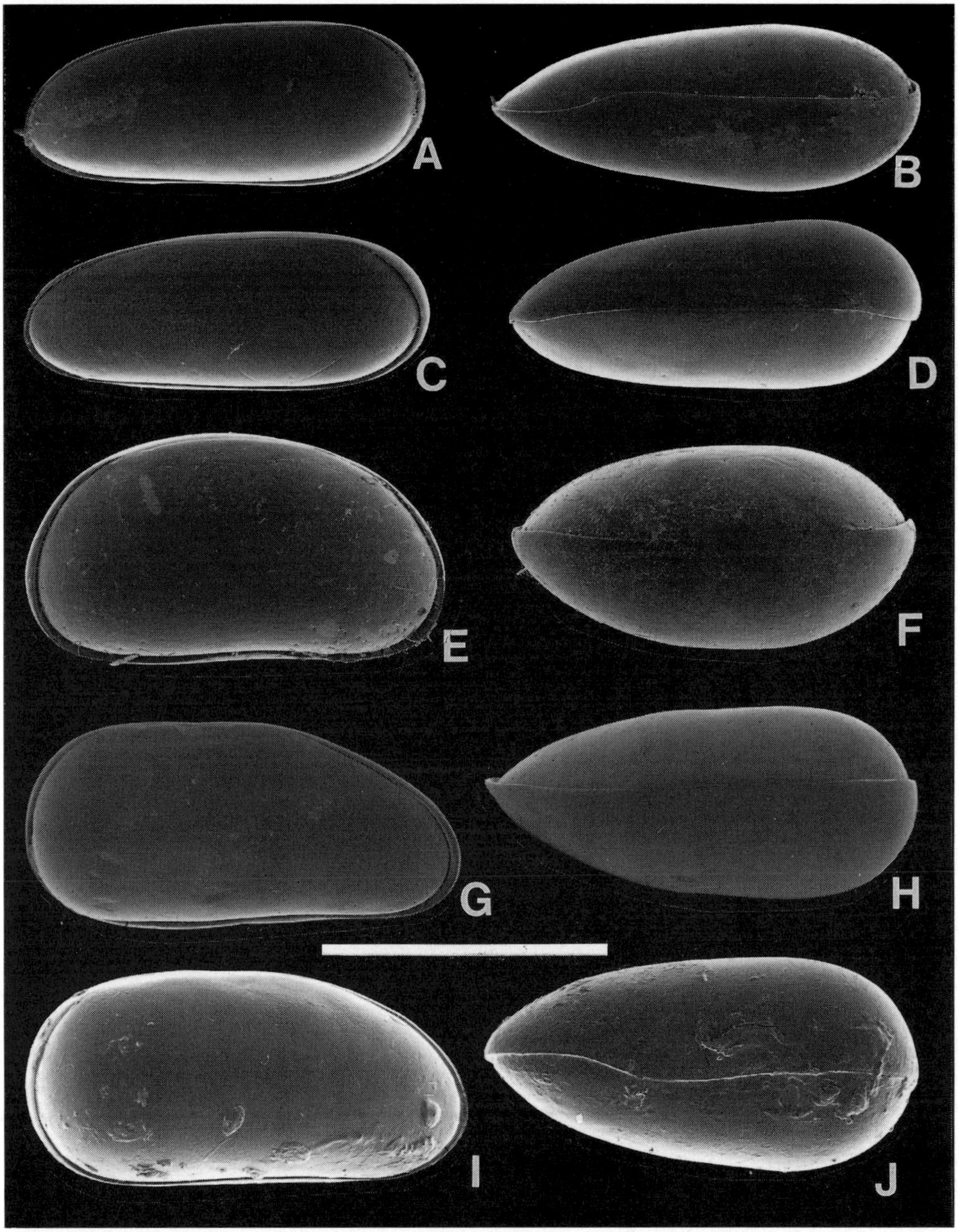

Fig. 4.2 Illustration of valves of extant genera in Darwinulidae, all females.
A-B. *Darwinula stevensoni* (Italy), A= Carapace left lateral; B= carapace ventral. C-D. *Alicenula inversa* (South Africa), C= carapace left lateral, D= carapace dorsal. E-F. *Microdarwinula zimmeri* (France), E= carapace right lateral, F= carapace dorsal. G-H. *Vestalenula molopoensis* (South Africa), G= carapace right lateral, H= carapace dorsal. I-J. *Penthesilenula brasiliensis* (France), I= carapace right lateral, J= carapace ventral. Scale= 417 μm for A,B; 373 μm for C,D; 313 μm for G-J; 238 μm for E,F. A,B from Rossetti & Martens (1996), others original by G.P. Rossetti & J. Cillis.

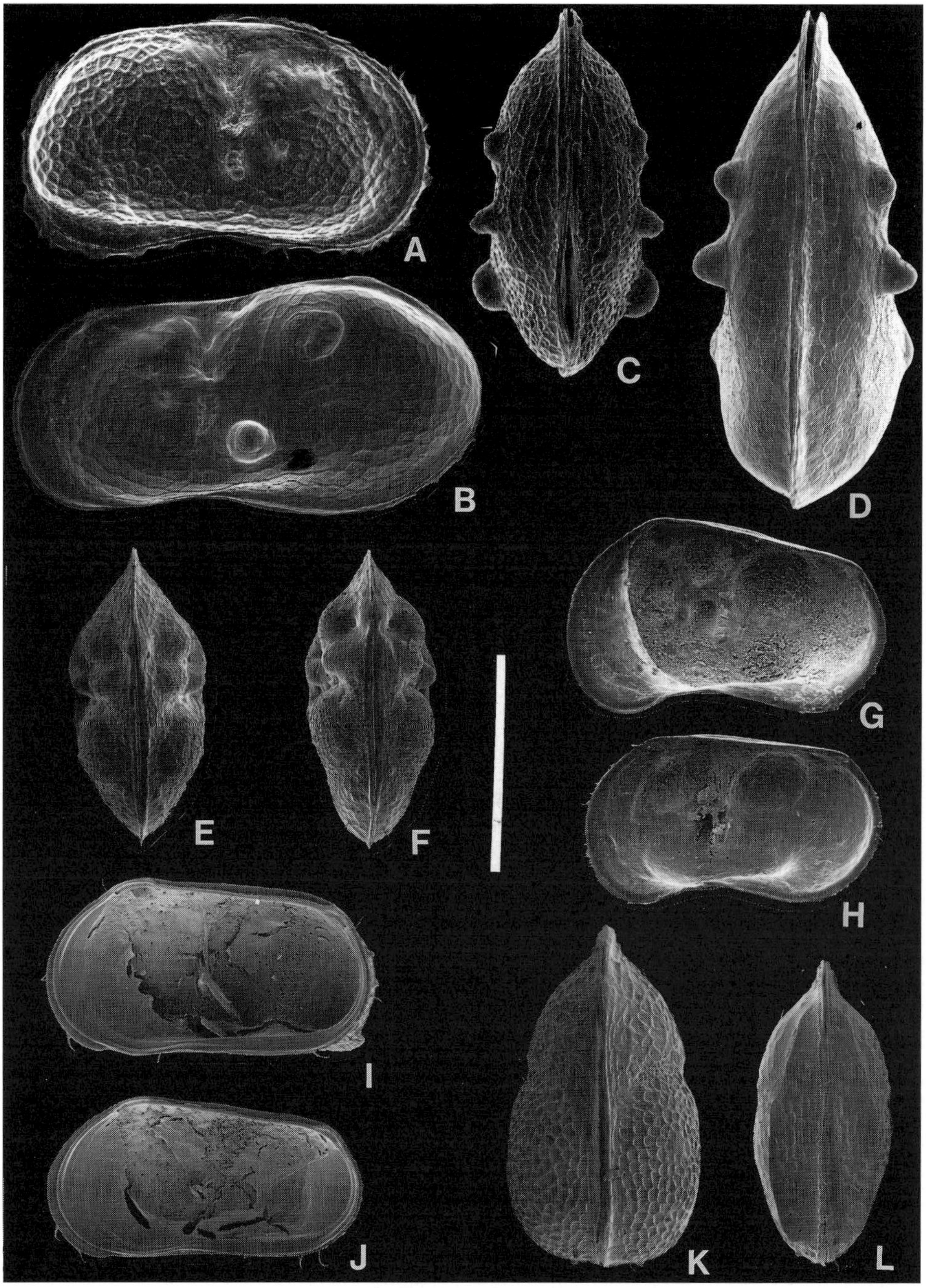

Fig. 4.3 Illustrations of valves and sexual dimorphism in 3 genera of Limnocytheridae. A-H= Limnocytherinae, I-L= Timiriaseviinae. A-D. *Limnocythere inopinata* (China, Lake Qinghai), A= Female, carapace right lateral, B= male, carapace left lateral, C= female, carapace ventral, D= male, carapace ventral. E-H. *Korannacythere devriesi* (South Africa), E= Female, carapace dorsal, F= male, carapace dorsal, G= female, right valve internal, H= male, right valve internal. I-L. *Gomphocythere capensis* (South Africa), I= Female, right valve internal, J= male, right valve internal, K= female, carapace dorsal, L= male, carapace ventral. Scale= 391 μm for I-L, 325 μm for A-D, 303 μm for E-H. E-H after Martens (1996), others original.

Fig. 4.4 Illustration of valves of selected genera in Cypridoidea. A-D= Notodromadidae; E-H= Candonidae, I,J= Ilyocyprididae, K-Q= Cyprididae
A-D. *Oncocypris mulleri* (South Africa), A= female, carapace dorsal, B= Female, right valve internal, C= male, right valve internal, D= male, carapace dorsal. E-F. *Physocypria capensis* (South Africa), E= female, carapace left lateral, F= carapace dorsal. G,H. *Fabaeformiscandona danielopoli*, female (China), G= carapace right lateral, H= female, carapace dorsal. I,J. *Ilyocypris gibba*, female (South Africa), I= carapace dorsal, J= carapace left lateral. K,L. *Cypris decaryi*, female (South Africa), K= carapace dorsal, L= right valve internal. M. *Pseudocypris triquetra*, female, (South Africa), carapace dorsal. N,O. *Herpetocypris reptans*, female (Belgium), N= carapace right lateral, O= carapace dorsal. P,Q. *Zonocypris tuberosa*, female (South Africa), P= carapace left lateral, Q= carapace dorsal.
Scale= 1880 µm for M, 1790 µm for N,O, 1389 µm for K, 1111 µm for L, 807 µm for G,H, 676 µm for P,Q, 625 µm for A-D,I,J, 417 µm for E-F. G,H after Yin & Martens (1997), K,L after Martens (1990a), N,O after Gonzalez Mozo *et al.* (1996), P,Q after Martens *et al.* (1996), others original.

limb is inserted posterior to this sternum. Kaufmann (1896) already suggested that the brush-like organs in cytherid males could be vestigial Mx2. Recently, Meisch (1996) and Smith & Martens (in press) provided evidence for the same, the former for the Myodocopida, the latter while describing the post-embryonal ontogeny of *Eucypris virens*. We thus maintain here that the fifth limb is the first thoracopod. Ostracoda thus have only four pairs of head appendages.

The sixth limb is the second thoracopod, and is a walking limb in all non-marine ostracods. It is pediform, with one large stout end claw.

The T3 can either be transformed into a cleaning limb with a distal pincer-organ (used to clean the natatory setae of at least the reflexed A1 – Cyprididae) or remain a walking limb (Cytheroidea, Darwinuloidea, other Cypridoidea).

The caudal rami in most Cypridoidea consist of a stout ramus, with two distal claws and setae; in other cypridinid groups (e.g. some Notodromadidae, Cypridopsinae) and in most Cytheroidea and Darwinuloidea, it is reduced to a short base and one or two setae in the female, and nearly always incorporated into the male hemipenis. In some darwinulids (e.g. in *D. stevensoni*), caudal rami are absent in the adult, but present in the last juvenile stages. Meisch (in prep.) argues that these caudal rami in ostracods are not homologous to the crustacean furca.

Reproductive organs

Cypridoidea

Males usually have 4 paired testicular tubes, which are spiralled in the carapace. Ostracods have, relatively speaking, the longest spermatozoa in the animal kingdom and a specimen 1 mm long can have spermatozoa of 10 mm and more in length (Wingstrand 1988, chapter 2). To handle these excessively long spermatozoa during insemination, the *ductus ejaculatorius* in Cypridoidea is modified to a muscular sperm pump, the *Zenker's organ*. There are two *hemipenes* and the morphology of these structures is highly diagnostic. They consist of (1) the peniferum, (2) the spermiduct, (3) the *bursa copulatrix* and (4) sclerotized rami and muscles (Danielopol 1969, Martens 1986) (Fig. 4.8).

The *peniferum* is a chitinous, transparent sheath, enveloping the internal structures and mostly forms two distal lobes: the *medial* (ms) and the *lateral* (ls) *shields*.

The *spermiduct* consists of an internal and an external part. The *internal spermiduct* has at least two parts: the labyrinth and a tubular part. The *labyrinth* is a solid, sclerotized structure with a number of distinct compartments. Danielopol (1969) distinguished four parts (named a-d or d1-d4): an ascending branch, a horizontal branch, a descending branch and a terminal, swollen compartment, from where the *tubular part* of the spermiduct runs to the *bursa copulatrix*. This part can display irregular loops and curves (for example in Cypridini, see Fig. 4.8, and Herpetocypridinae) and is distally surrounded by sclerotised structures.

The *bursa copulatrix* is a distally situated, bowl-shaped organ. It can be surrounded by elongated, sclerotized structures (named 'g' or 'M' in *Candona* by Danielopol (1969, 1978), latches 'l' in *Heterocypris* by Broodbakker (1982)). It is the actual copulatory process, the intromitting organ, which is pushed outside of the peniferum during erection.

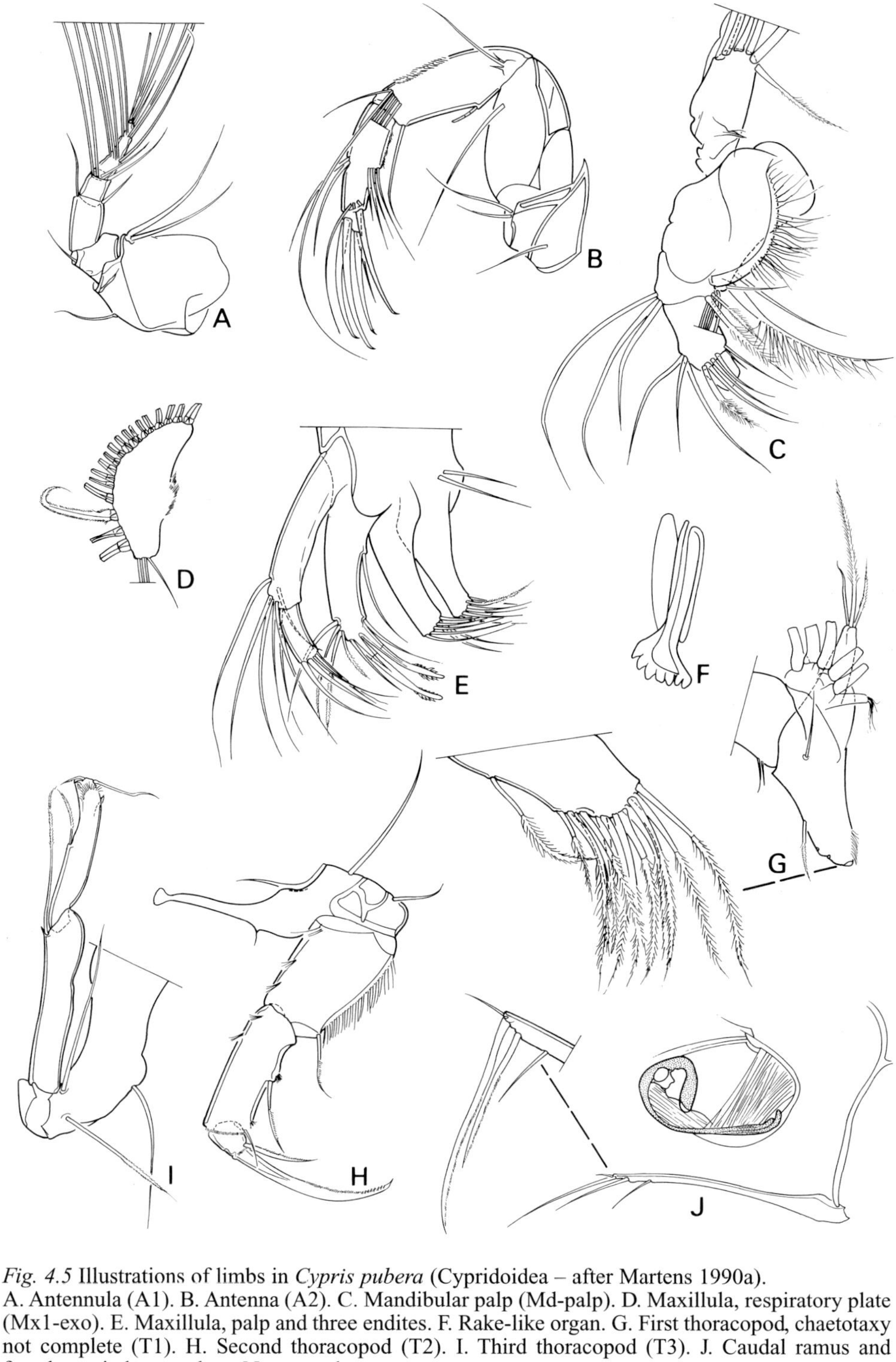

Fig. 4.5 Illustrations of limbs in *Cypris pubera* (Cypridoidea – after Martens 1990a).
A. Antennula (A1). B. Antenna (A2). C. Mandibular palp (Md-palp). D. Maxillula, respiratory plate (Mx1-exo). E. Maxillula, palp and three endites. F. Rake-like organ. G. First thoracopod, chaetotaxy not complete (T1). H. Second thoracopod (T2). I. Third thoracopod (T3). J. Caudal ramus and female genital operculum. Not to scale.

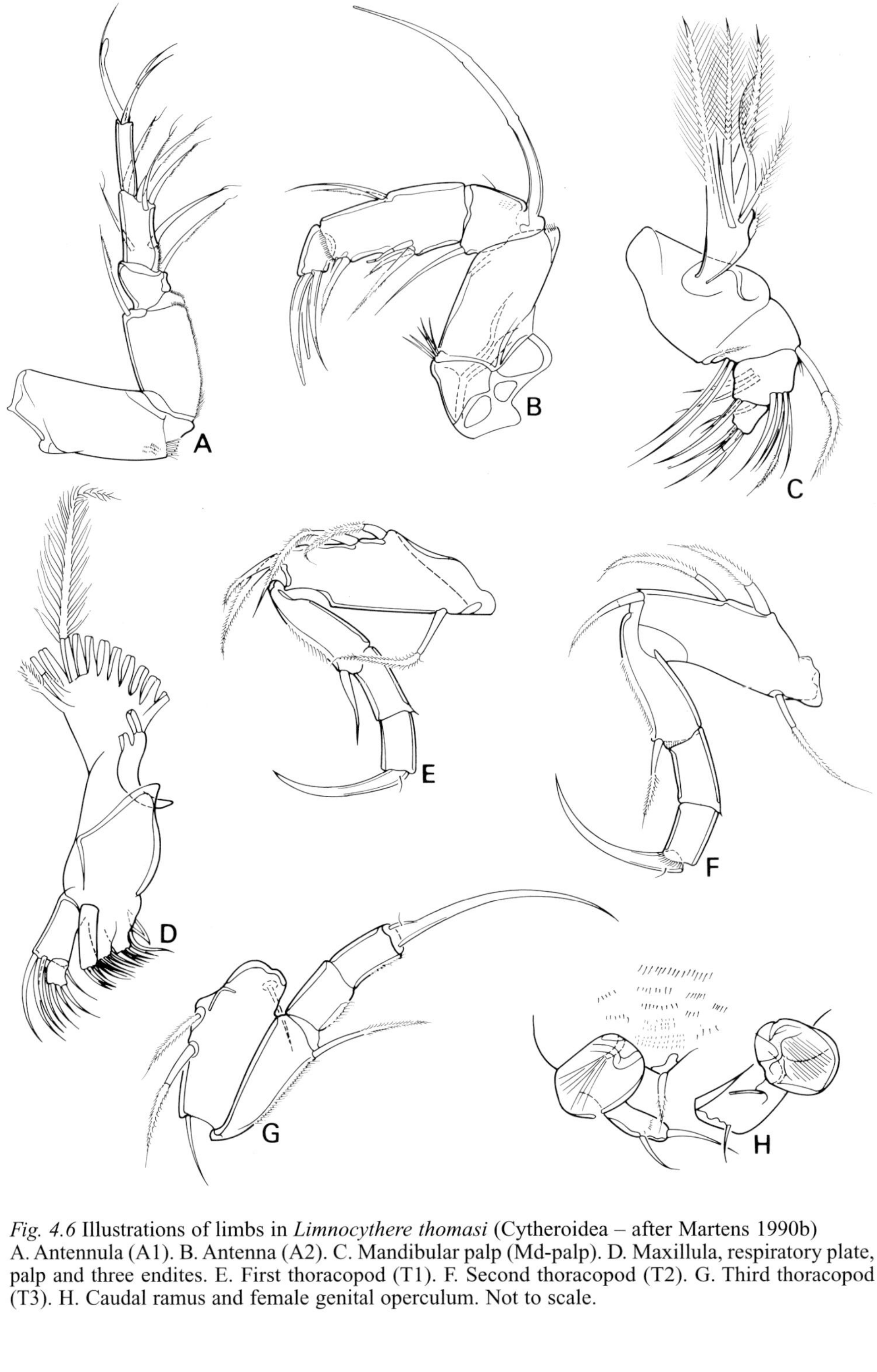

Fig. 4.6 Illustrations of limbs in *Limnocythere thomasi* (Cytheroidea – after Martens 1990b) A. Antennula (A1). B. Antenna (A2). C. Mandibular palp (Md-palp). D. Maxillula, respiratory plate, palp and three endites. E. First thoracopod (T1). F. Second thoracopod (T2). G. Third thoracopod (T3). H. Caudal ramus and female genital operculum. Not to scale.

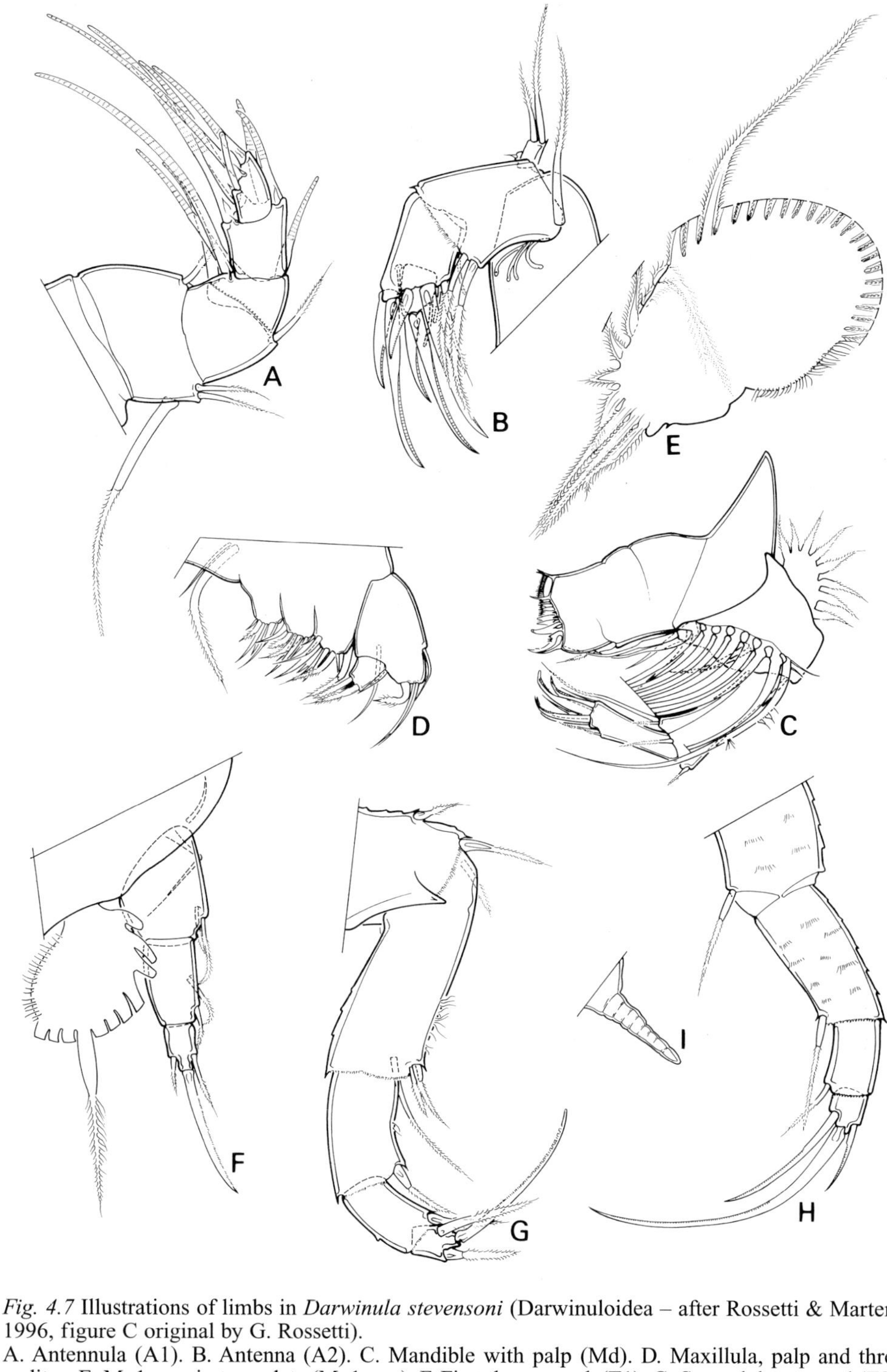

Fig. 4.7 Illustrations of limbs in *Darwinula stevensoni* (Darwinuloidea – after Rossetti & Martens 1996, figure C original by G. Rossetti).
A. Antennula (A1). B. Antenna (A2). C. Mandible with palp (Md). D. Maxillula, palp and three endites. E. Mx1, respiratory plate (Mx1-exo). F. First thoracopod, (T1). G. Second thoracopod (T2). H. Third thoracopod (T3). I. postabdomen. Not to scale.

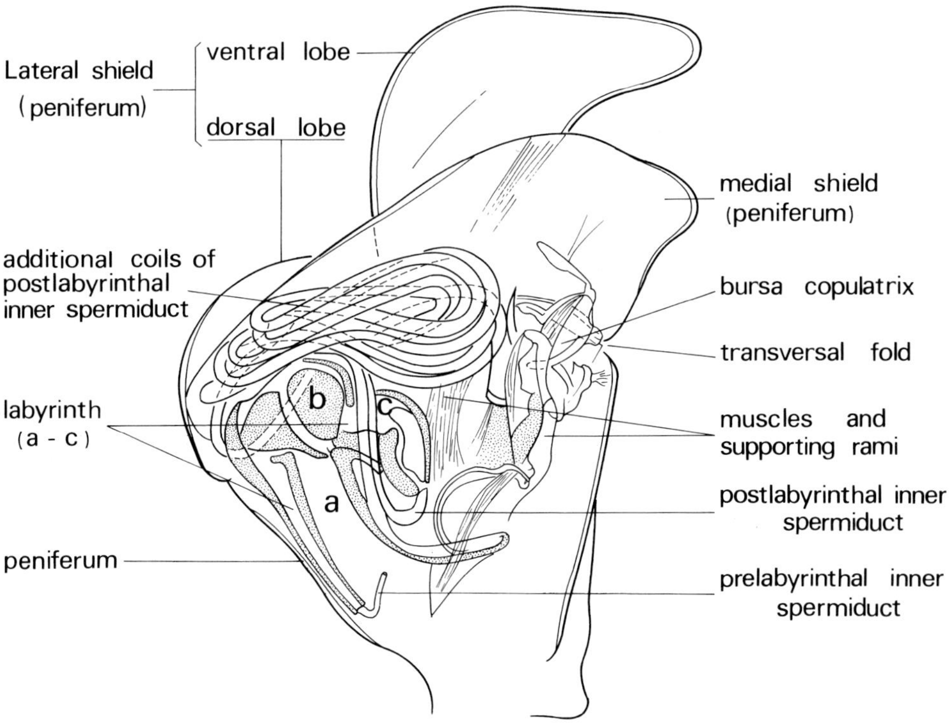

Fig. 4.8 General structure of Hemipenis in Cypridoidea; example of *Pseudocypris bouvieri* (Cyprididae, Cypridinae – after Martens 1990a).

Other *rami* and *muscles* attach the spermiduct to the peniferum and play an important role during erection. In Cypridoidea, erection involves two major movements: (1) displacement of both hemipenes, bringing them outside of the valves in a copulatory position (*external erection*) and (2) displacement of the internally situated copulatory complex (part 'e' in Martens 1986 among others), in a relaxed condition enveloped by the lateral and medial shield of the peniferum, to an external and free position (*internal erection*).

Females have fewer diagnostic features in their reproductive systems. The only characters used here are the shape of the *genital lobes* (situated at the postero-ventral part of the body) and of the lateral, paired *ovaria* (mostly visible in lateral view, through the transparent carapace).

This is the general pattern for Cypridoidea. Figs. 4.8-4.9 illustrate different modifications in the different families of the Cypridoidea, affecting especially the shape and position of the distal lobes of the peniferum and the number of additional coils in the postlabyrinthal inner spermiduct.

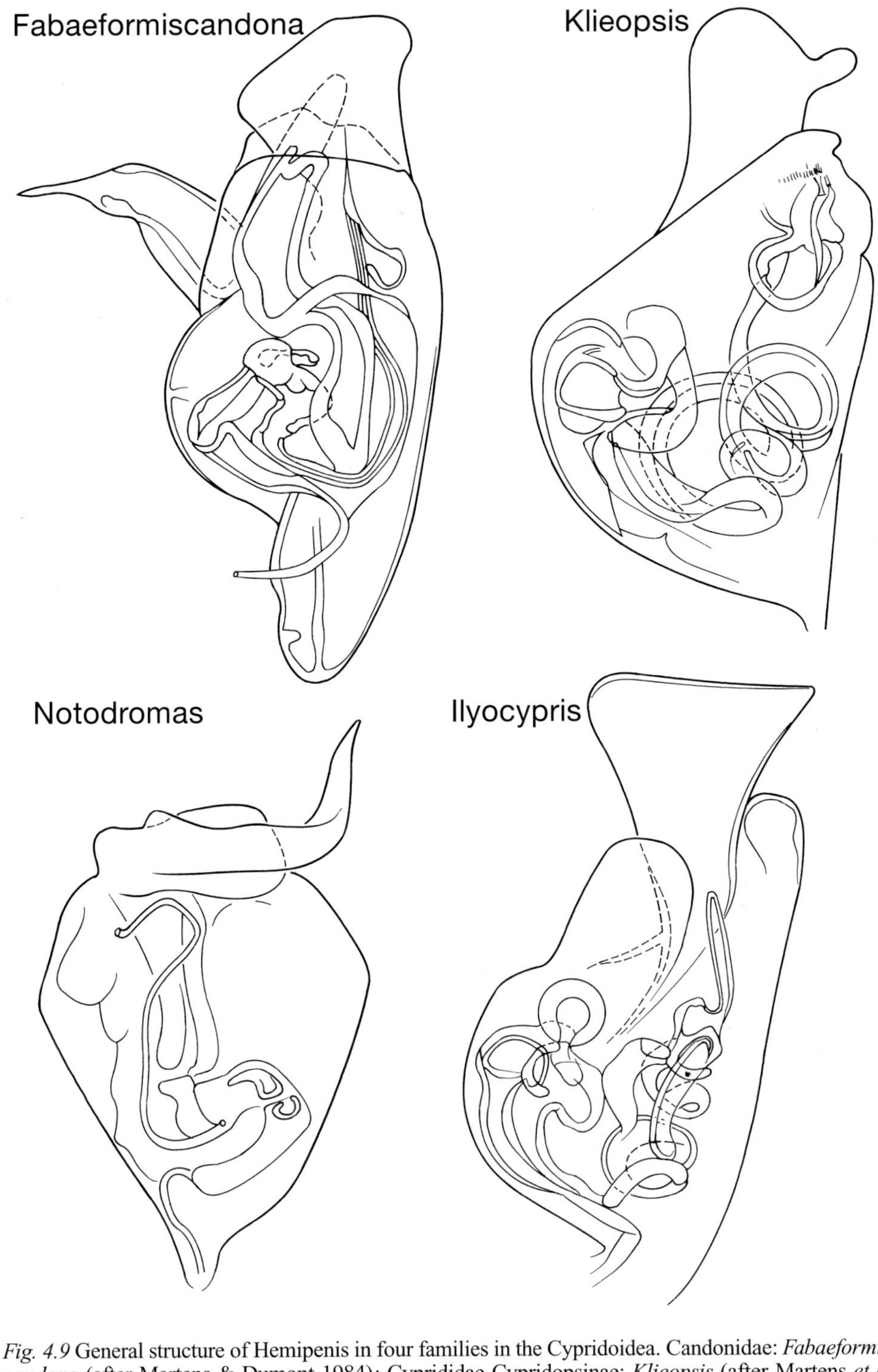

Fig. 4.9 General structure of Hemipenis in four families in the Cypridoidea. Candonidae: *Fabaeformiscandona* (after Martens & Dumont 1984); Cyprididae-Cypridopsinae: *Klieopsis* (after Martens *et al.* 1991). Notodromadidae: *Notodromas*. Ilyocyprididae: *Ilyocypris* (after Martens 1991a). Not to scale.

Cytheroidea

The hemipenes (Fig. 4.10) in this group have a built-in sperm-pump, which turns them into muscular organs. They are large, occupying c 1/4-1/3 of the male body cavity and consist mainly of the following parts: (1) the peniferum, in which different rami are incorporated, (2) a solid mass of different *muscles*, (3) an internal *labyrinth* and (4) the external *copulatory complex* (Martens 1990b).

The *peniferum*, a semi-transparent sheet, envelops the muscles and the labyrinth, but not the copulatory complex, and comprises the different rami. It has a frontal expansion, the *lateral lobe* and can have both a *distal* and a *proximal lobe*. The latter is present in all species, the former is pronounced in a few taxa only. Dorsally (and on the external, lateral, side) there is a *movable trabecule*, which has an uncertain function.

Several long and curved *ridges* fuse the medial and the lateral sheets of the peniferum and serve for the attachment of a number of muscles.

The *labyrinth* consists of 3 main parts, all sclerotised structures in which irregular cavities run, and at least one large muscle is attached to each of the three parts. These three muscles make up 50-70% of the total mass of muscles in the hemipenis.

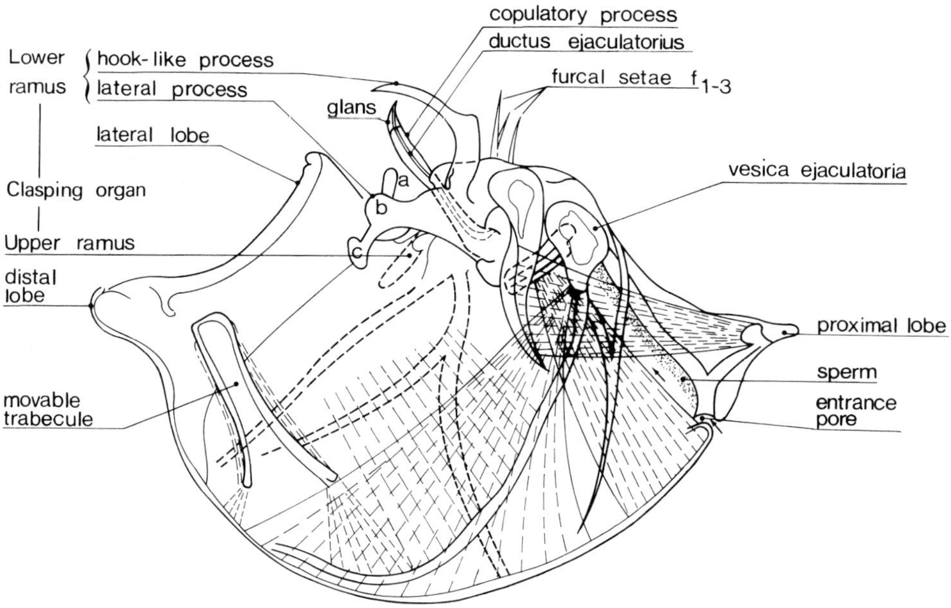

Fig. 4.10 General structure of Hemipenis in Cytheroidea, example of *Limnocythere inopinata* (Limnocytheridae – after Martens 1990b).

The *copulatory complex* typically consists of upper and lower ramus of the clasping organ, copulatory process and three setae of the caudal ramus, now incorporated in the hemipenis. The *clasping organ* can have different shapes and forms and can also be completely reduced. The *copulatory process*, the actual intromitting organ, is an elongated, sclerified trabecule in which a *ductus ejaculatorius* is clearly visible. The *glans*, the distal part, is somewhat separated from the main body of the copulatory process and can be of highly variable morphology, but nearly always ends in a distal hook, at the tip of which opens the *gonopore.* Three *setae of the caudal ramus* are present on each hemipenis. In Limnocytheridae, the copulatory complex is never completely enveloped by the peniferum, and there is therefore little or no need for an internal erection. Fig. 4.11 shows the difference in hemipenis structure in two genera of the two subfamilies of Limnocytheridae.

Again, females have far fewer diagnostic features in their reproductive organs.

Darwinuloidea

These are ancient asexuals and reproduce by obligate parthenogenesis. The group has reproduced asexually at least since the Cretaceous, probably since the end of the Triassic (see chapter 3) and no bisexual populations have ever been found in extant taxa.

From Recent faunas, putative males of *D. stevensoni* have been reported twice. In their original description of *D. stevensoni*, Brady & Robertson (1870) (see also Brady & Norman, 1889) reported and even illustrated hemipenes and carapace of a male of this species. Turner (1895) reported a male of *D. improvisa* (a synonym of *D. stevensoni* – Müller, 1912) from North America, but only illustrated the A1 and T3 of this presumed male and as these illustrations show no potential sexual dimorphism, this record cannot be validated on the evidence provided. The illustrations of Brady & Robertson (1870) and Brady & Norman (1889) (redrawn in Fig. 4.12), however, show a laterally compressed carapace in dorsal view and this morphology could be related to the absence of a brooding chamber in this specimen. The drawing of the hemipenis is difficult to interpret, although Maddocks (1973) attempted to homologise the different parts, and distinguished separate hemipenes and Zenker organs. It seems that at least the male found by Brady & Robertson (1870) cannot be disregarded. Rossetti & Martens (1996) and Schön *et al.* (1998) discussed this putative male and concluded that it was a non-functional atavistic specimen, useful from a morphological-taxonomical point of view, but of little importance for reproductive strategies (chapter 17).

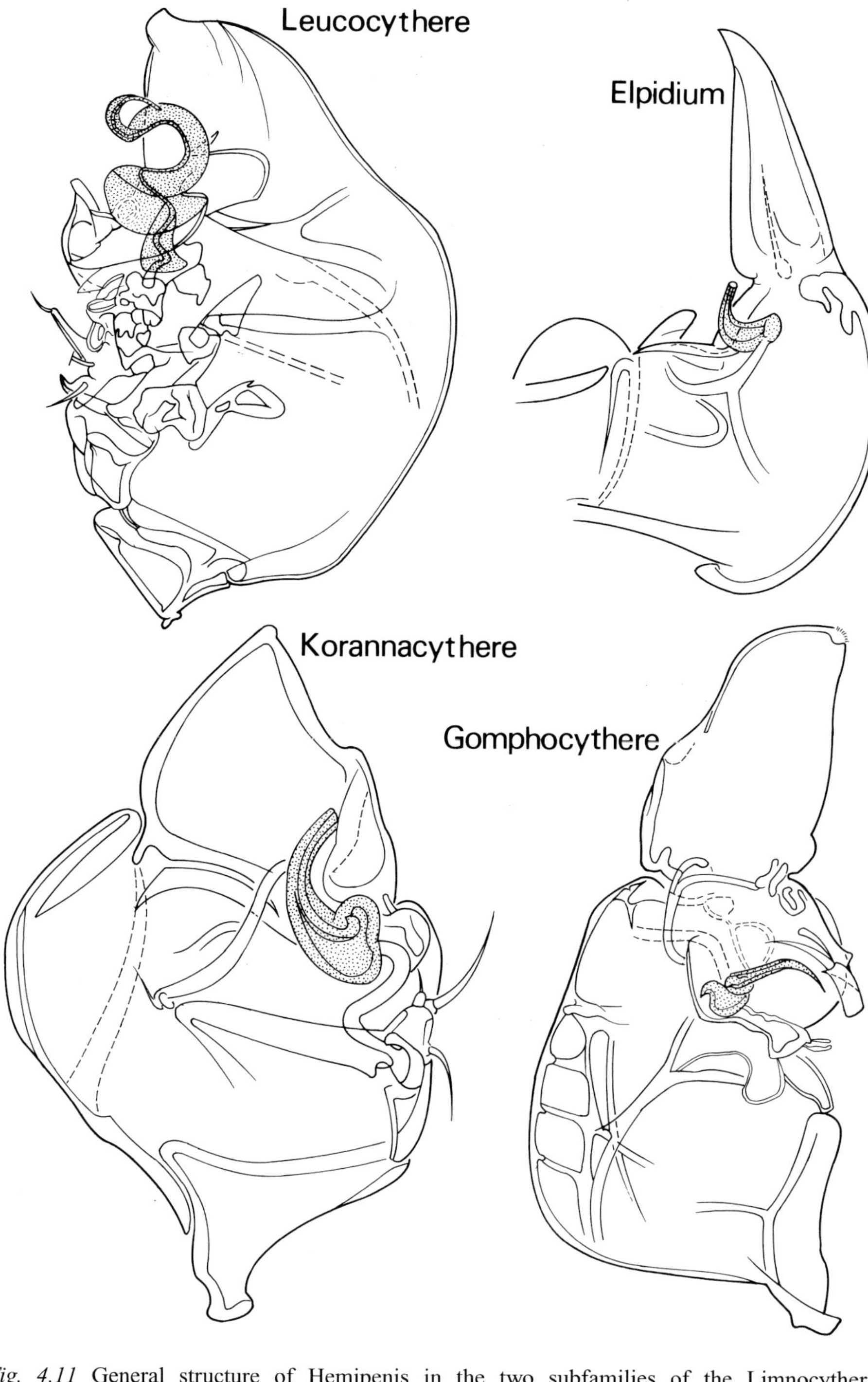

Fig. 4.11 General structure of Hemipenis in the two subfamilies of the Limnocytheridae. Limnocytherinae: *Leucocythere* (after Martens 1991b) and *Korannacythere* (after Martens 1996). Timiriaseviinae: *Elpidium* (after Danielopol, 1975) and *Gomphocythere* (after Martens 1993). Not to scale.

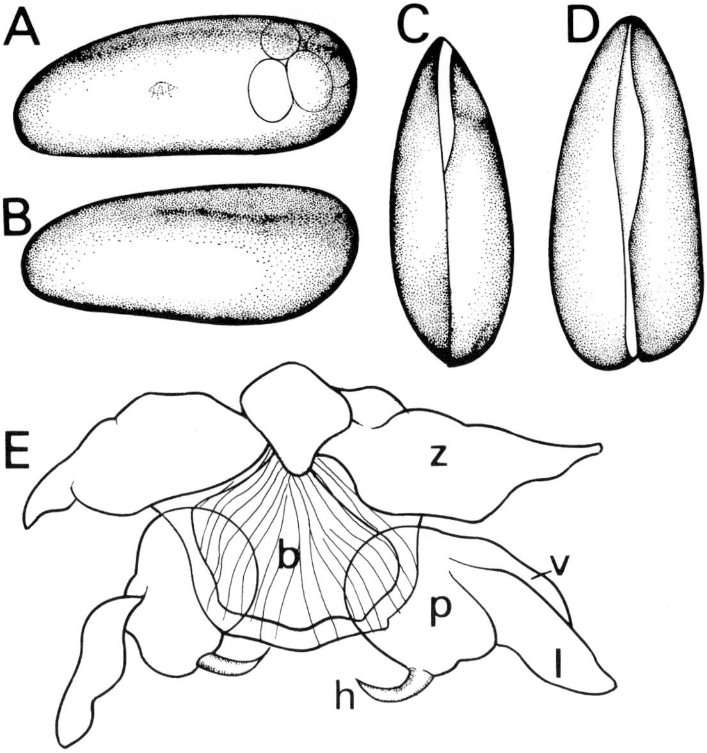

Fig. 4.12 Original illustrations of male and female *D. stevensoni*.
A, D= female. B, C= male. E= Hemipenis. (redrawn after Brady & Norman 1889). Symbols in E are interpretations following Maddocks (1973): b= posterior part of body, comprising testes; h= hook-like process; l= supporting lamella or lobe; p= body of hemipenis; v= vas deferens; z= Zenker's organ. Not to scale.

References

Athersuch, J., Horne, D.J.& Whittaker J.E. 1989. Marine and Brackish Water Ostracods. *Synopses of the British Fauna* (New Series), No.43: 359pp.

Brady, G.S. & Norman, A.M., 1889. A monograph of the marine and freshwater Ostracoda of the North Atlantic and of North-Western Europe. Section I: Podocopa. *Sci. Trans. R. Dublin Soc., ser.* **2, 4 (2):** 63-270.

Brady, G.S. & Robertson, D., 1870. The Ostracoda and Foraminifera of Tidal Rivers. *Ann. Mag. nat. Hist., Ser.* **4, 6:** 1-33, 307-309.

Broodbakker, N.W. & Danielopol, D.L. 1982. The Chaetotaxy of Cypridacea (Crustacea, Ostracoda) limbs: Proposals for a descriptive model. *Bijdr. Dierk.*, **52:** 103-120.

Broodbakker, N.W. 1982. The genus *Heterocypris* (Crustacea, Ostracoda) in the West Indies. Part I. Taxonomic characters. *Bijdr. Dierk.,* **52:** 207-227.

Danielopol, D.L. 1969. Recherches sur la morphologie de l'organe copulateur mâle chez quelques ostracodes du genre *Candona* Baird (Fam. Cyprididae Baird). *in*: Neale, J.W., *The taxonomy, morphology and ecology of recent Ostracoda*, Oliver & Boyd: 136-153.

Danielopol, D.L. 1975. Remarques sur la diversification morphologique de trois nouvelles espèces d'*Elpidium* (Ostracoda) à Cuba. *Bull. am. Paleontol.,* **65:** 47-60.

Danielopol, D.L. 1978. Über Herkunft und Morphologie der Süsswasser-hypogäischen Candoninae (Crustacea, Ostracoda). *Sitzungsber. Österr. Akad. Wissensch. Mathem.-naturw. Kl., Abr. I*, **187(1-5):** 162 pp.

Gonzalez Mozo, M., Martens, K. & Baltanás, A. 1996. A taxonomic revision of European *Herpetocypris* Brady & Norman, 1889 (Crustacea, Ostracoda). *Bull. k. belg. Instit. Natuurw. Biol.*, **66:** 93-132.

Hartmann, G. 1966. Ostracoda. *in*: *Dr H.G. Bronns Klassen und Ordnungen des Tierreichs*, 5. Arthropoda, 1. Crustacea, **2(4/1):** 1-216.

Hartmann, G., 1977. *in*: Neale, J. W., Discussion on ostracod terminology. *in*: Löffler, H. & Danielopol, D. (eds), *Aspects of ecology and zoogeography of Recent and fossil Ostracoda*, Junk, The Hague: 495-497

Kaufmann, A. 1896. Die Schweizerischen Cytheriden und ihre nächsten Verwandten. *Rev. suisse Zool.*, **4:** 313-384.

Kesling, R. V., 1951. The morphology of ostracod molt stages. *Illinois biol. Monogr.*, **21:** 1-26.

Klie, W. 1938. Ostracoda, Muschelkrebse. *in*: Dahl, F. (ed), *Die Tierwelt Deutschlands und der angrenzenden Meeresteile*, Gustav Fischer: 230 pp.

Maddocks, R. 1973. Zenker's organ and a new species of *Saipanetta* (Ostracoda). *Micropaleontology*, **19:** 193-208.

Maddocks, R. 1982. Part 4. Ostracoda. *in*: Abele, L.G. (ed), *The Biology of Crustacea. Vol. 1. Systematics, the Fossil Record and Biogeography*, Acad. Press: 221-239.

Martens, K. 1986. Taxonomic revision of the subfamily Megalocypridinae Rome, 1965 (Crustacea, Ostracoda). *Verhand. k. Acad. Wetensch., Lett. Schone Kunst., Kl. Wetensch.*, **48:** 81 pp + 64 figs.

Martens, K. 1990a. Revision of African Cypridini. I. The genera *Cypris* O.F. Müller, *Pseudocypris* Daday and *Globocypris* Klie. *Bull. k. belg. Inst. Natuurw. Biol.*, **60:** 127-172.

Martens, K. 1990b. Revision of African *Limnocythere* s.s. Brady, 1867 (Crustacea, Ostracoda) with special reference to the Eastern Rift Valley Lakes: morphology, taxonomy, evolution and (palaeo) ecology. *Arch. Hydrobiol., Suppl.*, **83:** 453-524.

Martens, K. 1991a. On a small collection of non-marine ostracods from Mongolia (Crustacea, Ostracoda). *Misc. zool. hung.*, **6:** 53-60.

Martens, K. 1991b. A new and enigmatic representative of *Leucocythere* Kaufmann, 1892 from the Eastern Cape Province (South Africa) (Crustacea, Ostracoda). *J. nat. Hist.*, **25:** 583-596.

Martens, K. 1993. On the taxonomy and zoogeography of the genus *Gomphocythere* Sars, 1924 (Crustacea, Ostracoda), with the description of a new species from the Nahal Dan (Israel). *Belg. J. Zool.*, **123:** 39-54.

Martens, K. 1996. On *Korannacythere* nov.gen. (Crustacea, Ostracoda) a new genus of temporary pool limnocytherids from southern Africa, with the description of three new species and a generic reassessment of the Limnocytherinae. *Bull. k. belg. Inst. Natuurw. Biol.*, **66:** 51-72.

Martens, K. & Dumont, H.J. 1984. The ostracod fauna (Crustacea, Ostracoda) of Lake Donk (Flanders): a comparison between two surveys 20 years apart. *Biol. Jb., Dodonaea,* **52:** 95-111.

Martens, K., Meisch, C. & Marmonier, P. 1991. On *Klieopsis* n.gen., with a redescription of *Cypridopsis horai* Klie, 1927 (Crustacea, Ostracoda). *Bull. k. belg. Inst. Natuurw. Biol.*, **61:** 55-64.

Martens, K., Davies, B.R., Baxter, A.J. & Meadows, M.E. 1996. Taxonomy and ecology of the Ostracoda (Crustacea) from the Verlorenvlei (Southwestern Cape, RSA). *S. afr. J. Zool.*, **31:** 23-36.

Matzke-Karastz, R. 1995. Aktuelle Gattungs- und Artmerkmale bei *Scottia, Cyclocypris, Psychrodromus* und *Mesocypris* (Ostracoda). *Geol. Univ. Köln, Sonderveröffentl.*, **97:** 285 pp. + 28 plts.

Meisch, C. 1996. Contribution to the taxonomy of *Pseudocandona* and four related genera, with the description of *Schellencandona* nov.gen., a list of Candoninae genera and a key to the European genera of the subfamily (Crustacea, Ostracoda). *Bull. Soc. Natur. Luxemb.*, **97:** 211-237.

Moore, R.C. (ed) 1961. *Treatise on Invertebrate Palaeontology. Part Q, Arthropoda 3, Crustacea, Ostracoda. Geol. Soc. Am.* Univ. Kansas Press.

Müller, G.W. 1912. Ostracoda. *in*: *Das Tierreich*, **31:** 434 pp.

Rossetti G. & Martens, K. 1996. Redescription and morphological variability of *Darwinula stevensoni* (Brady & Robertson, 1870) (Crustacea, Ostracoda). *Bull. k. belg. Inst. Natuurw. Biol.*, **66:** 73-92.

Schön, I., Butlin, R.K., Griffiths, H.I. & Martens, K. 1998. Slow evolution in an ancient asexual ostracod. *Proc. R. Soc. Lond. B,* **265:** 235-242.

Smith, R. & Martens, K. (in press). The ontogeny of the cypridid ostracod *Eucypris virens* (Jurine, 1820) (Crustacea, Ostracoda). *in*: Horne, D.J. & Martens, K. (eds), Proceedings of the XIII International Symposium for Ostracodologists. *Hydrobiologia*.

Turner, C.H., 1895. Fresh-water Ostracoda of the United States. *Minnes. geol. nat. Hist. Surv., zool. Ser.*, **2:** 277-337.

Van Morkhoven, F.P.C.M. 1962. *Post-Palaeozoic Ostracoda. Their morphology, taxonomy and economic use,* Elsevier: 204 pp.

Wingstrand, K.G. 1988. Comparative spermatology of the Crustacean Entomostraca. 2. Subclass Ostracoda. *Biol. Skr. Kong. Dansk. Vidensk. Selsk.,* **32:** 1-149.

Xianguang, H., Siveter, D.J., Williams, M., Walossek, D. & Bergström, J. 1996. Appendages of the arthropod *Kunmingella* from the early Cambrian of China: its bearing on the systematic position of the Bradoriida and the fossil record of the Ostracoda. *Phil. Trans. R. Soc. Lond. B*, **351:** 1131-1145.

Yin, Y. & Martens, K. 1997. On a new species of *Fabaeformiscandona* Krstić, 1972 (Crustacea, Ostracoda) from China, with a preliminary checklist of Recent non-marine ostracods. *Hydrobiologia,* **357:** 117-128.

CHAPTER 5

GEOGRAPHICAL DISTRIBUTION OF REPRODUCTIVE MODES IN LIVING NON-MARINE OSTRACODS

D.J. Horne, A. Baltanás and G. Paris

Abstract

The global distribution of non-marine ostracods is reviewed briefly and the limitations of available data are discussed. The geographical distribution of reproductive modes is discussed in general terms, with particular consideration of factors influencing our perception of the phenomenon of geographical parthenogenesis, including reproductive and dispersal strategies which may affect sex ratios. Geographical parthenogenesis is shown to occur in shallow marine ostracods, although not as commonly as in non-marine taxa. The European distribution of non-marine ostracods, with particular reference to reproductive modes, is illustrated and discussed using the NODE GIS database. Typical patterns of geographical parthenogenesis are found almost entirely in the Cyprididae (e.g., *Eucypris virens*, *Heterocypris incongruens*, *Herpetocypris* spp.) and only in one or two species belonging to other families (e.g. *Pseudocandona albicans*, *Limnocythere inopinata*).

Introduction

Although ostracods are well-represented in almost all kinds of non-marine aquatic habitats, they remain ill-investigated throughout the world, as compared to other micro-Crustacea. Martens & Behen (1994) suggest two reasons for this: (a) the skills required to dissect and hence to determine the taxonomic identity of ostracods, and (b) the scarcity of adequate taxonomic compilations including keys and figures to help newcomers in the field. These are but two of the large array of 'cultural' biases, like the boreal bias and the megafauna bias, known to exert a dramatic influence on our understanding of biogeographical patterns in the natural world.

Our knowledge of non-marine ostracod biogeography is biased towards temperate boreal regions (geography) and permanent water bodies (habitat). The geographical bias seems to be much more important for ostracods than for other groups and is not only 'boreal' but predominantly 'European'; this point can be illustrated by reference to some existing checklists. The ostracod list in *Limnofauna Europaea* (Löffler & Danielopol 1978), a classic checklist of all the animals inhabiting European inland waters (but also covering some areas of North Africa and Asia Minor) included 409 species for an area of about 11-12 x 10^6 km^2. Compared to estimates for other (sub-) continents (De Deckker 1983, for Australia; Martens

Sex and parthenogenesis: evolutionary ecology of reproductive modes in non-marine ostracods, edited by K. Martens, pp. 77–99

1984, for African inland waters; Delorme 1991, for North America; Martens & Behen 1994, for South America) there is a remarkable difference in species density (number of species per unit area) (Fig. 5.1). Of course, data in some of the above-mentioned checklists are somewhat obsolete and much new information has been gathered during the last decade from the least-known areas (e.g. China: see McKenzie 1986), but there remains a significant European bias in our current knowledge of non-marine ostracods. Considering the large area of North America and its diversity of habitats (from arctic to subtropical), our knowledge of its non-marine ostracod fauna is especially poor.

Global distribution of non-marine ostracods

Considered globally, a high degree of endemicity is seen at the species level, especially in the southern hemisphere (about 85% of African and South American species appear to be endemic: Martens 1984, Martens & Behen 1994, Martens 1998), but this decreases at the genus level (35% in Africa, 20% in South America) and is non-existent at family level. Australia seems to harbour a diverse halobiont fauna with many endemic genera (De Deckker 1983) and the endemic subfamily Ngarawinae. Africa has a diverse fauna of semi-terrestrial ostracods (*Terrestricandona*, *Mesocypris*, *Bryocypris*), while in South America (including Central America and the Caribbean) several species of *Elpidium* have been found living in bromeliads (Danielopol 1975). It is noticeable, nevertheless, that a majority of species (63%) known to occur in Africa have been reported in only a single publication (Martens 1984). Further field work is needed to determine whether or not such taxa are truly endemic.

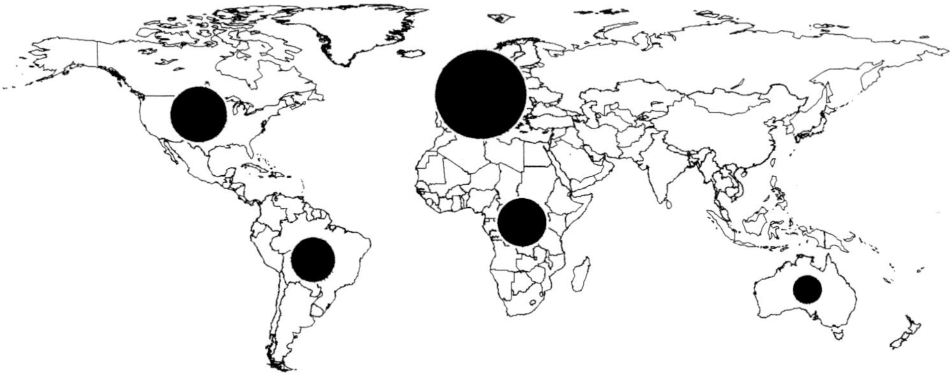

Fig. 5.1 Non-marine ostracod distribution at a global scale. Dot sizes are proportional to species density (species richness/area).

Attempts to explain the processes behind biogeographical patterns in non-marine ostracods have paid special attention to ecological factors (McKenzie 1971, 1986, Carbonel *et al.* 1988). The existence of latitudinal diversity gradients is difficult to assess although they are likely to occur. Species richness in Europe and Africa are similar, but research effort (evaluated as number of publications dealing with non-marine ostracods) is more than three times greater in the former region. This suggests that further attention to tropical areas will add significantly to the number of species recorded there, which is in line with the general trend of increasing species richness with decreasing latitude observed in most animal and plant groups (Myers & Giller 1988).

An increase in recorded global ostracod diversity is also anticipated from work in ancient lakes. These habitats are large lakes of tectonic origin and great antiquity, in which species-rich communities with large concentrations of endemics have developed (Martens 1994). The most important of these, Lake Tanganyika and Lake Baikal, together harbour more than 230 species and subspecies (although 400 are expected to be found), 90% of them endemic (Martens 1994). However, ancient lakes are but singularities in the wide array of continental habitats. The diverse ostracod faunas found in those lakes are the result of local processes acting over long periods of time, and their high proportion of endemics means that their actual influence on global distribution patterns (rather than simply global diversity) might be low.

General aspects of geographical distribution of reproductive modes in non-marine ostracods

Geographical distribution is closely linked to reproductive and dispersal methods in non-marine ostracods. McKenzie (1971) recognised three kinds of reproductive modes among non-marine ostracods: (1) reproduction wholly by parthenogenesis; (2) bisexual reproduction; and (3) geographical parthenogenesis (the occurrence of sexual populations in only a small, usually marginal, part of the range of the species). Although it is tempting to categorise each ostracod species according to the sexual/ parthenogenetic dichotomy (e.g. Chaplin *et al.* 1994, Griffiths & Butlin 1995), there are arguments against such a simple classification.

First, for a large number of taxa, reproductive mode seems to be a feature linked to the population level rather than the species level. For some species previously assumed to be strict parthenogens, males have eventually been found. *Potamocypris villosa* was considered to be entirely parthenogenetic until Martens & Meisch (1985) discovered a sexual population in northern Spain (with a sex ratio of approximately 1:1).

Second, a species is called parthenogenetic (or 'asexual') on the basis of negative evidence (no males have been found) which is heavily dependent on sampling effort. Among cypridids, males tend to be smaller than females and have shorter life spans. This could be attributed to a lack of mating competition between males and/or greater fecundity of larger females, neither of which has yet been demonstrated. Resource competition between the sexes is another possible explanation (Van Harten 1983), but there is no evidence for a relationship between degree of sexual dimorphism and reproductive success in ostracods. Although there are no actual data on intraspecific sex-ratio variability, it seems to be quite common at interspecific level in both marine (Abe 1990) and non-marine (Mikulic 1965) ostracods. But even if a 1:1 primary sex ratio is the rule, tertiary sex ratios can be heavily biased towards females due to differential male mortality. Thus, finding males is dependent not only on reproductive mode, but also on when (sampling season) and how (sample size) populations are studied (see also chapter 2).

Third, it has been shown that the presence of males need not imply the absence of parthenogenetic females, which may co-occur in the same water body (Chaplin 1993, see also chapter 15). If this is not an infrequent phenomenon, the distribution of parthenogenetic reproduction may have been underestimated, especially in rare species.

In spite of these problems, we are fairly certain that some species, like *Cypria ophtalmica*, reproduce almost exclusively syngamically, because hundreds of populations have been sampled and males have been recorded frequently. Unfortunately, the presence of males is not always reported in the literature; 75% of records for European populations of *C. ophtalmica* make no mention of the presence or absence of males. In most cases this is almost certainly because the ostracodologists concerned were either not interested in sex ratios or considered the presence of males to be trivial and not worth mentioning. Nevertheless, there are some intriguing records stating explicitly that no males were found in *C. ophtalmica* (Klekowski 1952, Petkovski 1977, Scharf 1982).

We are equally certain that some taxa are exclusively parthenogenetic, but once again there are enigmatic exceptions. The most striking case is that of *Darwinula stevensoni* (see Fig. 5.6), a species belonging to a lineage for which parthenogenesis has probably been the rule for more than 100 million years (see chapter 6), and for which only a single male (almost certainly non-functional) has been recorded reliably among hundreds of populations (Rossetti & Martens 1996, see chapter 4).

Between these extremes, wholly parthenogenetic and wholly sexual species, there is a continuum. The exact position of a species within that continuum may not be totally controlled by phylogeny, but there is a clear non-random distribution of sexuality and parthenogenesis among different taxonomic hierarchies. Parthenogenesis is fairly common in the Cypridoidea, less frequent in the Cytheroidea (see Fig. 5.15 and discussion later in this chapter). In several cypridid genera (*Bradleystrandesia*, *Cypridopsis*, *Eucypris*, *Potamocypris*) males are unknown for most species, whereas in some others (*Fabaeformiscandona*, *Cyclocypris*, *Candonopsis*, *Cypria*, *Mixtacandona*, *Pseudocandona*) syngamic populations are frequent and males have been described for most species.

Geographical parthenogenesis in non-marine ostracods

Overlapping the lineage-specific effect, there is a geographical effect: some conspecific populations and/or congeneric species display disjunct distributions for sexuals and parthenogens. This phenomenon, usually referred to as geographical parthenogenesis, has been hypothesised to be environmentally cued, mainly through temperature (McKenzie 1971, 1986, Tétart 1974, Neale 1983) but also through habitat stability (McKenzie & Moroni 1986). Sexually reproducing forms are assumed to occupy their native environments, whereas parthenogenetic forms disperse to peripheral areas and beyond (McKenzie 1971). The roles of passive dispersal and migrations are emphasised, thus focusing on those groups with desiccation-resistant eggs (Cypridoidea). The concept of 'centre of origin', a central tenet for dispersal biogeographers, is central to McKenzie's explanatory scheme. The genus *Isocypris* (Cyprididae, Isocypridinae) is a good example. Five species (*I. africana*, *I. devexa*, *I. nivea*, *I. perangusta* and *I. priomena*) are known to occur in Africa (South Africa, Kenya and Tanzania), some of them including males (McKenzie op.cit., Martens 1984). Species of this genus are also known in Europe (NODE database), North America (Delorme 1970) and South America (Martens & Behen 1994) but no males have ever been found. The presence of sexual reproduction is assumed to be the primitive condition and defines an area where 'such taxa evolved' (McKenzie 1971: 227) and from where the species migrated, occupying less favourable environments which impede the maintenance of sexual reproduction. Observed patterns of geographical parthenogenesis in Europe have been explained in terms of the recolonisation of new habitats after the ice retreated at the end of the last glaciation; parthenogens are thought to have spread rapidly northwards, while sexual populations are found in more stable refugia at lower latitudes (see chapter 6 for further discussion). Objections already raised with regard to dispersal biogeography are applicable here. The stability of large areas over long periods of time (the 'refuge' hypothesis) is debatable. There is growing evidence that even the tropical rain forests suffered from severe climate changes during Pleistocene glacial periods. Areas displaying the greatest diversity of species are not necessarily the 'centre of origin' for the group, as is shown in the remarkable case of Lake Victoria's cichlids (Johnson *et al.* 1996). Indeed, there is no agreement on criteria for determining 'centres of origin'. Finally, this approach completely ignores phylogenetic information. This is surprising, considering that phylogenetic analysis in ostracods is favoured by the wide array of morphological characters available (a feature common to most crustaceans) and their long fossil record (non-marine ostracods are certainly known from the Carboniferous onwards – see chapters 3, 6). Thus far, there has been only a single attempt to apply cladistic analysis to large-scale biogeography in non-marine ostracods (Martens & Coomans 1990).

Sex ratios, reproductive modes and dispersal mechanisms

The underlying genetic mechanisms of skewed sex-ratios (including rare male occurrences) are discussed in chapter 2; the functionality of the latter in chapter 17. However, it is also possible that at least some observations of skewed sex-ratios in non-marine ostracods may be explained by consideration of the relationships between reproductive modes and dispersal mechanisms. There are two main ways in which non-marine ostracods may achieve dispersal between separate aquatic habitats: (1) as desiccation-resistant eggs, either by wind or in mud stuck to a living intermediary such as a tetrapod or bird; (2) as adults or juveniles, possibly in a torpid state, in mud attached to animals (brood care may be advantageous but not essential to this type of dispersal) (De Deckker 1977). One possible explanation of an anomalous sex-ratio is that the population/ deme under consideration grew relatively recently in the sampled habitat. Such a population might have originated from a colonisation event that took place only a few generations ago. Whether dispersal was by eggs or by live transport of adults, it is highly probable that a small founding colony of ostracods would have a skewed sex ratio, given the genetic variation that might be expected to exist within the larger parent population. It could take many generations to achieve the equilibrium described above.

However, colonisation by a single individual is a risky business and likely to have a lower success rate than colonisation by many – by transplanting a fairly large sample of a population as adults, juveniles or eggs. So maybe we should expect to find strongly skewed sex ratios sometimes, but not very often. Perhaps this is why populations containing rare males are themselves rare. There are various ways in which a species might improve its prospects of multiple dispersal (a sample of a population) as opposed to solitary dispersal to a new site. One is by having high population densities (achieved either by high productivity, or by ecological restriction to a closely defined niche, or both) so that the dispersal medium is more likely to affect a lot of individuals at once. Aquatic birds wading or diving in the shallows of a lake might 'collect' a quite heterogeneous sample of an ostracod population to be transplanted when they fly to the next lake. Another strategy is to have fertilised females, or eggs, available for dispersal at a particular (and quite restricted) time of year, so as to take advantage of a particular dispersal medium (migrating birds, strong winds, etc.). Such temporal (seasonal?) maximisation of reproductive potential would also make it more likely that individuals from different populations would be dispersed simultaneously and be mixed in the new habitat. This may be the strategy of *Metacypris cordata* (see Danielopol & Horne 1996, Danielopol *et al.* 1996): the optimum time of year for dispersal would be autumn, when the transport of a single female containing fertilised eggs or developing juveniles in the brood space would be sufficient to colonise a new site. A sample collected in September, for example, from a square metre of lake margin vegetation (whether deliberately by a human with a net or unwittingly by a duck) would be likely to contain many, possibly hundreds of brooding females, thus greatly increasing the chances of a successful colonisation event.

This argument also has implications for the dispersal strategy of a parthenogenetic brooder, *Darwinula stevensoni.* This ostracod is often found with very high population densities in shallow water (around 1.5m deep) near the margins of lakes, which would facilitate the dispersal of brooding females encapsulated in wet mud on the feathers of diving birds. This in turn might explain why this species has such a well-defined optimum depth range. In a population in a single lake, individuals with a preference for deeper water would be less likely to get dispersed to other lakes, while those venturing into shallower water are more likely to experience unfavourable temperature extremes or even desiccation due to fluctuating lake levels. Selection and dispersal would thus define and maintain a fairly narrow optimum depth range for the species.

Asexual lineages can colonise new water bodies more easily than sexual ones (leaving aside considerations of actual dispersal mechanisms), since the minimum required to establish a new colony is a single egg, while in a sexual species at least one of each sex is needed to found a sexual population. For a species in which parthenogenesis is an option, therefore, the parthenogenetic component of the metapopulation will spread more easily than the sexual component; and this is indeed one of the main arguments advanced to explain geographical parthenogenesis in European non-marine ostracods (see chapter 13). Of course, colonisation involves not only dispersal, but also competition with other species, in which sexual reproduction may be advantageous (chapters 14, 17).

Geographical parthenogenesis in marine ostracods

During the course of our research on reproductive modes in European non-marine ostracods, we have often met with the view that parthenogenesis, while relatively common in non-marine ostracods, is rare in marine ostracods. In fact, we have found evidence of parthenogenesis in several marine and brackish water taxa, suggesting that it may not be such a rare phenomenon after all. Bearing in mind that this evidence is circumstantial, being based entirely on the absence of males from field collections, we are reasonably certain that parthenogenesis occurs in the cytheroidean genera *Aurila*, *Hemicythere* and *Callistocythere*, and in the cypridoidean genus *Paracypris*, at least in species living in European coastal waters.

The hemicytherid genus *Aurila* has a wide distribution in the northern hemisphere (Atlantic and Pacific) with most species inhabiting warm temperate shallow marine waters. Three European species offer evidence of geographical parthenogenesis (Athersuch *et al.* 1989). *Aurila arborescens* has sexual populations in the Mediterranean, but only females have been recorded in the only known English population on the coast of SW Wales. *Aurila woutersi* lives around the coast of southern England and probably also in the Mediterranean; no males have yet been recorded, in spite of a study of its population dynamics in the Bristol Channel, based on frequent sampling spanning 18 months (Horne 1980). In the same study, the population dynamics of *A. convexa* were also monitored and no males were ever found; in the western Mediterranean all-female populations are found, but in the eastern Mediterranean sexual populations have been described from Cyprus. There is an isolated record of a male *A. convexa* from interglacial deposits in the southern North Sea (Horne *et al.* 1990), suggesting that the distribution of sexual and parthenogenetic populations has shifted over time, possibly in response to climate change. Parthenogenetic *Aurila arborescens* extended as far north as Denmark (57°N) during the postglacial temperature maximum (approximately 6000-7000 years ago) (Vork & Thomsen 1996) but the most northerly living population known today is on the Welsh coast (52°N).

Another hemicytherid, *Hemicythere villosa*, a very common littoral to shallow sublittoral marine species on the coasts of NW Europe, is found in both sexual and asexual populations within a few km of each other in at least two different areas in the UK. In the Bristol Channel in normal marine salinities, no males were ever recorded during an 18-month long study of its life-cycle, yet in slightly reduced salinities in the mouth of the Severn Estuary less than 20 km distant, populations with 1:1 sex-ratios were recorded during the same period (Horne 1980). A similar phenomenon has been noted in the vicinity of the Fleet in Dorset (Whittaker 1972).

The Leptocytheridae also seem to include parthenogenetic representatives in Europe: the genus *Leptocythere* always seems to be represented by sexual populations, but two out of the three British species of the genus *Callistocythere* appear to exhibit parthenogenesis (Athersuch *et al.* 1989). *Callistocythere littoralis* is common on littoral and shallow sublittoral marine algae around Britain, Ireland, Norway, the Atlantic coast of France and in the Mediterranean; no males have ever been found. *Callistocythere badia* is another marine, littoral species that is widespread around British coasts, where only all-female populations are recorded; males have been described, however, from Cyprus in the eastern Mediterranean. Parthenogenetic *C. badia* occur in the Holocene of Denmark during the postglacial thermal maximum (Vork & Thomsen 1996). *C. murrayi*, on the other hand is a brackish water species known only from the British Isles, where it occurs as sexual populations in tidal creeks and estuaries.

Most marine cypridoideans in European coastal waters are represented by sexual populations (e.g., the genera *Pontocypris* and *Propontocypris*; Pontocyprididae), but males are unknown in *Paracypris polita* (Candonidae).

On the evidence available we can conclude that parthenogenesis occurs in less than 5% of the total British coastal marine and brackish water ostracod fauna, compared with over 40% of European non-marine species. We know of no examples of parthenogenesis in marine or brackish water ostracods in other parts of the world, but this may be partly due to inadequate sampling, as relatively few seasonal studies of ostracod population dynamics have been undertaken outside European seas. Bearing in mind the global, or at least northern hemispherical distribution of the four marine genera in which parthenogenesis has been observed, we consider it probable that other examples will be recognised in the future. Parthenogenesis in marine ostracods is thus comparatively rare, but not unknown.

European distribution of non-marine ostracods – the NODE database

As Europe has the most complete record of living non-marine ostracods (Fig. 5.1), it is the best area in which to test existing hypotheses concerning the distribution of taxa and their reproductive modes, particularly the question of geographical parthenogenesis. To this end, a database of the geographical and ecological distribution of living and fossil (Quaternary) ostracods in Europe has been compiled, based mainly on data abstracted from published literature (Horne *et al.* 1998). It was decided from the outset to develop a GIS (Geographical Information System) database using Unix-based Oracle and Arc/Info software, to ensure the greatest flexibility with regard to both data analyses and future expansion and applications; subsequently, the acquisition of pc-based ArcView GIS has facilitated data manipulation and map production (see Box 5.1). The existing database (named NODE – Non-marine Ostracod Distribution in Europe) contains more than 10,000 records of ostracods. Of course, Europe represents only a part of a much larger geographical area, the Palaearctic, but there are relatively few ostracod records available for the rest of the area. Coverage in NODE is best in central and western Europe, sparse in eastern, northern and southern areas, for several reasons, including our decision to take a very broad geographical view of Europe (including North Africa), the paucity of data from some regions such as northern Scandinavia, and the difficulty of abstracting data and identifying localities in foreign-language literature, especially in certain eastern-European countries. The most complete dataset for any country is that of Poland, resulting from the outstanding efforts of one of our associates, Tadeusz Namiotko. While it might have been quicker and easier to achieve the short-term aims of the project by simply marking records on paper maps, such a traditional approach would have been a dead end. As well as allowing flexibility for analyses, the GIS approach facilitates taxonomic corrections and continued development, with future potential applications in biodiversity studies, biomonitoring of water quality and Quaternary palaeoclimatology and palaeoecology.

In the sense that future growth and refinement of NODE will undoubtedly improve the dataset, the analyses that follow are preliminary, but this should not detract from their current value and validity.

NODE currently contains 10179 records of living and fossil ostracods abstracted from more than 500 publications (and including some of our own unpublished records). The majority of the fossil records were abstracted from a single recent compilation (Griffiths 1995). Before preparing the maps presented and discussed in this chapter, however, a number of records of marine/brackish taxa were removed; these were entered originally because they occur alongside non-marine taxa in slightly saline coastal waters (e.g., the Gulf of Bothnia), but they are not considered to be part of the European non-marine fauna. Following this adjustment, the Recent (living) dataset comprises 7968 records of 309 species (60 genera) from 1997 localities in 43 countries (Fig. 5.2); the Fossil (Quaternary) dataset comprises 1916 records representing 168 species (46 genera) from 198 localities in 20 countries (Fig. 5.3).

Box 5.1 NODE- a GIS database of non-marine ostracod distribution in Europe

NODE (Non-marine Ostracod Distribution in Europe) is a GIS (Geographical Information System) application developed at the University of Greenwich by members of the EU network project: *The evolutionary ecology of reproductive modes in non-marine Ostracoda.* It comprises two main components, which can be used either jointly or independently: (1) a relational geographical database, and (2) a graphic user interface (GUI). A third component is pc-based ArcView GIS software which we have used since the end of 1997 for most of our data manipulation and analyses, although NODE continues to 'live' on the Unix system.

1. Unix-based software: the database

The core of the application is a geographical database developed using the relational database management system ORACLE, running on a Unix platform. The database has been (and continues to be) compiled mainly with data from published literature, but also contains unpublished records from sampling carried out by members of the network laboratories during their research programmes. It is itself divided into two separate sub-databases: at the time of writing, RECENT contains more than 8000 records of living ostracods, FOSSIL nearly 2000 records of Quaternary ostracods (Pleistocene and Holocene). The structure of the two subsets is very similar, the main difference being that the RECENT database has fields for environmental information such as type of waterbody, while the FOSSIL database holds stratigraphical information. The basic unit for each sub-database is the occurrence of an ostracod species at a certain location, expressed as latitude and longitude coordinates. Data are input via a form; while data entry is in progress, key strokes can provide lists of possible values (e.g., generic and specific names) which have already been entered for other records, a feature which helps to maintain taxonomic consistency and is particularly useful when entering many records of the same species. Each record is automatically given a unique reference number (not shown on the data entry forms) and must contain certain essential information (generic and specific name, latitude and longitude) to be accepted. The geographical limits of the database extend from 25°W to 48°E and from 25°N to 72°N, representing an area considerably larger than the politically-defined Europe. A 'standard' water body classification system turned out, in practice, to have serious flaws, so we developed our own environmental classification of water bodies which required a considerable investment of time from various network members.

2. Unix-based software: the Graphic User Interface

The GIS GUI has been developed in Arc/Info for Unix platforms. The application allows the user to retrieve information from the database and create maps of ostracod distribution, either over the whole European study area or in smaller subsets. The user may choose to draw maps at different levels of taxonomic classification, from species up to subfamilies and families.

Box 5.1 Continued

Map displays may also distinguish between records for which males were reported and those for which males were stated not to be present or were simply not mentioned. Fossil (Quaternary) and living distributions for individual species may be compared. Points displayed on the map may be queried for locality information or species lists. The customized GUI allows all operations to be performed easily by point and click, selecting items from pop-up menus. The user therefore does not need to be a GIS expert to benefit from the database. The GUI now represents the first-stage interface with NODE for most users; those who need to work more intimately with the database are likely to use a copy of the dataset with pc-based software.

3. PC-based software: ArcView GIS
The acquisition of ArcView, a pc-based (and more user-friendly) version of Arc/Info, has made use of the database much easier. Working datasets are copied from ORACLE and can be manipulated in ArcView without risk of compromising the integrity of NODE itself. The examples of distribution maps presented herein were produced with ArcView. Some potential users do not require the whole dataset, or only need the data in tabular form; in one such case, data for a single species (*Candona candida*, approx. 400 records) were abstracted from NODE and emailed to the enquirer as an Excel file. Corrections or additions to the database can be fed back to the 'editor' (DJH) and, if considered acceptable, entered directly into NODE.

Establishing NODE has brought many advantages: the data are stored in one place in a structured, standardised way, information from very different sources can be integrated and interconnected, data can be used by different people for different purposes, and it is relatively easy to verify and correct data once they have been entered. It must be appreciated, however, that the development of NODE required a substantial investment of time and effort, not only in planning its structure and dealing with programming difficulties, but also (and most particularly) in abstracting data from the literature. The required data are difficult to find in many published papers since they are presented in many different forms, or not at all. Many authors simply do not give adequate locality data, so that data recovery from the literature is unlikely ever to be complete. Sometimes a sample site cannot be located precisely, either due to ambiguity (there may be several places with the same name in a country), name changes (many older Scandinavian and eastern European records suffer from this problem), or other reasons. In many cases a (temporary) solution is to give approximate coordinates, since on a Europe-wide scale a degree of latitude or longitude either way will not make much difference; in future, though, some users may wish to work at higher resolution, on individual countries or regions, in which case some record coordinates will need to be checked and refined. Suprisingly, perhaps, it is often the records of common species that cannot be entered: authors tend to be precise about where they collected a rare or unusual species, but for a ubiquitous ostracod like *Cypria ophtalmica* may simply state that it was found 'in lakes and ponds throughout the region' without actually naming even one locality. These problems are compounded when dealing with literature in languages other than English. It requires someone with a working knowledge of ostracods as well as the language concerned to abstract data; within our network we have been able to deal adequately with papers in French, German, Spanish and Italian, but have had to enlist external assistance for Polish, for example.

NODE is the first European ostracod database to be implemented in a GIS environment. Although at present it cannot be regarded as complete, it is nevertheless the best available compilaton of such data, with enormous potential for future development and applications. It will continue to be developed and managed under the control of an editorial board comprising members of each of the original network laboratories. Although it 'lives' at the University of Greenwich, it can be accessed remotely by anyone with appropriate facilities (subject to the agreement of the editorial board); an alternative is to be given a copy of the current dataset for use in one's own computer environment (many of the analyses discussed in this chapter were done in this way, using PC-based software, rather than in NODE itself). Eventually we intend to make it available for study at a Web site. GIS is difficult and time-consuming in the short-term, but in the long term it is worth the effort.

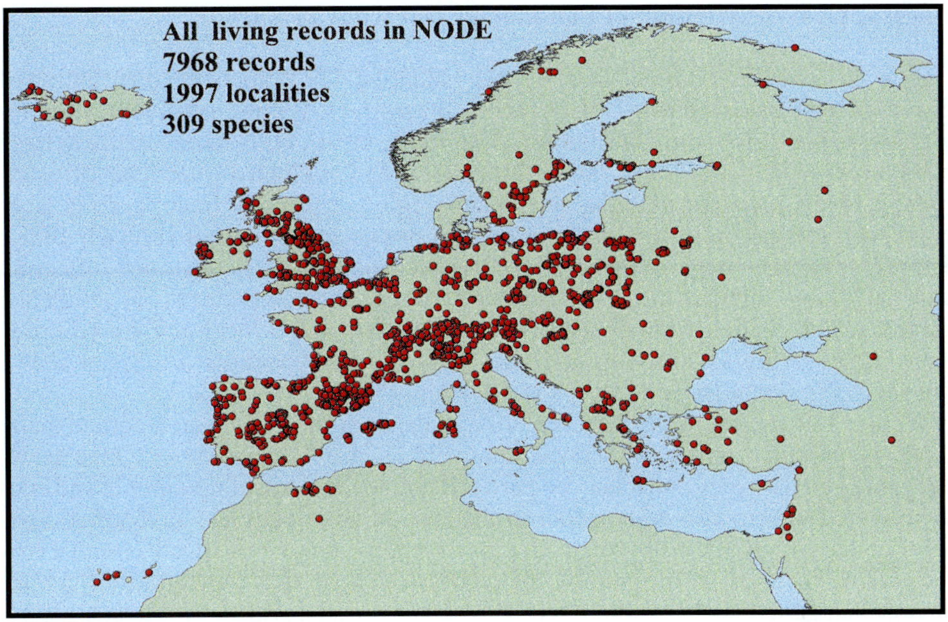
All living records in NODE
7968 records
1997 localities
309 species

Fig. 5.2

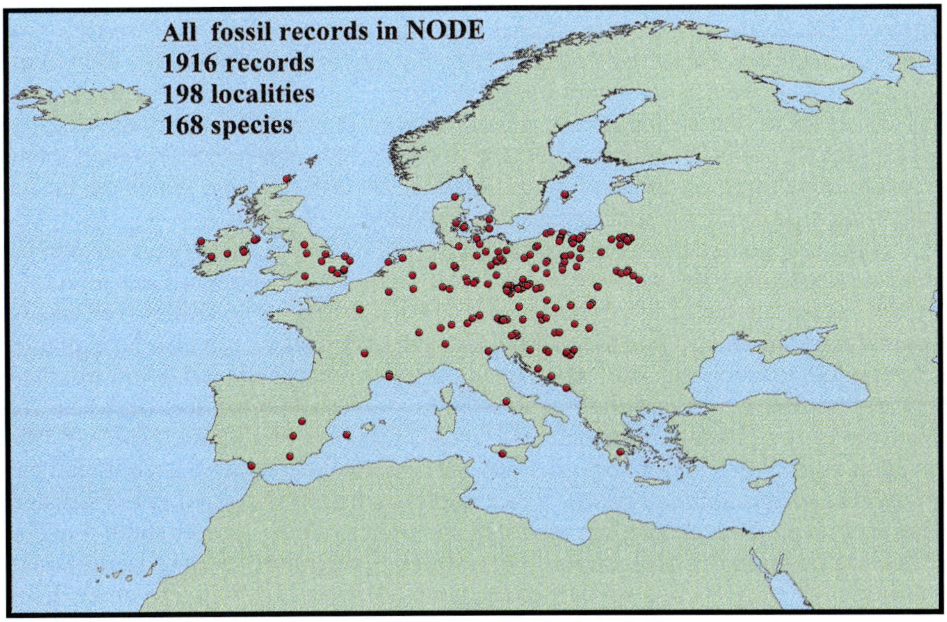
All fossil records in NODE
1916 records
198 localities
168 species

Fig. 5.3

Geographical distribution of non-marine ostracods in Europe

One of the first observations we can make is that NODE reflects the distribution of ostracodologists as well as of the ostracods themselves. Within individual countries we can see a non-random distribution of sampling effort, with some relatively large blank areas (Fig. 5.2). Given the high diversity and ecological plasticity of ostracods, it seems inconceivable that such blank regions truly lack them; even in large cities we can find them living in ornamental ponds and fountains (Meisch 1988). The exceptionally large number of records in Poland (2899) does not reflect an unusually large number of localities or species, but the completeness of the Polish dataset which includes much seasonal data, so that some localities are recorded many times. After Poland, the countries with the most records are Spain (844), France (633), Great Britain (595) and Germany (537).

Of the 309 species in the Recent (living) dataset, some have been recorded many times, while others appear only once (about 30%), or only a few times. Once again, sampling and data entry bias may be partly to blame, but lack of taxonomic and ecological precision is also responsible in the case of some species. Taxonomic revisions of certain groups will probably reduce the total number of species (with a corresponding increase in the number of records of those that remain valid), as has already happened with genera such as *Potamocypris*, revised by Meisch (1984, 1985) and *Herpetocypris*, revised by Gonzalez Mozo *et al.* (1996). The most frequently-recorded species are *Cypria ophtalmica* (573 records), *Cyclocypris ovum* (403), *Candona candida* (373), *Cypridopsis vidua* (368) and *Heterocypris incongruens* (332). In the Quaternary dataset the commonest species are *Candona candida* (112 records), *Candona neglecta* (94), *Cytherissa lacustris* (78), *Darwinula stevensoni* (71) and *Limnocythere inopinata* (69). The differences between the two datasets illustrate the strong preservational bias of the fossil record: lake-dwellers like *C. lacustris* and *D. stevensoni* have a much greater preservation potential than ostracods inhabiting temporary ponds, such as *H. incongruens*, which so far has only 10 records in the Quaternary dataset of NODE.

The geographical density of records is also influenced by ecological factors, but once again we must be cautious in our interpretation of low record densities. If a species has been recorded in only one or a few neighbouring localities we can suggest that it is endemic to that area, although such occurrences can also be explained as taxonomic or sampling artefacts. The majority of records in NODE are from lentic water bodies, while sampling of lotic and hypogean habitats has been less frequent and geographically more patchy. Nevertheless, in cases where the density and spread of records are reasonably good we can observe interesting distributional patterns. *Fabaeformiscandona protzi*, for example, is mainly restricted to northeastern Europe, with an isolated English record at a similar latitude; further south it occurs at higher altitude in the Alps, suggesting a strong temperature/climate control on its distribution (Fig. 5.4). Two species of the genus *Cypris* (Fig. 5.5) also show distributions suggestive of a latitudinal/climatic influence: *C. pubera* is widespread across central and northern Europe, while *C. bispinosa* is found mainly in the southwest, although the two have overlapping ranges in France and northern Spain. Records of *Darwinula stevensoni* are scattered throughout Europe (Fig. 5.6) and show no particular pattern, as might be expected of a cosmopolitan species;

Cytherissa lacustris (another inhabitant of lakes), on the other hand, shows a restricted distribution similar to that of *F. protzi*, living only in high-altitude lakes at the southern limits of its distribution (Fig. 5.7).

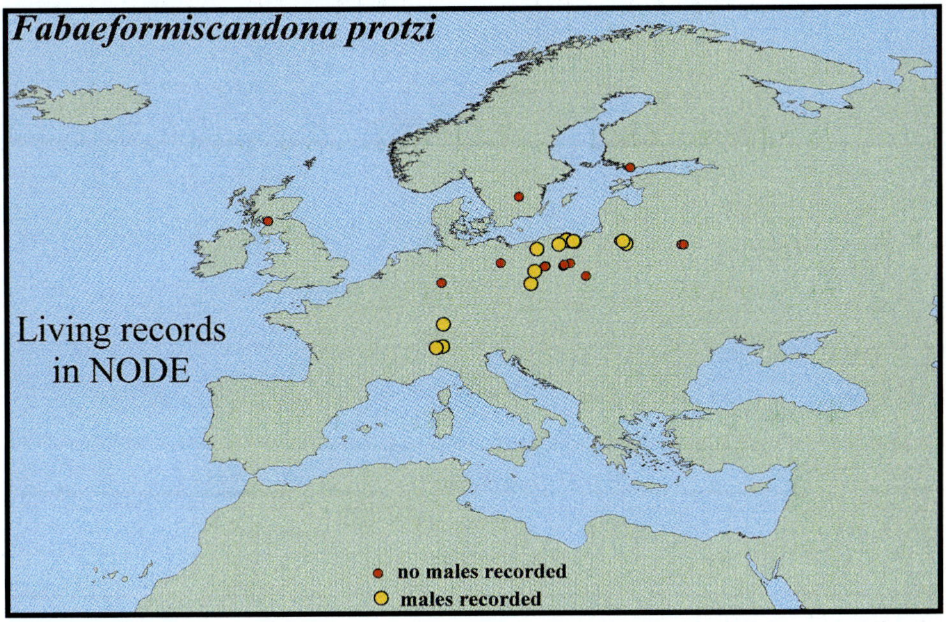

Fig. 5.4

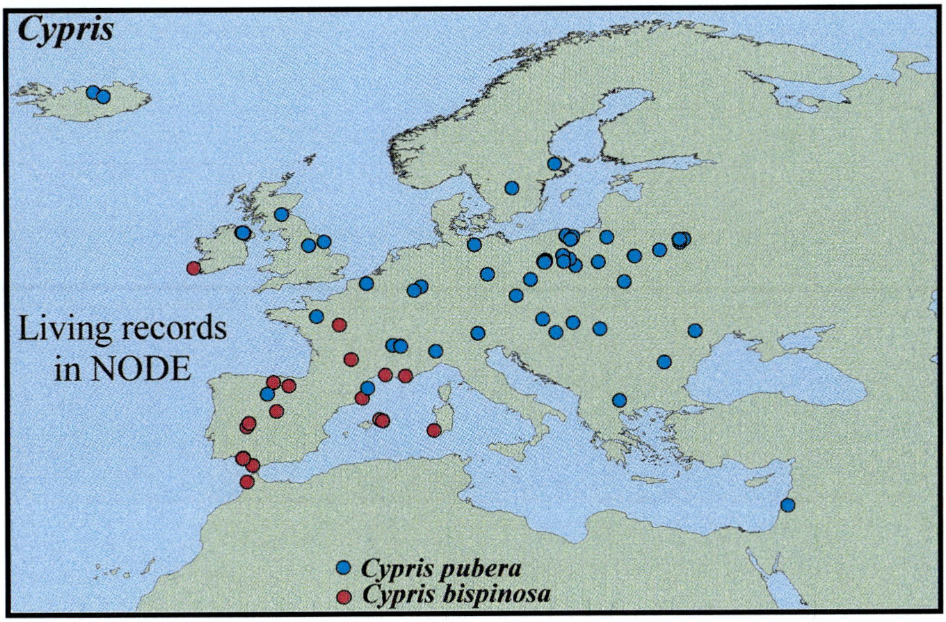

Fig. 5.5

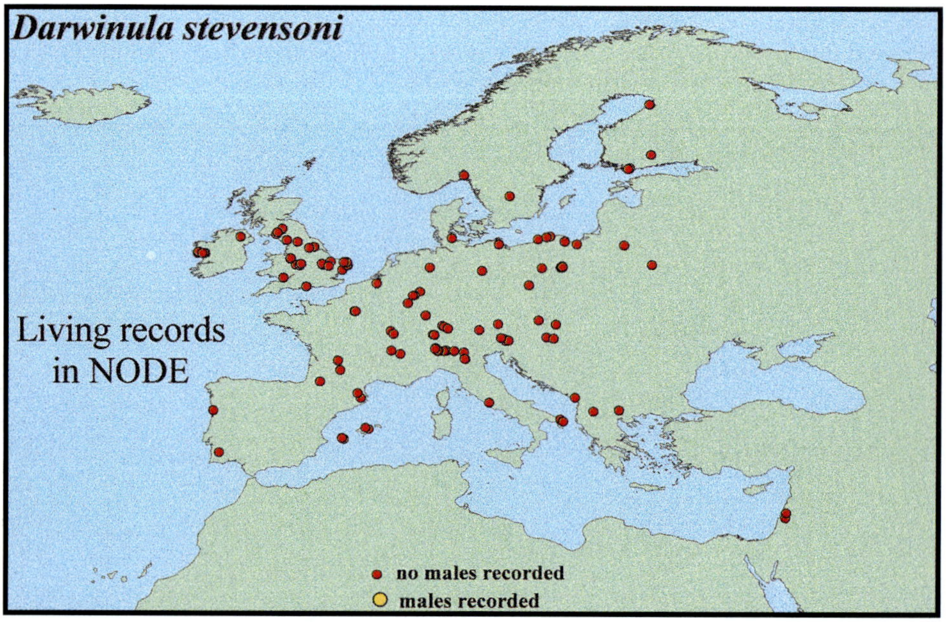
Darwinula stevensoni
Living records in NODE
no males recorded
males recorded

Fig. 5.6

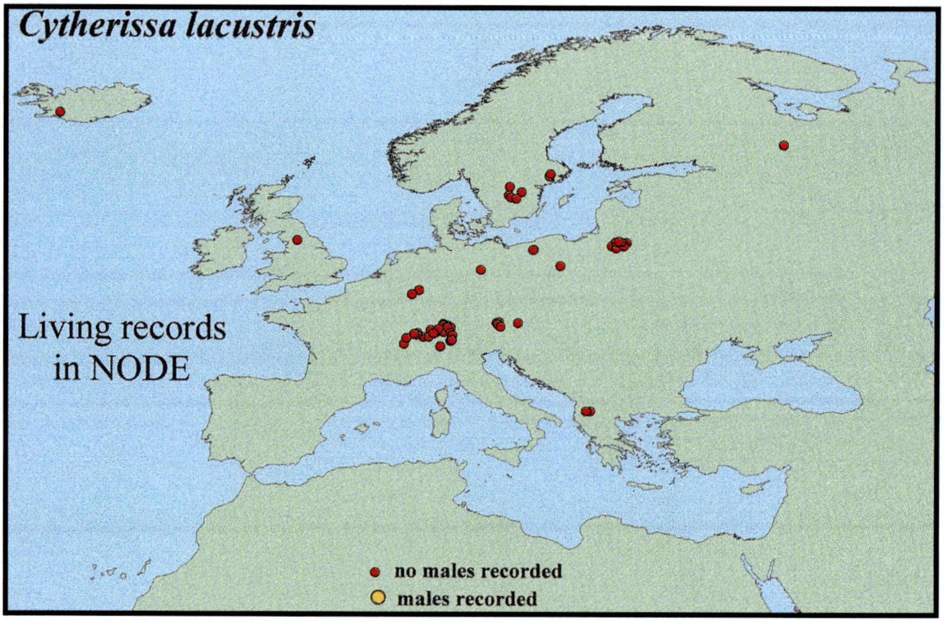
Cytherissa lacustris
Living records in NODE
no males recorded
males recorded

Fig. 5.7

The taxonomic and geographical distribution of reproductive modes in Europe

The distribution of sexual reproduction has been examined at several taxonomic levels. The superfamilies Cypridoidea and Cytheroidea both have at least some sexually-reproducing representatives in Europe; the Darwinuloidea do not (see Fig. 5.15). One of the primary objectives of the NODE project was to test hypotheses concerning geographical parthenogenesis in European non-marine ostracods. Using NODE, we can immediately refine the general view that sexual populations are to be found in southern and eastern locations (e.g., Spain, Sicily, Turkey) which were unaffected by ice cover or permafrost during the last glaciation, while parthenogenetic populations are widespread in Europe, largely representing the colonisation of new habitats which became available during the Late-glacial and Holocene (see chapter 6 for further discussion). No clear pattern of geographical parthenogenesis is seen in the Cypridoidea, the superfamily with the greatest proportion of apparently parthenogenetic taxa (see Fig. 5.15). At family level, neither the Candonidae (Fig. 5.8), Ilyocyprididae nor the Notodromadidae show a pattern of geographical parthenogenesis. Most species in the Candonidae show an apparently random distribution of male records and it is doubtful that those where males were not recorded actually represent parthenogenetic populations (e.g., *Candona neglecta*, Fig. 5.10). Of 136 candonid species, only one, *Pseudocandona albicans* (Fig. 5.11), shows the 'classic' pattern of geographical parthenogenesis.

In contrast, a much clearer pattern is displayed by the distribution of the family Cyprididae (Fig. 5.9). There are 20 cypridid species with at least one male record (compared to 100 with none – see Fig. 5.15), the great majority of which occur south of approximately latitude 50°N; most of those further north belong either to the subfamily Scottiinae (genus *Scottia*, for which there are no certain instances of parthenogenesis) or represent unexpected and interesting sexual populations of *Heterocypris incongruens*, one of the 'typical' geographical parthenogens; there are isolated records of males of *Cypris pubera* and *Potamocypris zschokkei* in Poland. Cypridids showing geographical parthenogenesis include *Eucypris virens*, *Tonnacypris lutaria*, *Prionocypris zenkeri*, *Plesiocypridopsis newtoni*, *Cypridopsis werneri* and species of *Heterocypris*, *Herpetocypris* and *Psychrodromus*.

Heterocypris incongruens is a common and widespread inhabitant of temporary waterbodies (Fig. 5.12), demonstrating the effectiveness of desiccation-resistant eggs as a dispersal mechanism for parthenogenetic lineages. Sexual populations are known from southern, circum-Mediterranean localities, which, according to the post-glacial invasion hypothesis of geographical parthenogenesis (see chapter 6), represent refugia far enough south to have been unaffected by glaciation or permafrost. The small number of male records which exist further north suggest that although parthenogenetic forms spread quickly after the last glaciation and have enjoyed success for several thousand years, sexual forms are slowly but surely spreading into Europe again. An alternative explanation, that parthenogenesis is currently widespread because relative climatic stability in Europe over the past 11,500 years has permitted parthenogens to out-compete sexual forms (Horne & Martens in press), permits a different interpretation, however. The sporadic occurrence of sexual populations of *Heterocypris incongruens* in Europe might not represent the gradual invasion of sexuals from the south, as suggested above, but the

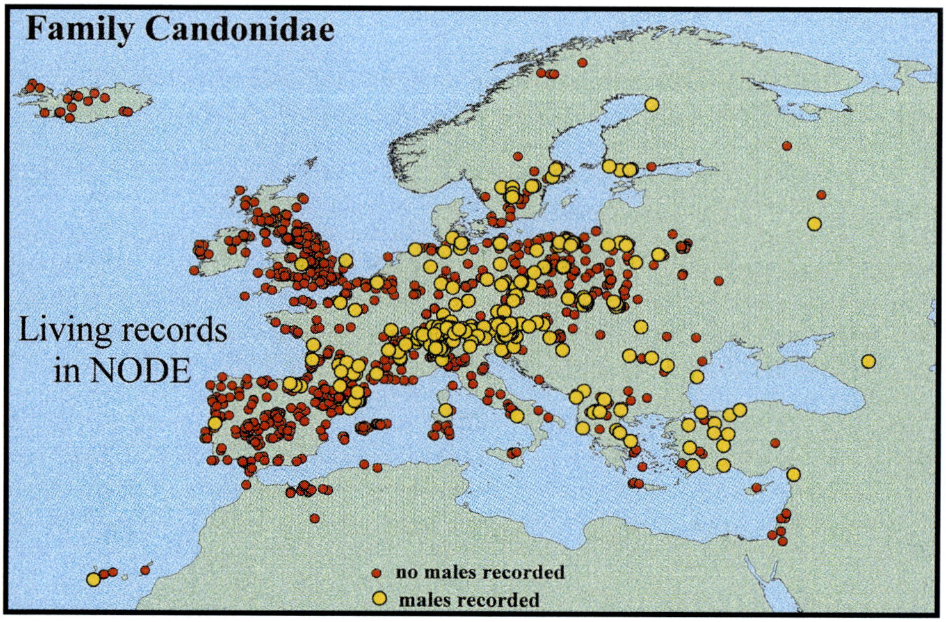
Family Candonidae
Living records in NODE
no males recorded
males recorded

Fig. 5.8

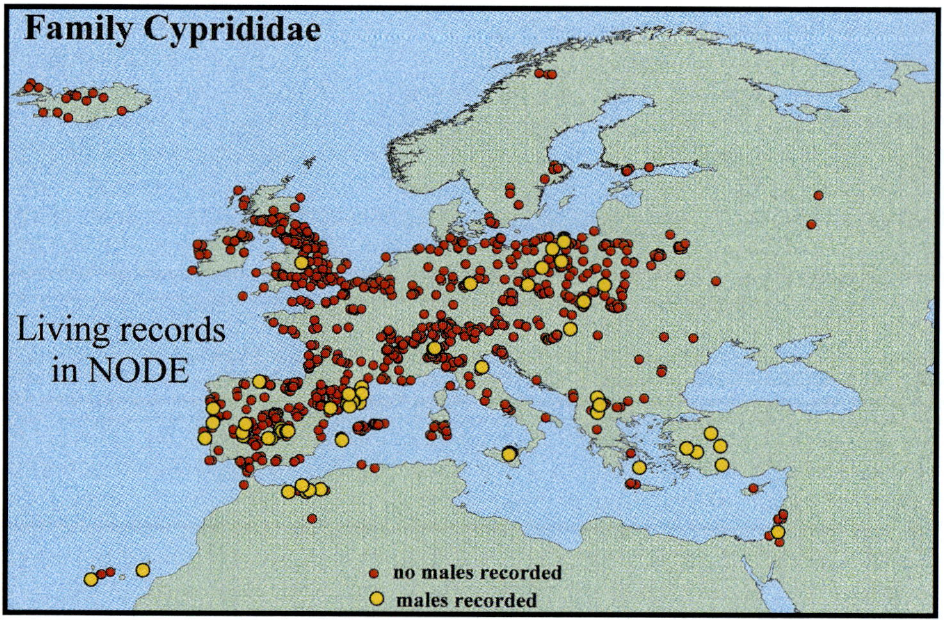
Family Cyprididae
Living records in NODE
no males recorded
males recorded

Fig. 5.9

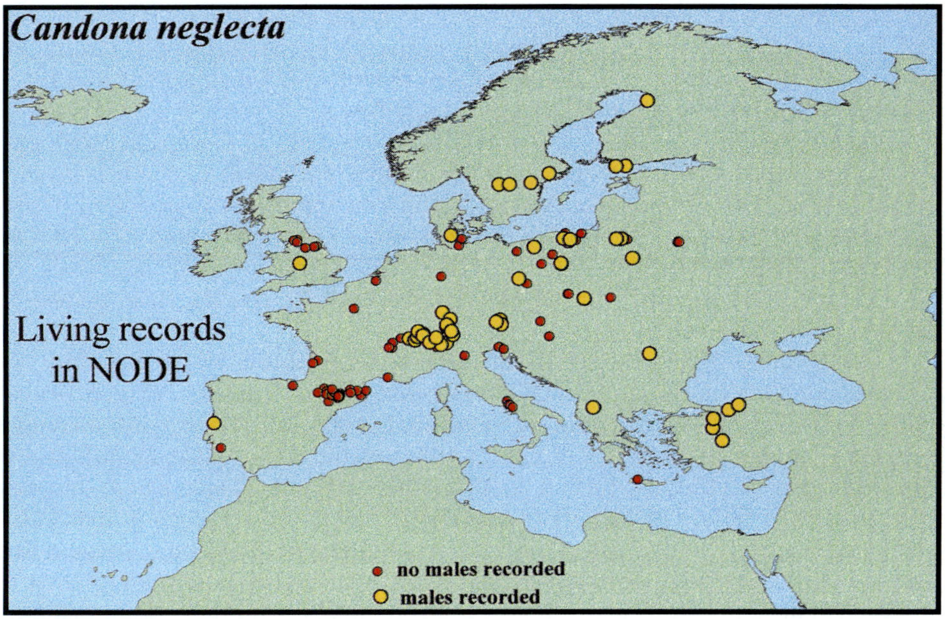

Fig. 5.10

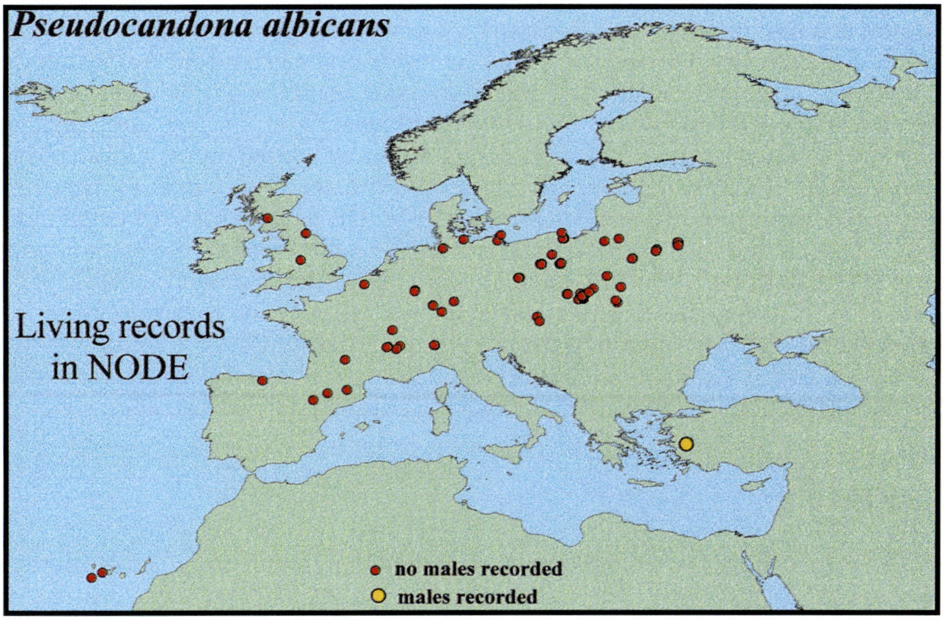

Fig. 5.11

last remnants of a once much larger sexual metapopulation which was established after the last glaciation but was largely replaced by parthenogens during the Holocene. Unfortunately, the very poor fossil record of *H. incongruens* alluded to above renders it useless for testing the two hypotheses.

Another cypridid, *Eucypris virens*, is also widespread in European temporary waterbodies, but the only known sexual populations are in Spain, southern France and Sicily (Fig. 5.13), once again demonstrating a clear pattern of geographical parthenogenesis. It, too, has a poor fossil record – only 8 Quaternary records in NODE, compared with 228 living records. Patterns of geographical parthenogenesis are also apparent in the distributions of several other cypridid species, most notably in *Herpetocypris brevicaudata.*

In the Cytheroidea, geographical parthenogenesis seems to be rare, but two examples are particularly interesting. *Limnocytherina sanctipatricii* (Fig. 5.14) has a fairly wide distribution in central and northern Europe, but male records are either at higher latitudes (> 53°N) or higher altitudes in the Alps, while records where males were not noted occur further south at lower altitudes. Once again, this may be an artefact (a failure of authors to observe or record the presence of males in a species which may be exclusively syngamic), but the pattern could suggest that there are parthenogenetic populations, more easily dispersed and more tolerant of environmental extremes than their sexual relatives. Another limnocytherid, *Limnocythere inopinata* displays 'typical' geographical parthenogenesis, with sexual populations confined to south-eastern extremities (see chapters 2, 6, Fig. 6.4). One occurrence of 'rare males' (see chapter 6 for further discussion) is recorded in Poland; another, in Denmark (B. Scharf, pers. comm.) is not included in NODE, neither are such records from laboratory cultures originating from Austrian lakes (Yin 1997). In support of the post-glacial invasion hypothesis it has been claimed that sexual populations of *L. inopinata* were present in Europe until the late Pleistocene, but only parthenogenetic populations recolonised the area after the last glacial retreat (McKenzie 1986, Chaplin *et al.* 1994, see also chapter 6). However, Quaternary records of males of this species in NODE include some from the Late-glacial and Holocene, suggesting that sexual (as well as parthenogenetic) forms were early colonisers of the new habitats (Griffiths & Butlin 1995, Griffiths 1995). One record of males, from Neusiedlersee in Austria, may date from within the last two centuries (Löffler 1990); at present it is not possible to decide whether this represents a sexual population that has persisted throughout the Holocene, a relatively recent colonisation or a sequence of incidences of rare male occurrences.

Conclusions

The taxonomic distribution of geographical parthenogenesis in Europe is more restricted than previously thought. The classic pattern, with sexual populations confined to southern or eastern refugia, is almost exclusive to the Cyprididae, also the family with by far the greatest proportion of apparently totally parthenogenetic taxa (83%; see Fig. 5.15). In other groups, geographical parthenogenesis is seen only in a very few species, such as *Pseudocandona albicans* in the Cypridoidea (Candonidae, Candoninae) and *Limnocythere inopinata* (Cytheroidea, Limnocytheridae, Limnocytherinae).

Geographical parthenogenesis remains undemonstrated in the Darwinuloidea, the cypridoidean families/subfamilies Notodromadidae, Ilyocyprididae and Cyclocypridinae (Candonidae), and the Timiriaseviinae (Cytheroidea, Limnocytheridae).

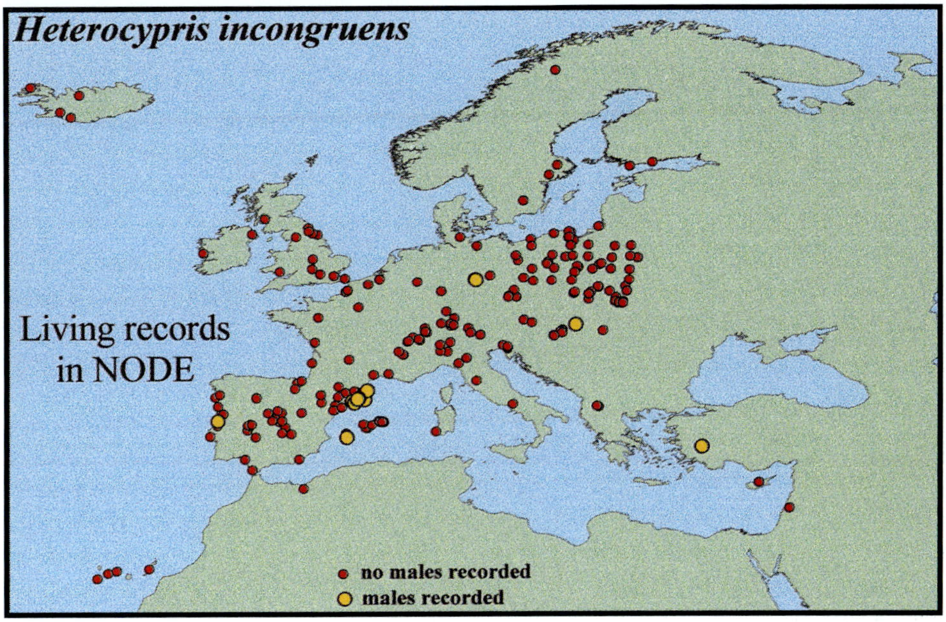

Fig. 5.12

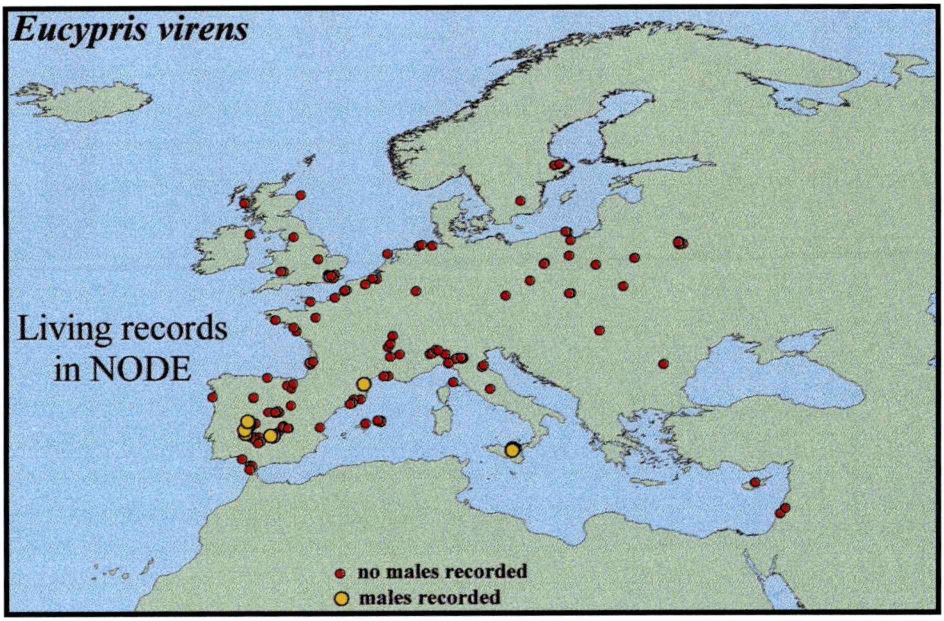

Fig. 5.13

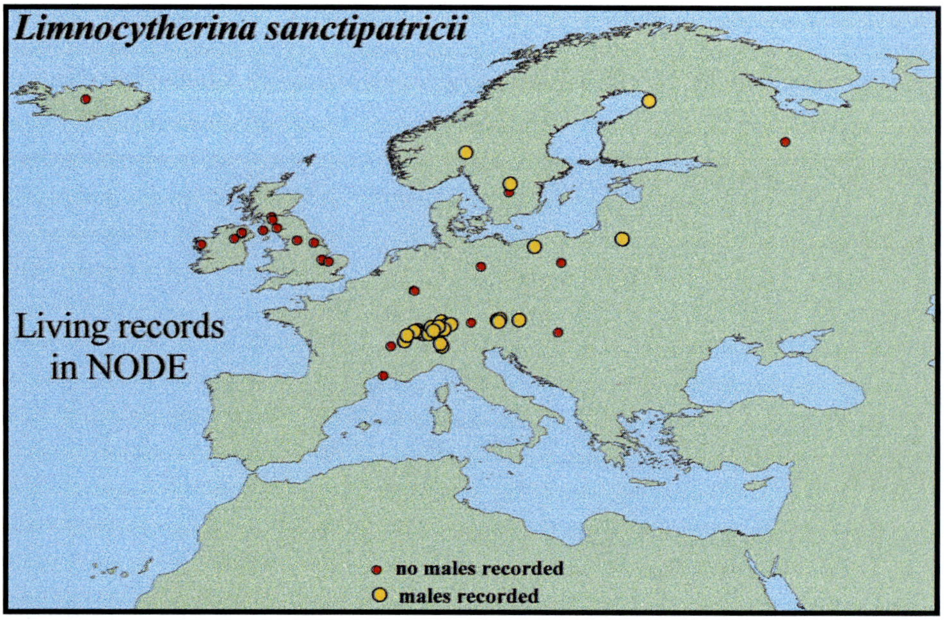

Fig. 5.14

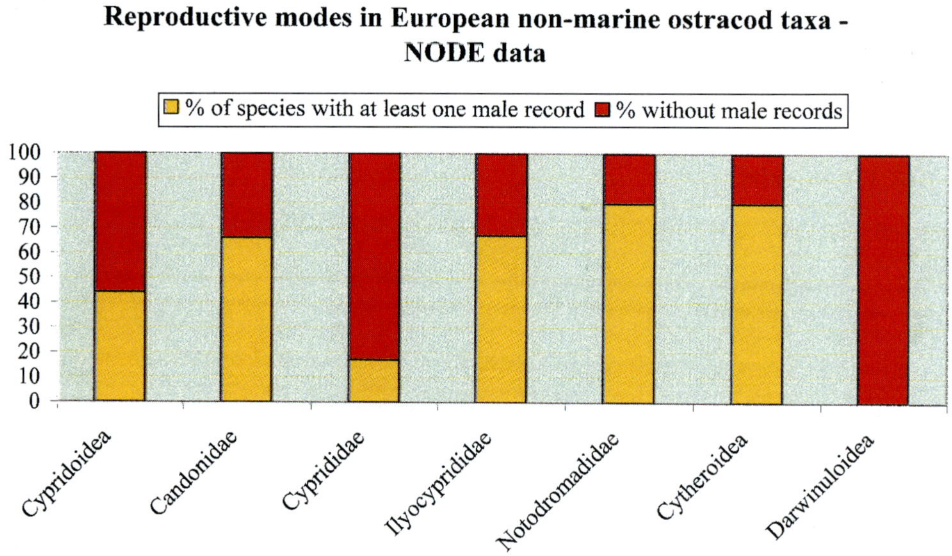

Fig. 5.15 Taxonomic distribution of reproductive modes in European non-marine ostracods. For each taxonomic group, the chart shows the percentage of species for which there is at least one male record (for which sexual reproduction is therefore at least possible, even if there is geographical parthenogenesis) against those with no male records (for which reproduction is assumed to be exclusively parthenogenetic).

References

Abe, K. 1990. What the sex ratio tells us: a case from marine ostracods. *in*: Whatley, R. & Maybury, C. (eds), *Ostracoda and Global events*, Chapman and Hall: 175-185.

Athersuch, J., Horne, D.J. & Whittaker, J.E. 1989. *Marine and brackish ostracods. Synop. Brit. Fauna (N. Ser.)*, **43:** E. J. Brill, 343 pp, 7pls.

Carbonel, P., Colin, J.-P., Danielopol, D.L., Löffler, H. & Neustrueva, I. 1988. Paleoecology of limnic ostracodes: a review of some major topics. *Palaeogeogr., Palaeoclimatol., Palaeoecol.*, **62:** 413-461.

Chaplin, J.A. 1993. The local displacement of a sexually reproducing ostracod by a conspecific parthenogen. *Heredity*, **71:** 259-268.

Chaplin, J. A., Havel, J. E. & Hebert, P. D. N. 1994. Sex and ostracods. *Trends Ecol. Evol.*, **9:** 435-439.

Danielopol, D.L. 1975. Remarques sur la diversification morphologique de trois nouvelles espèces d'*Elpidium* (Ostracoda) à Cuba. *Bull. Amer. Paleont.*, **65:** 47-60.

Danielopol, D.L. & Horne, D.J. 1996. Some observations on sexual behaviour and the morphology of the male copulatory appendages in *Metacypris cordata* (Ostracoda, Timiriaseviinae). *in*: Keen, M.C. (ed): *Proceedings of the 2nd European ostracodologists' Meeting, Brit. Micropalaeontol. Soc., Lond.*: 171-174.

Danielopol, D.L., Horne, D.J. & Wood, R.N. 1996. Notes on the ecology of *Metacypris cordata* (Ostracoda, Timiriaseviinae): why does it not colonise groundwater habitats? *in*: Keen, M.C. (ed), *Proceedings of the 2nd European ostracodologists' Meeting, Brit. Micropalaeontol. Soc., Lond.*: 175-180.

De Deckker, P. 1977. The distribution of the 'giant' ostracods (family: Cyprididae Baird, 1845) endemic to Australia. *In:* Löffler, H. & Danielopol, D. (eds), *Aspects of ecology and zoogeography of recent and fossil Ostracoda*, Junk, The Hague: 285-294.

De Deckker, P. 1983. Notes on the ecology and distribution of non-marine ostracods in Australia. *Hydrobiologia,* **106:** 223-234.

Delorme, L.D. 1970. Freshwater Ostracoda from Canada. Part I. Subfamily Cyprididae. *Can. J. Zool.,* **48:** 153-168.

Delorme, L.D. 1991. Ostracoda. *in: Ecology and Classification of North American Freshwater Invertebrates,* Acad. Press: 691-722.

Hartmann, G. 1968. Ostracoda (3. Lieferung). *Klassen und Ordnungen des Tierreichs,* **5** *(Arthropoda),* 1. Abt. Crustacea, 2. Buch, IV. Teil: 409-568.

Gonzalez Mozo, M., Martens, K. & Baltanás, A. 1996. A taxonomic revision of European *Herpetocypris* Brady and Norman, 1889 (Crustacea, Ostracoda). *Bull. k. belg. Inst. Natuurw. Biol.,* **66:** 93-132

Griffiths, H.I. 1995. European Quaternary freshwater Ostracoda: a biostratigraphic and palaeobiogeographic primer. *Scopolia*, **34:** 1-168.

Griffiths, H.I. & Butlin, R.K. 1995. A timescale for sex versus parthenogenesis: evidence from subfossil ostracods. *Proc. R. Soc. Lond.* B, **260:** 65-71.

Horne, D.J. (MS) 1980. *Recent ostracods from the Severn Estuary and Bristol Channel.* PhD Thesis, Univ. Bristol: xi + 320 pp, 19 pls.

Horne, D.J., Lord, A.R., Robinson, J.E. & Whittaker, J.E. 1990. Ostracods as climatic indicators in glacial deposits, or, On some new and little-known British Quaternary Ostracoda. *Cour. Forsch. Inst. Senckenb.*, **123:** 129-140.

Horne, D.J., Cocks, F., Paris, G. & Connaghan, D. 1998. The trials of NODE (Non-marine Ostracod Distribution in Europe): practical problems in the development of a GIS application. *(electronic publication).* http://www.bris.ac.uk/Depts/Geol/gig/conf/giseng/horne.htm

Horne, D.J. & Martens, K. (in press). Geographical parthenogenesis in European non-marine ostracods (Crustacea, Ostracoda); post-glacial invasion or Holocene stability? *Hydrobiologia.*

Johnson, T.C., Scholtz, C.A., Talbot, M.R., Kelts, K., Ricketts, R.D., Nagobi, G., Beuning, K., Ssemmanda, I. & McGill, J.W. 1996. Late Pleistocene desiccation of Lake Victoria and rapid evolution of cichlid fishes. *Science* **273:** 1091-1093.

Klekowksi, R. 1952. Etude des ostracodes (Ostracoda) des eaux continentales salées et sulfuriques. *Soc. sci. lodz., Sect.* 3, **19:** 1-36 (in Polish).

Löffler, H. 1990. Paleolimnology of Neusiedlersee, Austria. I. Succession of ostracods. *Hydrobiologia*, **214:** 229-238.

Löffler, H.H. & Danielopol, D.L. 1978. Ostracoda (Cyprididae). *in:* Illies, J. (ed), *Limnofauna Europaea,* Gustav Fischer Verlag: 196-208.
McKenzie, K.G. 1971. Palaeozoogeography of freshwater Ostracoda. *in*: Oertli, H.J. (ed), *Paléoécologie des Ostracodes, Pau 1970. Bull. Centre Rech. Pau-SNPA Suppl.*, **5:** 207-237.
McKenzie, K.G. 1986. Ostracoda: new aspects of their biogeography. *in*: Gore, R.H. & Heck, K.L. (eds), *Crustacean biogeography*, Balkema: 257-277.
McKenzie, K.G. & Moroni, A. 1986. Man as an agent of crustacean passive dispersal via useful plants – exemplified by Ostracoda 'ospiti esteri' of the Italian ricefield ecosystem – and implications therefrom. *J. crustacean Biol.,* **6:** 181-198.
Martens, K. 1984. Annotated checklist of non-marine ostracods (Ostracoda, Crustacea) from African inland waters. *Zool. Bijdr., K. Mus. mid. Afr., Tervuren,* **20:** 51 pp.
Martens, K. 1994. Ostracod speciation in ancient lakes: a review. *in*: Martens, K., Goddeeris, B. & Coulter, G. (eds), *Speciation in ancient lakes, Adv. Limnol.,* **44:** 203-222.
Martens, K.1997. Speciation in ancient lakes. *Trends Ecol. Evol.,* **12:** 177-182.
Martens, K. 1998. Diversity and endemicity of recent non-marine ostracods (Crustacea, Ostracoda) from Africa and South America: a faunal comparison. *Verhandl. int. Verein. Limnol.,* **26:** 2093-2097.
Martens, K. & Meisch, C. 1985. Description of the male of *Potamocypris villosa* (Jurine, 1820) (Crustacea, Ostracoda). *Hydrobiologia,* **127:** 9-15.
Martens, K. & Coomans, A.C. 1990. Phylogeny and historical biogeography of the Megalocypridinae Rome, 1965 with an updated checklist of this subfamily. *in*: Whatley, R.C. & Maybury, C. (eds), *Ostracoda and Global events, Proceedings of the 10th international symposium on Ostracoda*, Chapman & Hall: 545-556.
Martens, K. & Behen, F. 1994. A checklist of the recent non-marine ostracods (Crustacea, Ostracoda) from the inland waters of South America and adjacent islands. *Trav. sci. Mus. nat. hist. nat. Luxemb.*, **22**: 1-84.
Meisch, C. 1984. Revision of the recent Western Europe species of the genus *Potamocypris* (Crustacea, Ostracoda). Part I. Species with short swimming setae on the second Antenna. *Trav. sci. Mus. Hist. nat. Luxemb.,* **3:** 55 pp.
Meisch, C. 1985. Revision of thc recent West European species of the genus *Potamocypris* (Crustacea, Ostracoda). Part II. Species with long swimming setae on the second Antenna. *Trav. sci. Mus. Hist. nat. Luxemb.,* **6:** 95 pp.
Meisch, C. 1988. Ostracodes récoltés à Paris. Avec une clef pour la détermination des espèces européennes du genre *Ilyocypris* (Crustacea, Ostracoda). *Bull. Soc. Nat. Luxemb.,* **88:** 145-163.
Mikulic, F. 1965. Syngamy and parthenogenesis in *Candona* species in Lake Ohrid. *Arkh. biol. Nauk.,* **17:** 11-12.
Myers, A.A. & Giller, P.S. 1988. *Analytical biogeography. An integrated approach to the study of animal and plant distributions,* Chapman & Hall: 578pp.
Neale, J. 1983. The Ostracoda and uniformitarianism. I. The later record: Recent, Pleistocene and Tertiary. *Proc. Yorkshire geol. Soc.,* **44:** 305-326.
Petkovski, T.K. 1977. Bemerkenswerte Entomostraken aus Jugoslavien. *Acta Mus. Mac. Sci. nat.,* **9:** 147-182.
Rossetti G. & Martens, K. 1996. Redescription and morphological variability of *Darwinula stevensoni* (Brady & Robertson, 1870) (Crustacea, Ostracoda). *Bull. k. belg. Inst. Natuurw. Biol.,* **66:** 73-92.
Scharf, B.W. 1982. Muschelkrebse aus dem Murnauer Moos (Crustacea, Ostracoda). *in:* Maddocks, R.F. (ed), *Applications of Ostracoda, Proceedings of the 8th International Symposium on Ostracoda:* 255-263.
Tétart, J. 1974. Les Entomostracées des milieux peu profonds de la Vallée du Rhône. Essai d'Etude écologique: composition des associations et répartition des espèces. *Trav. Lab. Hydrobiol. Piscicult. Grenoble,* **64/65:** 109-245.
Van Harten, D. 1983. Resource competition as a possible cause of sex ratio in benthic ostracodes. *in*: Maddocks, R.F. (ed), *Applications of Ostracoda, Proceedings of the 8th International Symposium on Ostracoda: 568-580.*
Vork, K.A. & Thomsen, E. 1996. Lusitanian/Mediterranean ostracods in the Holocene of Denmark: implications for the interpretation of winter temperatures during the postglacial temperature maximum. *The Holocene*, **6:** 423-432.

Whittaker, J.E. (MS) 1972. The taxonomy, ecology and distribution of Recent brackish and marine Ostracoda from localities along the coast of Hampshire and Dorset (Christchurch Harbour, The Fleet, Weymouth Bay). PhD Thesis, Univ. Wales, Aberystwyth. 2 vols, 643 pp, 68 pls.
Yin, Y. 1997. Contribution to the morphology and ecology of ostracoda Limnocytheridae and Candonidae: Comparative studies between Austrian and Chinese Crustacea. *PhD Thesis. Univ. Vienna*, 163 pp.

CHAPTER 6

FOSSIL DISTRIBUTION OF REPRODUCTIVE MODES IN NON-MARINE OSTRACODS

H. I. Griffiths and D. J. Horne

Abstract

The excellent fossil record of ostracods makes them a valuable group for studying the evolution of reproductive modes. Many lineages exhibit distinctive sexual dimorphism of the carapace, facilitating the recognition of fossil sexual and parthenogenetic taxa, although there are problems with the interpretation of some groups. Over geologically short time-scales, the study of Quaternary non-marine ostracods permits detailed consideration of the origin of the geographical parthenogenesis observed in certain modern distribution patterns. Two hypotheses to explain these distributions are discussed: the first argues that parthenogens were able to colonise newly-available habitats more quickly than sexuals after the last glaciation, the second suggests that the relative stability of Holocene climate has conferred a competitive advantage on parthenogenetic populations. On longer time-scales (millions of years), the evidence of reproductive modes in Mesozoic non-marine ostracods is examined, with particular reference to 'Wealden' palaeoenvironments, and the antiquity of obligate parthenogenesis in Darwinuloidea is discussed. A new hypothesis for the origin of parthenogenetic lineages through interspecific hybridisation is proposed, with particular reference to the Neogene – Recent genus *Tonnacypris*.

Introduction

Although the fossil record has potential advantages for the study of the evolution of parthenogenesis, few of the groups in which parthenogenesis is common are well represented in ancient non-marine or marine sedimentary successions. However, because of their calcareous shells, Ostracoda have a detailed fossil record, so that sufficient taxonomic detail and temporal resolution are available for fossil reproductive strategies to be studied. Some authors (e.g., Whatley 1992) have attributed the successful invasion of Mesozoic non-marine habitats by the Cypridoidea to the acquisition of parthenogenetic reproduction, as well as their ability to produce desiccation-resistant eggs, both of which are advantageous to survival and wide dispersal in 'unstable', temporary non-marine waters (cf. Williams 1987). Justification for this approach comes from studies of modern faunas, in which parthenogenesis (unisexuality) is common in non-marine waters, whilst the overwhelming majority of species inhabiting 'stable' marine environments reproduce sexually (bisexuality) (Cohen & Morin 1990, but see also chapter 5).

Sex and parthenogenesis: evolutionary ecology of reproductive modes in non-marine ostracods, edited by K. Martens, pp. 101–118

Recognising fossil sexuality and parthenogenesis

There are problems with the identification of reproductive mode within extinct populations, and not only because of the effects of differential and/or selective preservation. For example, many modern freshwater ostracod species (mainly Cyprididae) are known to exhibit geographical parthenogenesis (Vandel 1928), and reports of male-bearing populations in species that had been thought previously to be parthenogenetic are not uncommon (e.g. Gülen 1985, Järvekülg 1959, Martens & Meisch 1985, Petkovski & Meisch 1995). Moreover, it is not unusual for geographically isolated sexual populations to be accorded (sub-)specific ranking: *Cytherissa lacustris* is basically a Holarctic parthenogen (Danielopol *et al.* 1990), but with a sexual isolate (*C. l. baikalensis*) endemic to the Siberian lakes Baikal and Hovsogul (Mazepova 1990), while the western Eurasian *Eucypris lilljeborgi* has a sexual isolate (*E. l. turcica*) in eastern Anatolia (Hartmann 1964). Clearly, geographical parthenogenesis is unlikely either to be preserved or recognised in the fossil record, even when subject to extensive spatial or temporal averaging. Some efforts have been focused on the Quaternary fossil record in attempts to determine relationships between reproductive mode (sexual or parthenogenetic), dispersal mode (as resting eggs or as live individuals) and geographical distribution in non-marine ostracods. Sex was undoubtedly the ancestral reproductive mode of the Ostracoda; distinctive sexual dimorphism is a feature of many Lower Palaeozoic marine ostracods.

The simplest means of determining reproductive mode in an extinct taxon is to be able to identify morphologies within it that seem consistent with bisexuality, *i.e.* the presence of sexually dimorphic adults in a fossil population. These dimorphs may be identified through size and shape changes in late ontogeny, often associated with the development of male reproductive organs or a female brood pouch. Frequency distributions in the larger size classes of sexually dimorphic species should be bimodal (see Fig. 6.1, v). Thus, claims that modern darwinuloids represent one of the oldest parthenogenetic lineages known, are based on the assumption that, if males should occur in ancient assemblages, their valve morphology would be recognisable because most modern darwinuloids have an externally visible brood pouch (Butlin & Griffiths 1993, Martens *et al.* 1997). One would expect males to be smaller and to lack a brood pouch, causing carapaces to be less wide in dorsal view and then to have the adductor muscle scars situated approximately in the middle of the valves, as in *Microdarwinula* which has no externally visible brood pouch (the adductor muscle scars are situated forward of mid-length in females with a brood pouch). However, juveniles have similar features, while other so-called males could simply be different species. If, in fact, sexual dimorphism was negligible, then interpretations of the age of parthenogenesis in *Darwinula* would be weakened (but see Griffiths & Butlin 1994). If two apparent adult morphotypes are present within a particular fossil assemblage, interpretations regarding reproductive mode stand or fall upon whether these represent two sexual dimorphs, two different ontogenetic stages, or two different species. Conversely, in a monomorphic assemblage it may be unknowable whether this represents a parthenogenetic species, or a sexual species that exhibits no dimorphism.

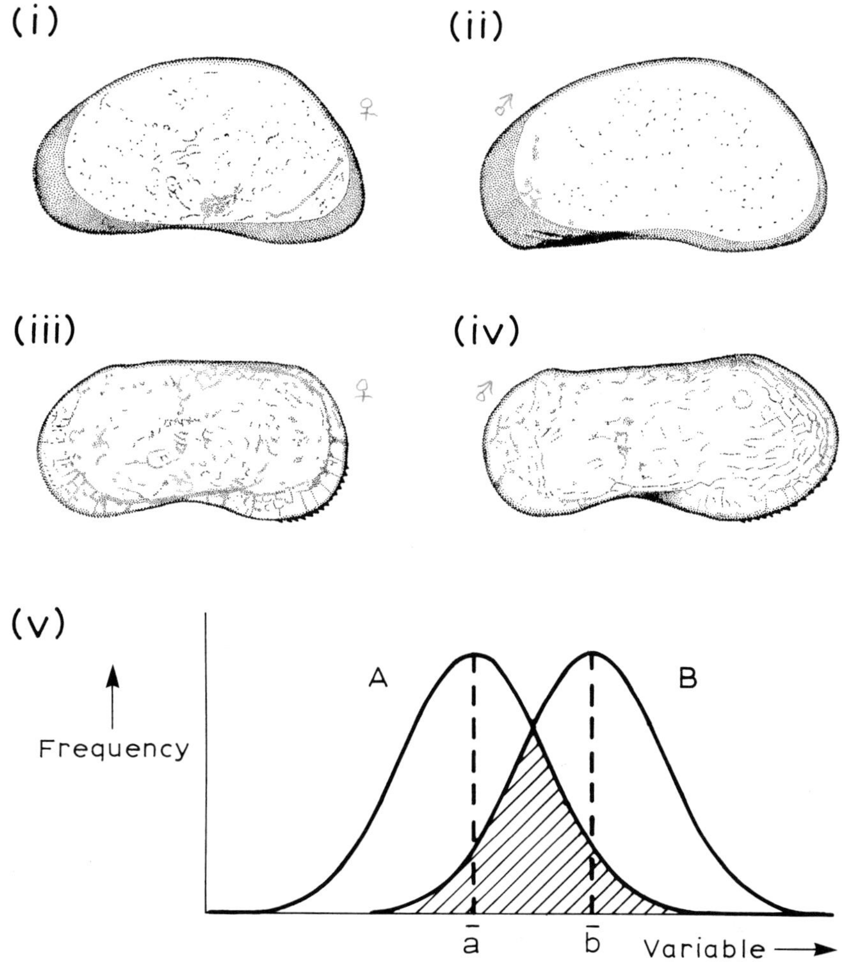

Fig. 6.1 Sexual dimorphism in the valves of non-marine ostracods: (i) female *Candona candida* (left valve), (ii) male *Candona candida* (left valve), (iii) female *Limnocythere inopinata* (left valve), (iv) male *Limnocythere inopinata* (left valve) (all redrawn from Sywula & Pietrzeniuk 1989), (v) size distribution frequencies in a dimorphic fossil population.

It is often possible to extrapolate probable reproductive mode by comparisons with living relatives. Thus, comparison with the extant species *Metacypris cordata* allows the identification of males and females in other, fossil, Timiriaseviinae (e.g. *Theriosynoecum*, *Timiriasevia* and *Frambocythere*) as far back as the Jurassic. In carapaces of Timiriaseviinae, sexual dimorphism is particularly obvious from the inflated posterior brood-pouch of the female. Most fossil timiriaseviines show this type of sexual dimorphism, but at least one genus, *Kovalevskiella* (which is found in European benthic non-marine environments in the Tertiary and Quaternary, but exclusively in groundwaters today), appears to have adopted parthenogenesis (Carbonel *et al.* 1986). Furthermore, extant members of the related modern group

Limnocytherinae have no brood care, but neverthless display quite marked shell sexual dimorphism. Sexual dimorphism is not always well expressed in the shape of the shell, however. In *Eucypris virens*, for example, males are often (but not always) smaller than females, but differ from them only slightly in shape; in fact, the shape variation between male and female is well within the overall range of shape variation observed in all-female populations (see chapter 8). Recognising males of *E. virens* is easy when they are alive (presence of hemipenes, testes, etc.) but extremely problematical when only fossil shells are available.

In many Cypridoidea, if fossil preservation is good, males and females may be distinguishable from the imprints of the gonads on the inside of the valves. Thus Janz (1995) was able to identify the shells of males (with four narrow imprints in each valve) and females (with two broader imprints in each valve) of *Strandesia risgoviensis* from the Miocene lake deposits of the Ries meteorite crater (southern Germany). Of course, such clear examples of fossil sexual status are not so common, even when careful analyses are carried out. Staplin (1963a) was able to recognise both sexes in several dimorphic species of *Candona s.l.* in Pleistocene freshwater deposits from Illinois (USA), but unable to separate the sexes in *Cyclocypris* spp. from the same sediments (Staplin 1963b). *Cyclocypris* is invariably sexual, and the males and females are of similar shape (although the former are slightly smaller). Similarly, Bless & Pollard (1975) analysed shape variation in the Carboniferous non-marine species *Carbonita humilis*, but were unable to decide whether the dimorphic variation observed represented sexual reproduction, parthenogenesis but with a high degree of intraspecific variation, or some environmentally-determined mixture of the two.

In extinct taxa without close living relatives (e.g. the Palaeozoic *Carbonita*) the problems of recognising reproductive mode are greater. Dimorphism is characteristic of marine Silurian Beyrichioidea (Palaeocopida), but dispute over which sex the two morphs represented persisted for many years. Only the discovery of juveniles within the crumina (= brood chambers) showed that the cruminate morph was the female (see chapter 10), and even this could be contested if one allows for the possibility of male pouch-brooding (although this is presently unknown in ostracods). Finally, it must be noted that the expected form of dimorphism may be reversed. In many groups (e.g. the marine/brackish Cytheruridae), the male morph is larger and more posteriorly inflated because of the relatively large size of the copulatory appendages. If this lineage was extinct, fossil males could be mistaken for pouch-brooding females. A final complication is that some populations of some species may produce occasional males that do not appear to be sexually functional; such is apparently the case of *Limnocythere inopinata* in two Austrian lakes (Yin 1997), although it should be noted that the living males in question have been found in laboratory cultures originating from the lakes, and not in the lakes themselves (D. Danielopol pers. comm.). Bearing in mind these considerations and observing that the recognition of reproductive mode in fossil ostracods is not as straightforward as some have claimed, the remainder of this chapter will be devoted to exploring the evidence and implications of some examples from the fossil record, on geologically short (thousands of years), intermediate (millions of years) and long (hundreds of millions of years) timescales.

Short timescales: temporal and geographical parthenogenesis in the Quaternary

20,000 years ago, at the height of the last glaciation, much of Europe was covered by ice or permafrost. The Scandinavian ice-sheet extended as far south as 52°N and others advanced outwards from the Pyrenees and the Alps (West 1977). About 18,000 years ago the ice began to melt and retreat. As the climate grew warmer, new lakes, ponds, rivers and other non-marine habitats became available. Much of the non-marine fauna of the waters of present-day central and northern Europe derives from colonisation events during the Late-glacial and Holocene (*i.e.* during the last *c.* 15,000 years) (Fryer 1980, Fryer & Joyce 1981, Roberts 1998, Segerstråle 1957). The last Pleistocene glaciation temporarily wiped out many European non-marine ostracod populations, which then recolonised their former ranges and entered newly-created habitats as the climate warmed and the ice retreated. This colonisation occurred in distinct waves, presumably as temperate species moved outwards from refugial areas in the circum-Mediterranean region and Asia Minor, and cold-stenothermal species retreated northwards (Absolon 1973, Griffiths & Evans 1995, Griffiths *et al.* in press). As parthenogenetic lineages should be able to disperse more quickly than sexual ones, this has often been invoked to explain the presence of sexual populations in areas such as Iberia and the Balkans, which were not glaciated during the Pleistocene (e.g. Martens 1994, Willis 1994), compared to the parthenogenetic ones which spread into formerly glaciated areas; this is the 'post-glacial invasion' hypothesis of geographical parthenogenesis.

The use of ostracod-bearing Quaternary deposits for the study of both the pattern and process of evolution is comparatively recent (Evans & Griffiths 1993), even though a small number of authors had previously sought to examine evolutionary trends within specific sediment sequences (e.g. Benson & Macdonald 1963, Stout 1975). Griffiths & Butlin (1995) used early Holocene (between *c* 11,500 and 7,000 Calendar years Before Present) ostracod assemblages from 34 sediment profiles from a wide variety of freshwater environments to analyse the effect of reproductive mode on colonisation success and ecological persistence. Results suggest that the time-scale over which parthenogenesis offers an advantage over sexual reproduction (at least in such 'unstable' environments) is limited (Griffiths & Butlin 1995). There was no evidence that parthenogenetic species had colonised water bodies any faster or more successfully than sexual ones. Moreover, a further analysis based upon nineteen individual species (rather than on species assemblage data) suggests that although there were no significant differences between the two strategies, there was some tendency for sexual species to be more persistent, and for parthenogens to respond more to environmental fluctuation (Griffiths *et al.*, in press a). These findings are clearly of interest in the light of the 'general purpose genotype' view of parthenogenesis (e.g. Lynch 1984, Weider 1993). An alternative hypothesis is proposed by Horne & Martens (in press b), who argue that the current predominance of parthenogens in Europe results from the unusual climatic stability of the Holocene, compared to the frequent and strong oscillations of the preceding interglacial/glacial cycle. The presence of sexual populations of some non-marine ostracods in the Late-glacial and early Holocene of northern Europe refutes the post-glacial invasion hypothesis; on the other hand, under relatively stable climatic conditions, clonal populations are likely to be competitively superior to conspecific sexual lineages.

At a coarser scale, where different populations display different reproductive modes (including the occasional production of males), the Quaternary record offers information on the occurrence of the different types. This is particularly the case for well-known European faunas, and most notably for common, well-preserved, and clearly-recognisably sexually dimorphic species such as *Candona candida* and *Limnocythere inopinata* (see Fig. 6.1).

Candona candida is one of the most common modern freshwater ostracods. It was, for example, present in 113 of almost 900 Yorkshire sites sampled by Fryer (1993). Reports in the literature give no indication whether this species is a facultative or a geographical parthenogen, is occasionally sexual, or has some other strategy. The only detailed examination of the reproductive biology of *C. candida* is Semenova's (1979) study at Rybinsk Reservoir in southern Russia. Here, *C. candida* is sexual, but males are comparatively short-lived (present between August and December), whilst females are both more abundant and occur throughout the winter. Elsewhere, most records of *C. candida* are based on 'spot-sampling', which may have led to the under-representation of reports of the shorter-lived males. Of 373 living records in NODE (see chapter 5), only 16 (*c* 5%) include males, most of these being scattered through central and northern Europe from the Alps to Finland (Fig. 6.2). For fossil populations, Griffiths (1995) lists 222 records of *C. candida* in the European Quaternary, including 99 from the Upper Pleistocene and Holocene. Of these, only nine (*c* 7%) are known to contain males (but this is certainly an under-representation, as many reports give few details other than a species list). The geographical distribution of known male occurrence in *C. candida* in the Quaternary of Europe is shown in Fig. 6.3. There does not appear to be any clear correlation between male occurrence and any obvious environmental or geographical parameter, other than that all male-bearing fossil sites known are situated north of the Alps. The living distribution of *Candona candida* shows no clear pattern of geographical parthenogenesis either, and it might be concluded that in spite of recognisable sexual dimorphism, the apparent absence of males in many populations (both living and fossil) could simply be because authors have not bothered to distinguish the sexes.

Living *Limnocythere inopinata* shows a well-defined pattern of geographical parthenogenesis (see Martens 1994), at least in Europe (Fig. 6.4), but sexual populations seem to have been more widespread during the Pleistocene. Amongst modern populations, those from North America (generally known as *Limnocythere sappaensis*) seem always to be sexual (Dobbin 1941, Delorme 1971, Staplin 1963b), whilst those from the former Soviet Union (USSR) and Africa are parthenogenetic (Bronshtein 1947, Martens 1990). Sexual populations are also known from Chinese lakes (Yin 1997). In the Western Palaearctic, in northern and central European living populations, males are known only from occasional individuals from cultures taken from Austrian lakes (Yin 1997) and single lakes in Denmark (B. Scharf pers. comm.) and Poland (NODE). In contrast, in Greco-Macedonian Lake Dojran, and the Grecian lakes Vegoritis and Valtonera, males are almost as abundant as females (Petkovski 1959, Griffiths & Reed, in press, T.K. Petkovski, pers. comm.).

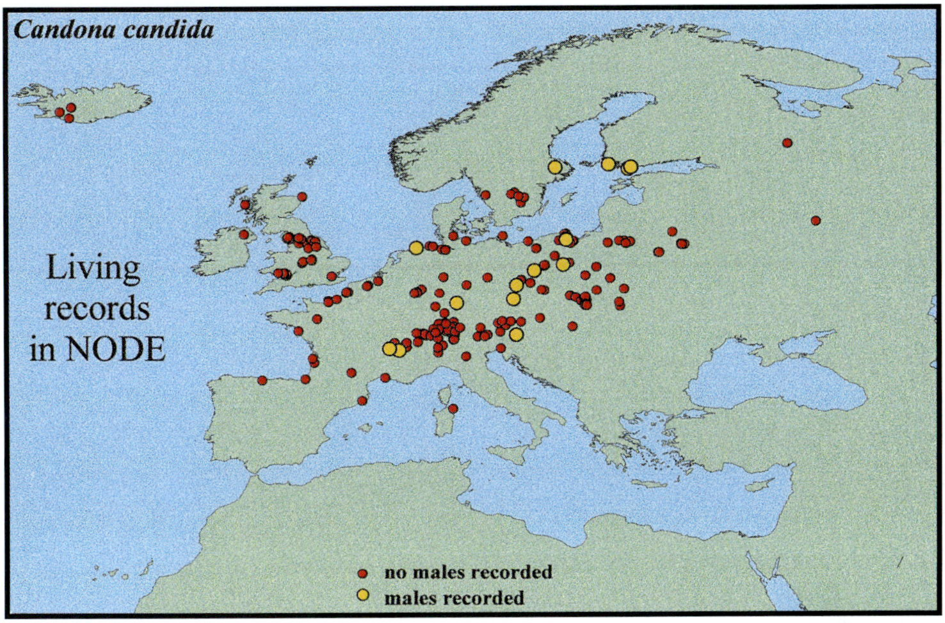

Fig. 6.2 Living distribution of presumed sexual and parthenogenetic populations of *Candona candida* in Europe.

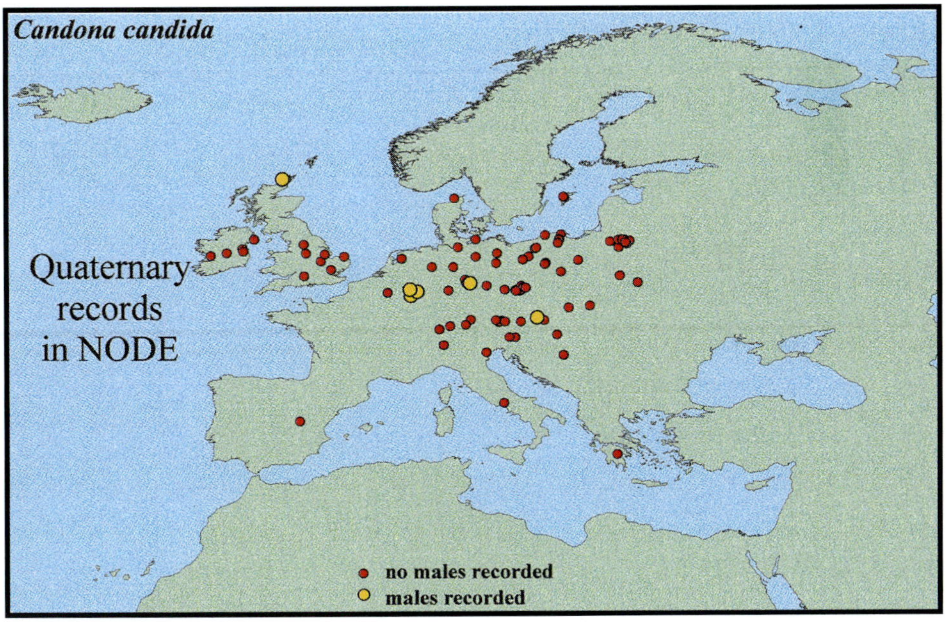

Fig. 6.3 Fossil distribution of presumed sexual and parthenogenetic populations of *Candona candida* in the Quaternary of Europe.

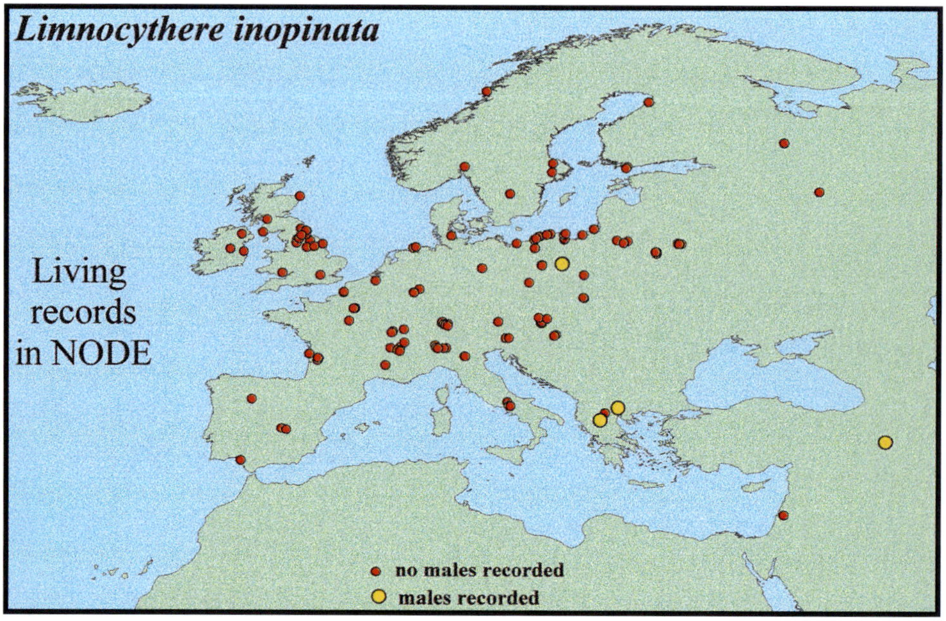

Fig. 6.4 Living distribution of presumed sexual and parthenogenetic populations of *Limnocythere inopinata* in Europe.

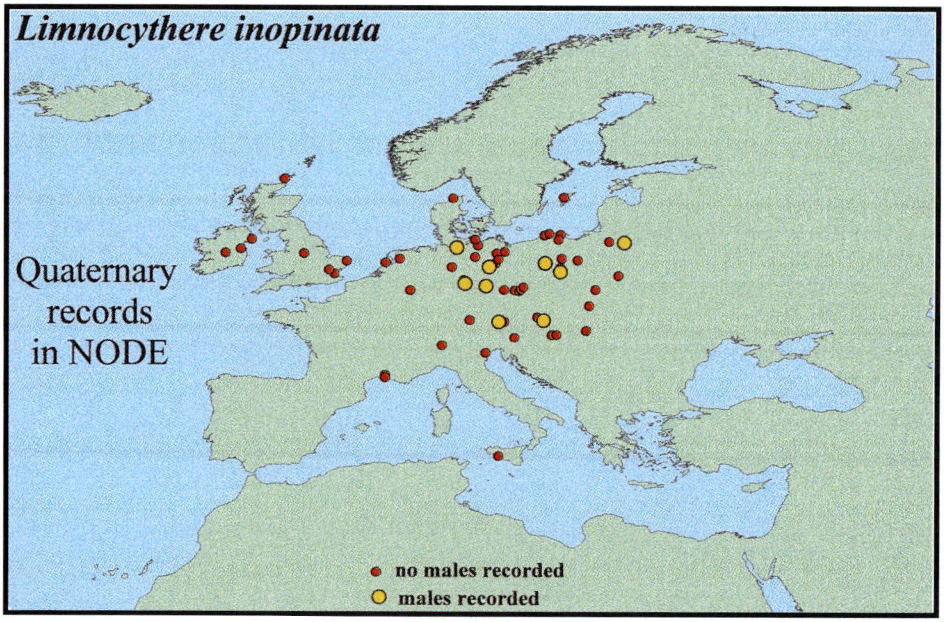

Fig. 6.5 Fossil distribution of presumed sexual and parthenogenetic populations of *Limnocythere inopinata* in the Quaternary of Europe.

Moreover, a fossil/historical 'rare occurrence' of males has been noted in the Neusiedlersee on the Austro-Hungarian border (Löffler 1990: 232). Here males are known only in short, isolated 'bursts' within two of the 18 lake cores studied, and in sediments dating (approximately) up to the 1930s (Löffler 1990).

The status of European Quaternary populations of *L. inopinata* is unclear. In north-eastern Europe the abundance of males may be low, as at the Holsteinian (Middle Pleistocene) sites at Elze and Paludinenbank, or quite high, as at various Eemian (Upper Pleistocene) localities (see Diebel & Pietrzeniuk 1975a). In southern Europe the situation is further aggravated by division of fossil *L. inopinata* into different subspecies (Krstić 1987). *Limnocythere inopinata* cf. *postconcava* Negadaev-Nikonov, 1955 *sensu* Krstić (1987) includes the population from modern Lake Dojran and Quaternary populations from Moldavia, whilst *Limnocythere inopinata pleistocenica* Krstić, 1987 includes populations from Germany, Moldavia, Croatia and Serbia (see also Griffiths 1995). Both clearly include males, but no data are available on relative male abundances within these taxa, nor is it clear whether both are invariably sexual.

The European distribution of Upper Pleistocene and Holocene male-bearing populations of *L. inopinata s.l.* is shown in Fig. 6.5. As in *C. candida*, no clear pattern can be seen, although again most Upper Pleistocene and Holocene populations have been reported from north of the Alps (sexual populations from sites in the former Yugoslavia usually are from the Middle or Lower Pleistocene; Griffiths 1995). The comparative abundance of sexual populations of both *C. candida* and *L. inopinata* may simply be an artefact caused by the comparative intensity of ostracod palaeontological work in Austria, Germany and the British Isles, but in the case of *L. inopinata* it is significantly different from the living distribution. Perhaps, as suggested by Gomez-Cely (1997) this distribution may reflect some response to unusually intense selective pressures created in an area lying between two dynamic, fluctuating ice-sheets (the Alpine and Scandinavian). According to the 'Holocene stability' hypothesis (Horne & Martens in press b), both sexual and parthenogenetic populations could have become established following the last glaciation, but the parthenogens have subsequently outcompeted the sexuals under relatively stable conditions, leaving only occasional rare males which may go mainly undetected. Further palaeontological study is required to test both the post-glacial invasion and Holocene stability hypotheses.

Löffler (1990) suggests that the presence of males of *L. inopinata* in sediments from the Neusiedlersee is related to increased solute content (more specifically, that males appear when alkalinity increases; see also discussion in chapter 12). A similar effect appears to have occurred within a population from Lake Süleymanhaci, Turkey (Roberts *et al.* in press). In this succession, males appear in the last glaciation (just below a thermoluminescence date of 22,528 ± 3988 BP), and are associated with sediment chemistry and diatom assemblages suggesting increased salinity (J.M. Reed, pers. comm.). In the case of the best documented extant European sexual population, data on Lake Dojran reviewed by Griffiths *et al.* (submitted), also show high levels of alkalinity, here through magnesium and bicarbonate enrichment. Lastly, although little is known of the ecology of the North American *L. sap-*

paensis, Dobbin's report of the presence of males at Alkali Lake, Chelan Co. (Washington State) notes that the lake is 'very alkaline' (Dobbin 1941: 186). All this evidence is somewhat apocryphal and circumstantial, and it remains unclear whether a link does exist between male-production and hydrochemistry. Certainly, Martens (1990) found African sexual populations of *Limnocythere* spp. to inhabit waters of a wide range of salinities. It is difficult to see how solute chemistry could relate to reproductive strategy, but it is not impossible that European sexual populations (both fossil and modern) originate from as yet undiscovered relict populations, and thus have been able to successfully colonise certain types of waters, perhaps carried by migrating water birds (De Deckker 1977).

Intermediate timescales: reproductive modes in Mesozoic non-marine ostracods

The earliest convincing non-marine ostracods are Carboniferous Darwinuloidea (e.g. *Darwinula s.l.*, *Whipplella*) accompanied by Carbonitoidea (e.g. *Carbonita*), Cytheroidea (e.g. *Tomiella*) and possibly Cypridoidea (Neustrueva in Carbonel *et al.* 1988; see also discussion in chapter 3). Following the diversification of these groups during the Permian, the only non-marine group to survive into the Triassic seems to have been Darwinuloidea. Post-Palaeozoic non-marine cytheroids (represented by Limnocytheridae) radiated early in the Mesozoic (the Triassic) and are important in the Jurassic and Cretaceous. Possible non-marine cypridoids of doubtful affinity are known from the Lower Permian (Damotte *et al.* 1992) but it was not until the latest Jurassic, with the development of 'Wealden' palaeoenvironments that ostracods, mostly non-marine Cypridoidea, began their major diversification, largely through the rapid evolutionary expansion of the *Cypridea* lineage. Thus cypridoids, which dominate in non-marine waters today, achieved this after a late start. What gave them an advantage over other groups then already established in freshwaters?

'Wealden'-type non-marine basins and their characteristic ostracod faunas occur in the latest Jurassic and Cretaceous in many parts of the world including Europe, North and South America, Africa and China. The classic 'Wealden' association of Cypridoidea (*Cypridea* in particular), darwinuloids (*Darwinula*) and cytheroids (*Theriosynoecum, Timiriasevia*) is first evident in the Late Jurassic (Kimmeridgian) of the USA and became most widespread in the Early Cretaceous, but persisted until the Late Cretaceous, mainly in China. Modern non-marine faunas with cypridoids other than *Cypridea* began to appear in about the mid-Cretaceous; the development and diversification of many modern non-marine genera (e.g. *Cypridopsis*, *Cypris*, *Herpetocypris* and *Ilyocypris*) began in the Late Cretaceous (Babinot *et al.* 1996, Colin *et al.* 1996, Musacchio & Simeoni 1989).

The palaeoecological investigation of Purbeck-Wealden ostracods is controversial (see reviews by Neale 1984, Colin in Carbonel *et al.* 1988, Horne 1995). They are perhaps best regarded as being entirely non-marine, although including both temporary and permanent waters, and saline and freshwater lakes. Alternations of *Cypridea*-dominated faunas with others in which genera such as *Darwinula* and

Theriosynoecum are prominent (F.W. Anderson's C-phase and S-phase faunicycles) may relate to the influence of climatic variation on the relative abundance of permanent and temporary water bodies in the Wealden landscape (Horne 1988, 1995). Most species of the commonest Purbeck-Wealden genera (*Darwinula*, *Theriosynoecum*, *Mantelliana*, *Cypridea*) probably inhabited fresh or slightly saline waters. The reproductive strategies of the first two genera (a darwinuloid and a cytheroid, respectively) involved female brood care; such ostracods are confined to permanent water bodies. Cypridoids such as *Mantelliana* and *Cypridea*, on the other hand, are believed to have laid desiccation-resistant eggs, enabling them to colonise ephemeral pools, survive periods of habitat desiccation, and effect world-wide dispersal (Horne & Martens, in press a).

Whatley (1992) attributes the success and diversification of Cypridoidea to the evolution, late in the Jurassic, of parthenogenesis and desiccation-resistant eggs. These, combined with the ability to swim, gave cypridoids an advantage over other non-marine lineages. Neither Limnocytheridae nor Darwinuloidea, he argues, have resting eggs. Horne & Martens (in press a) present evidence against Whatley's hypothesis. Firstly, parthenogenesis is now well-documented in some modern limnocytherids. In Timiriaseviinae, parthenogens include both Oligocene-Recent *Kovalevskiella* spp. and *Frambocythere tumiensis*; the latter are dominant in the latest Cretaceous to earliest Tertiary limnic sediments of the Pyrenees – being initially represented by both sexes but becoming parthenogenetic after the K-T boundary (Tambareau *et al.* 1997). Furthermore, it is now known that the modern southern African limnocytherids *Ovambocythere* and *Korannacythere* both produce desiccation-resistant eggs (Martens 1989a, 1996). Another problem is that much of the cypridoid expansion is due to the diversification of *Cypridea* (and close relatives), a genus with distinctive shell morphology but which is now extinct. The English Purbeck-Wealden alone (duration approx. 26 million years) has more than sixty species of *Cypridea* and the world-wide Cretaceous diversity of the genus runs to many hundreds of species, even at a conservative estimate. It is now apparent that representatives of extant cypridoidean lineages which also have resting eggs (Candonidae, Cyprididae, Ilyocyprididae, Notodromadidae) were contemporaries of the *Cypridea* group in the early Cretaceous (Horne & Martens in press a). The apparent dominance of *Cypridea* may be at least partly due to preservation effects: other abundant and widespread cypridoids of the time (e.g. *Mantelliana*) had thin, fragile shells that are rarely recovered from sediments; they, like *Cypridea*, often occur as dense concentrations of valves on bedding planes. As other contemporary cypridoids presumably had desiccation-resistant eggs, we must seek an alternative explanation for the success of *Cypridea*. Also, if *Cypridea* was exclusively parthenogenetic, how did it proliferate and diversify to such an extent in only 80 million years, or are we simply unable to recognise sexual dimorphism in the genus? Both Sohn (1969) and Kilenyi & Neale (1978) point out that there is no evidence either for or against sexual reproduction in *Cypridea*, and that sexual dimorphism is not apparent. In fact, sexual dimorphism has been described in species of *Longispinella* and *Paracypridea* (which many authors consider subgenera or synonyms of *Cypridea*) (Krommelbein 1961, Sohn 1969). Kilenyi & Neale (1978) discuss the possibility, first suggested by Sylvester-Bradley (1947), that high speciation rates in

Cypridea may have been due to polyploidy following hybridisation, combined with high selection pressures imposed by constantly changing Purbeck-Wealden environments. Reproductive and dispersal modes are surely of key importance to our interpretation of the evolution and palaeoecology of Mesozoic non-marine Ostracoda, but it is apparent that many questions remain to be answered.

Long timescales: sex and parthenogenesis in Darwinuloidea

Darwinuloids predominate in the earliest well-developed, non-marine ostracod faunas (Carboniferous and Permian). The infra-order is (possibly) known from marine sediments in the mid-Ordovician (e.g. genus *Predarwinula*; see Sohn 1979) and may have entered non-marine environments in the Devonian (Whatley 1992) (although debate continues over whether these were truly 'fresh' waters). It is established that many early darwinuloids were sexual, but that the lineage has been obligately parthenogenetic since some time in the Mesozoic (McKenzie 1971, Griffiths & Butlin 1994). Such a statement must be examined critically in view of its evolutionary significance.

Darwinula is often cited as an example of a genus showing long-term parthenogenesis, in apparent defiance of evolutionary models (Butlin & Griffiths 1993, Griffiths & Butlin 1994, see chapter 1). Leaving aside the debate over the reliability of reports of occasional males in living forms, the exact time at which sexual reproduction was lost from Darwinuloidea is a source of debate. The apparent high diversity and high incidence of sexual dimorphism in Permian darwinuloids (Sohn 1987) also requires critical reappraisal as both may eventually be shown (at least in part) to be taxonomic artefacts. Despite this, there is evidence suggesting sexual reproduction in the genus in the Early Cretaceous: possible males of *D. leguminella* have been identified in 'Purbeck' (late Jurassic-early Cretaceous) assemblages on Bornholm (Christensen 1963) and in southern England (Horne & Martens unpublished data), albeit in a highly female-biased sex ratio. Because of this, it is difficult to state precisely when darwinuloids lost sexual reproduction. The best guess is perhaps mid-Cretaceous (*c* 100 million years ago), although searches for males in other well-preserved fossil assemblages should be pursued. The data do suggest, however, that as far as can be determined at present, darwinuloids perhaps have been parthenogenetic for more than 100 million years.

Modern darwinuloids are not a diverse group, consisting only of *Darwinula* itself and the monospecific genus *Microdarwinula* (but see Rossetti & Martens 1998). Within *Darwinula*, Sohn (1987) lists 24 described modern species and subspecies. Although some of these are junior synonyms, the list has been augmented by the description of several new species, and some still await formal description (e.g. Martens & Rossetti 1997, Rossetti & Martens 1998). However, as there is no sex, and thus no gene exchange or recombination, speciation within Darwinuloidea can occur only through mutation. Certainly there is mounting evidence that some of the extant darwinuloids are very ancient (Martens *et al.* 1997; Schön *et al.* 1998, chapter 17).

Origin of parthenogenetic lineages: a hypothesis

One possible explanation for the origin of certain parthenogenetic lineages is through interspecific hybridisation, as this has been documented for a variety of parthenogenetic vertebrate lineages (e.g. Avise 1994, Vrijenhoek 1993). Such processes are also known to produce temporally and morphologically-stable hybrid taxa in other microcrustacean groups such as Cladocera (e.g. Leider 1991, Taylor & Hebert 1993) and this has been identified in cladoceran successions within fossil sediment sequences (e.g. Hofmann 1991). The first suggestion that this may also occur in Quaternary ostracod faunas was by Stout (1975). Stout re-examined Benson & Macdonald's (1963) report of a Holocene transition between taxa termed *Candona caudata* and *Candona novacaudata*, contending that this actually represented a hybridisation event rather than a genuine evolutionary transition. More recently, Turgeon (1993) used electrophoretic and cytological markers to investigate relationships between various North American cypridoids: '*Cyprinotus*' (= *Heterocypris*) and *Cypricercus* (= *Bradleystrandesia*). Both are well-known freshwater taxa, with males being reported in certain species of both genera (see also Chaplin *et al.* 1994, Turgeon & Hebert 1994).

Turgeon (1993) concludes that lineages of certain sexually-reproducing species have hybridised with parthenogenetic congeners to produce a triploid with a novel and, more importantly, non-intermediate, stable morphology. Thus, interspecific mating between parthenogenetic females from one taxon and males (or perhaps even occasionally-produced males) from another, may provide a means whereby a third, morphologically divergent form is generated. Should such a hybrid population be found within a fossil assemblage, it is highly likely that this would be determined as a new taxon. Certainly, a number of predominantly parthenogenetic modern cypridoid genera (e.g. *Ilyocypris*) are rich in Pleistocene chronoendemic species (see Griffiths 1995) – hybridisation provides a means whereby 'speciation' could occur in such taxa.

Another genus in which Pleistocene speciation may have occurred through hybridisation is *Tonnacypris*, a Holarctic ostracod which usually inhabits temporary water bodies (see Griffiths *et al.* 1998). The genus was erected by Diebel & Pietrzeniuk (1975 b) for German Pleistocene species, other species being added by Martens (1989b), Freels (1980) and Griffiths *et al.* (op. cit.). *Tonnacypris* now includes the Miocene *T. turcica*, the Pleistocene *T. loessica* and *T. convexa*, and three extant species: *T. lutaria* (from temperate western Eurasia), the Holarctic *T. glacialis*, and a localised Baltic crenophile, *T. estonicus*. All appear to be usually parthenogenetic, but a male of *T. lutaria* was reported at the southern edge of the species' range in Anatolia by Gülen (1985), and Bronshtein (1947) reports a putative male of *T. glacialis* from Novaja Zemlja, Russia (again near the southern edge of the species' range).

Griffiths *et al.* (loc. cit.) propose that the ranges of the extant *Tonnacypris* spp. reflect the results of last glaciation biogeography, with different species 'tracking' different thermal regimes (cf. Fryer & Joyce 1981). The result is that *Tonnacypris* comprises a cold-stenothermal form isolated within the Arctic Circle itself (*T. glacialis*), another cold stenotherm in sub-Arctic groundwater-associated habitats (*T.*

estonicus), and a third more eurythermal form in the temperate zone (*T. lutaria*). Species distributions did change within the Pleistocene, and there are cold-stage records of the Arctic *T. glacialis* from the Orkney Islands (UK) and southern Germany (similarly, the Holarctic *Cytherissa lacustris* is known from as far south as Mexico, during the Wisconsin Glaciation, R.A. Forester pers.comm.). Under conditions in which species ranges are (constantly) shifting, formerly isolated taxa may be brought into contact, providing potential conditions for hybridisation (Hewitt 1993, 1995). If this was to result in, for example, a sexual (or occasionally sexual) allopatric population from a refugial area coming into contact with a (usually) more northerly allopatric congeneric parthenogen, and if these should mate and hybridise, then this would produce a new morphospecies (Turgeon 1993, Turgeon & Hebert 1994). This may be locally persistent, or even have some adaptive advantage and disperse outwards. This may explain the generation of new taxa within predominantly parthenogenetic lineages, as well as the geographically- or temporally-restricted ranges seen in many Pleistocene chronoendemics (Griffiths 1995) and hence their utility in biostratigraphy.

Conclusions

It is clear that the investigation of fossil faunas, and particularly those that are well preserved and which offer good stratigraphic detail (both in time and space), offers great potential in the study of the evolution of sex and the adaptive value of parthenogenesis. Few other animal groups have much utility in this respect. Ostracoda offer one of the best fossil records of any invertebrate group inhabiting non-marine waters, augmented by an excellent marine record. This, then, is the strength of Ostracoda – that their stratigraphical range and excellent preservation potential allow their application to a wide variety of evolutionary questions. Further work will help to develop an understanding of how sex and parthenogenesis interact both geographically and at different timescales, and at the level of the population, the species assemblage, and broader-scale phylogenetic lineages.

References

Absolon, A. 1973. Ostracoden aus einigen Profilen spät- und postglazialer Karbonatablagerungen in Mitteleuropa. *Mitt. Bayer. Staatssammlung Paläontologie histor. Geol.*, **13:** 47-94, pl. 3.

Avise, J.C. 1994. *Molecular markers, natural history and evolution,* Chapman & Hall.

Babinot, J.-F., Colin, J.-P. & Tambareau, Y. 1996. Late Cretaceous non-marine ostracods from Europe: biostratigraphy, palaeobiology and taxonomy. *Cretaceous Research*, **17:** 151-167.

Benson, R.H. & MacDonald, H.C. 1963. Post-glacial (Holocene) ostracodes from Lake Erie. *Univ. Kansas Paleontological Contributions (Arthropoda)*, **4:** 1-26.

Bless, M.J.M. & Pollard, J.E. 1975. Quantitative analysis of dimorphism in *Carbonita humilis* (Jones & Kirby). *in*: Swain, F.M. (ed), Biology and Paleobiology of Ostracoda. *Bull. amer. Paleontology*, **65:** 109-127.

Bronshtein, Z.S. 1947. *Freshwater Ostracoda. Fauna of the USSR. Crustaceans. Volume II, Number I.* [English ed. (1988). Nayar, I. (trans.).], Amerind Publ., 470 pp.

Butlin, R.K. & Griffiths, H.I. 1993. Ageing without sex? *Nature*, **364:** 680.

Carbonel, P., Colin, J.-P., Danielopol, D.L. & Londeix, L. 1986. *Kovalevskiella* (Ostracoda, Timiriaseviinae), genre à mode de vie benthique depuis l'Oligocène, son adaptation à la vie interstitielle. *Géobios*, **19:** 677-687.

Carbonel, P., Colin, J.-P., Danielopol, D.L., Löffler, H. & Neustrueva, I. 1988. Paleoecology of limnic ostracodes: a review of some major topics. *Palaeogeogr., Palaeoclimatol., Palaeoecol.*, **62:** 413-461.

Chaplin, J.A., Havel, J.E. & Hebert, P.D.N. 1994. Sex and ostracods. *Trends Ecol. Evol.*, **9:** 435-439.

Christensen, O.B. 1963. Ostracods from the Purbeck-Wealden beds in Bornholm. *Danmarks Geologiske Undersogelse, 2. Räkke*, **86:** 1-58, pls.1-5.

Cohen, A.C. & Morin, J.G. 1990. Patterns of reproduction in ostracodes: a review. *J. Crustacean Biol.*, **10:** 184-211.

Colin, J.-P., Tambareau, Y. & Krasheninnikov, V.A. 1996. Ostracodes limniques et lagunaires dans le Crétacé Supérieur du Mali (Afrique de l'Ouest): systématique, paléoécologie et affinitiés paléobiogéographiques. *Rev. micropaléontologie*, **39:** 211-222.

Damotte, R., Lethiers, F., Corbin, J.-C., Renard, M. & Debriette, P. 1992. Premières données sur la faune d'ostracodes du Permien Inférieur de Souvigny (Nord du Massif Central, France). *Rev. micropaléontologie*, **35:** 253-270.

Danielopol, D., Olteanu, R., Löffler, H. & Carbonel, P. 1990. Present and past geographical-ecological distribution of *Cytherissa* (Ostracoda, Cytherideidae). *Bull. Inst. Geol. Bassin d'Aquitaine, Bordeaux*, **47:** 97-118.

De Deckker, P. 1977. The distribution of the 'giant' ostracods (Family: Cyprididae Baird, 1845) endemic to Australia. *in*: H. Löffler & D. Danielopol (eds), *Aspects of ecology and zoogeography of recent and fossil Ostracoda,* Junk, The Hague: 285-294.

Delorme, L.D. 1971. Freshwater ostracodes of Canada. Part V. Families Limnocytheridae, Loxoconchidae. *Can. J. Zool.*, **49:** 43-64, pls.1-19.

Diebel, K. & Pietrzeniuk, E. 1975a. Mittel- und jungpleistozäne Ostracodenfaunen des Raums Potsdam – Brandenburg – Rathenow in stratigraphischer und ökologischer Sicht. *Z. geol. Wiss.*, **3:** 1197-1233.

Diebel, K. & Pietrzeniuk, E. 1975b. Neue Ostracoden aus dem Pleistozän von Burgtonna (Bezirk Erfurt). *Tonnacypris* gen. n. und *Amplocypris tonnenis* sp. n. *Z. geol. Wiss.*, **3:** 87-97.

Dobbin, C.N. 1941. Fresh-water Ostracoda from Washington and other western localities. *Univ. Washington Publ. Biol.*, **4** (3): 175-245, 14 pls.

Evans, J.G. & Griffiths, H.I. 1993. Holocene mollusc and ostracod sequences: their potential for examining short-timescale evolution. *in*: Lees, D.R. & Edwards, D. (eds), *Evolutionary patterns and processes. Linn. Soc. Lond. Symp. Series* **14**, Acad. Press: 125-137.

Freels, D. 1980. Limnische Ostrakoden aus Jungtertiär und Quartär der Türkei. *Geol. Jb. B*, **9:** 3-169.

Fryer, G. 1980. The copepod *Salmincola edwardsii* as a parasite of *Salvelinus alpinus* in Britain, and a consideration of the so-called relict fauna of Ennerdale Water. *J. Zool.*, **193:** 253-268.

Fryer, G. 1993. *The freshwater Crustacea of Yorkshire: a faunistic and ecological survey.* Yorkshire Naturalists' Union and the Leeds Philosophical and Literary Society.

Fryer, G. & Joyce, A. 1981. The distibution of some freshwater copepods and its bearing on the history of the fauna and flora of the British Isles. *J. Biogeogr.*, **8:** 281-291.

Gomez-Cely, D.M. 1997. *Changes in the distribution of fresh-water ostracods in Europe during Quaternary – training exercise on application of geographic information systems in biodiversity studies.* Diploma thesis, Univ. Hull.

Griffiths, H.I. 1995. European Quaternary freshwater Ostracoda: a biostratigraphic and palaeobiogeographic primer. *Scopolia,* **34:** 1-168.

Griffiths, H.I. & Butlin, R.K. 1994. *Darwinula stevensoni*: a review of the biology of a persistant parthenogen. *in*: Horne, D.J. & Martens, K. (eds), *The Evolutionary Ecology of Reproductive Modes in Non-marine Ostracoda,* Univ. Greenwich Press: 27-36.

Griffiths, H.I. & Butlin, R.K. 1995. A timescale for sex *vs.* parthenogenesis – evidence from Holocene subfossil ostracods. *Proc. R. Soc. Lond. B*, **260:** 65-71.

Griffiths, H.I. & Evans, J.G. 1995. The Late-glacial and early Holocene colonisation of the British Isles by freshwater ostracods. *in*: Riha, J. (ed), *Ostracoda and Biostratigraphy,* Balkema: 291-302.

Griffiths, H.I. & Reed, J.M. (in press). Environmental change at Lake Dojran: preliminary data from fossils and sediments. *in*: Petkovski, P., Griffiths, H.I. & Haberl, W. (eds), *BIOECCO 2: Biodiversity, Ecology and Conservation of Balkan Fauna*, Macedonian Museum Nat. Hist.

Griffiths, H.I., Pietrzeniuk, E., Fuhrmann, R., Lennon, J.J., Martens, K. & Evans, J.G. 1998. *Tonnacypris glacialis* (Ostracoda, Cyprididae): taxonomic position, (palaeo-) ecology and zoogeography. *J. Biogeogr.,* **25**: 515-526.

Griffiths, H.I., Butlin, R.K. & Geiger, W. (in press a). Specific considerations of time-scale effects on reproductive mode in European Holocene non-marine Ostracoda. *in*: Crasquin, S., Braccini, E. & Lethiers, F. (eds), Proceedings of the 3rd European Ostracodologists Meeting. *Bull. Centr. Rech. Explor.-Prod. Elf-Aquit.*

Griffiths, H.I., Reed, J.M., Krstić, N., Petkovski, S., Stojanovski, P., Levkov, Z. & Ratajac, R. (in press b). The recent history and decline of Macedonian-Greek Lake Dojran (Dojransko ezero): a review. *Limnologica.*

Gülen, D. 1985. Bisexual Ostracoda (Crustacea) populations in Anatolia. *Istanbul Üniversiteti Fen Fakultesi Mecmuasi, Seri B,* **50:** 65-80.

Hartmann, G. 1964. Asiatische Ostracoden. Systematische und zoogeographische Untersuchungen. *Intern. Revue ges. Hydrobiol. (System. Beiheft)*, **3:** 1-155.

Hewitt, G.M. 1993. Post-glacial distribution and species substructure: lessons from pollen, insects and hybrid zones. *in*: Lees, D.R. & Edwards, D. (eds), *Evolutionary Patterns and Processes. (Linn. Soc. Symp. Ser. 14)*, Academic Press: 97-123.

Hewitt, G.M. 1995. Some genetic consequences of ice ages, and their role in divergence and speciation. *Biol. J. Linn. Soc.*, **58:** 247-276.

Hofmann, W. 1991. The Late Glacial/Holocene and recent *Bosmina* (*Eubosmina*) fauna of Lake Constance (Untersee) (F.R.G.): traces of introgressive hybridization. *Hydrobiologia*, **225:** 81-85.

Horne, D.J. 1988. *Cretaceous Ostracoda of the Weald. (BMS Field Guide no. 4).* British Micropalaeontological Society.

Horne, D.J. 1995. A revised ostracod biostratigraphy for the Purbeck-Wealden of England. *Cretaceous Res.*, **16:** 639-663.

Horne, D.J. & Martens, K. (in press a). The importance of resting eggs for the evolutionary success of non-marine Ostracoda (Crustacea). *In*: Brendonck, L., DeMeester, L. & Hairston, N. (eds), Diapause in the Crustacea. *Adv. Limnol.* **52:** 549-561.

Horne, D.J. & Martens, K. (in press b). Geographical parthenogenesis in European non-marine ostracods (Crustacea, Ostracoda); post-glacial invasion or Holocene stability? *Hydrobiologia.*

Janz, H. 1995. Neubeschreibung von *Strandesia risgoviensis* nov. comb. (Sieber, 1905) (Crustacea, Ostracoda) aus dem Nordlinger Ries (Miozän, Süddeutschland). *Stuttgarter Beitr. Naturk. B (Geologie und Paläontologie)*, **233:** 1-17.

Järvekülg, A.A. 1959. [Sluchay nahozhdeniya samtsa *Ilyodromus olivaceus* (Brady & Norman) (Ostracoda Cypridae)]. *Eesti NSV Teaduste Akadeemia Toimetised (Bioloogiline Seeria)*, **9:** 242-243. (In Russian).

Kilenyi, T.I. & Neale, J.W. 1978. The Purbeck/Wealden. *in*: Bate, R.H. & Robinson, J.E. (eds), A stratigraphical index of British Ostracoda. *Geol. J. Spec. Issue*, **8:** 299-324.

Krommelbein, K. 1961. Über Dimorphismus bei Arten der Ostracoden-Gattung *Paracypridea* Swain (Cyprideinae) aus dem NE-brasilianischen 'Wealden'. *Senckenbergiana lethaia*, **42:** 353-375.

Krstić, N. 1987. [*Limnocythere inopinata* species and its subspecies]. *Radovi Geoinstituta*, **20** (1986)**:** 213-221, pls I-III. (In Serbian with English summary).

Leider, U. 1991. The *Bosmina kessleri*-like morphotype of *Eubosmina* in Lake Muskoka, Ontario, Canada, as putative interspecific hybrids. *Hydrobiologia*, **225:** 71-80.

Löffler, H. 1990. Palaeolimnology of Neusiedlersee, Austria. I. The succession of ostracods. *Hydrobiologia*, **214:** 229-238.

Lynch, M. 1984. Destabilizing hybridization, general-purpose genotypes, and geographical parthenogenesis. *Quart. Rev. Biol.*, **59:** 257-290.

McKenzie, K.G. 1971. Palaeozoogeography of freshwater Ostracoda. *in*: Oertli, H.J. (ed), Paléoécologie des Ostracodes, Pau 1970. *Bull. Centre Rech. Pau-SNPA Suppl.*, **5:** 207-237.

Martens, K. 1989a. *Ovambocythere milani* gen. n. spec. n. (Crustacea, Ostracoda, Limnocytheridae) an African limnocytherid reared from dried mud. *R. zool. Afric.*, **103:** 379-388.

Martens, K. 1989b. On the systematic position of the *Eucypris clavata*-group, with a description of *Trajancypris* gen. nov. (Crustacea, Ostracoda). *Arch. Hydrobiol. Suppl.*, **83:** 227-251.

Martens, K. 1990. Revision of African *Limnocythere s.s.* Brady, 1867 (Crustacea, Ostracoda), with special reference to the Rift Valley Lakes; morphology, taxonomy, evolution and (palaeo-) ecology. *Arch. Hydrobiol. Suppl.*, **83:** 453-524.

Martens, K. 1994. A European network on the evolutionary ecology of reproductive modes in non-marine Ostracoda: background and objectives. *in*: Horne, D.J. & Martens, K. (eds), *The Evolutionary Ecology of Reproductive Modes in Non-marine Ostracoda,* Univ. Greenwich Press: 3-7.
Martens, K. 1996. On *Korannacythere* gen. nov. (Crustacea, Ostracoda), a new genus of temporary pool limnocytherids from southern Africa, with the description of three new species and a generic reassessment of the Limnocytheridae. *Bull. k. belg. Inst. Natuurw. Biol.*, **66:** 51-72.
Martens, K. & Meisch, C. 1985. Description of the male of *Potamocypris villosa* (Jurine, 1820) (Crustacea, Ostracoda). *Hydrobiologia*, **127:** 9-15.
Martens, K. & Rossetti, G. 1997. On two new species of *Darwinula* Brady & Robertson, 1885 (Crustacea, Ostracoda) from South Africa. *Bull. k. belg. Inst. Natuurw. Biol.*, **67:** 57-66.
Martens, K., Rossetti, G. & Fuhrmann, R. 1997. Pleistocene and Recent species of the family Darwinulidae Brady & Norman, 1889 (Crustacea, Ostracoda) in Europe. *Hydrobiologia*, **357:** 99-116.
Mazepova, G.F. 1990. [*Rakuschkovye Ratschki (Ostracoda) Baikala.*] Novosibirsk; Akademia Nauk SSSR/Sibirskoe Otdelenie Limnologicheskii Institut. (In Russian).
Musacchio, E.A. & Simeoni, M. 1989. Cretaceous non-marine cypridacean Ostracoda from central and northern Argentine Patagonia. *Cour. Forschungsinst. Senckenberg*, **113:** 77-88.
Neale, J.W. 1984. The Ostracoda and uniformitarianism. 2. The earlier record: Cretaceous to Cambrian. *Proc. Yorks. geol. Soc.*, **44:** 443-478.
Petkovski, T.K. 1959. Beiträge zur Kenntnis der Ostracoden-Fauna Jugoslawiens (V). *Istanbul Üniversitesi Fen Fekültesi Hidrobiologi Arastirma Enstitüsü Yayinlarindan (Seri B)*, **4:** 158-165.
Petkovski, T.K. & Meisch, C. 1995. Interesting freshwater Ostracoda (Crustacea) from Macedonia. *Bull. Soc. Natural. Luxembourg.*, **96:** 167-183.
Roberts, N. 1998. *The Holocene. An environmental history,* (2nd Edition), Blackwell: 316 pp.
Roberts, N., Black, S., Boyer, P., Eastwood, W., Griffiths, H.I., Leng, M., Parish, R., Reed, J., Twigg, D. & Yigitbasioglu, H. (in press). Chronology and stratigraphy of Late Quaternary sediments in the Konya Basin, Turkey: results from the KOPAL project. *Quart. Sci. Rev.*
Rossetti, G. & Martens, K. 1998. Taxonomic Revision of the Recent and Holocene representatives of the Family Darwinulidae (Crustacea, Ostracoda), with a description of three new genera. *Bull. k. belg. Inst. Natuurw. Biol.*, **68.**
Schoen, I., Martens, K. & Rossi, V. 1996. Ancient asexual scandals. *Trends Ecol. Evol.*, **11:** 296-297.
Schön, I., Butlin, R.K., Griffiths, H.I. & Martens, K. 1998. Slow molecular evolution in an ancient asexual ostracod. *Proc. R. Soc. Lond. B*, **265:** 235-242.
Segerstråle, S.G. 1957. On the immigration of the glacial relicts of Northern Europe, with remarks on their prehistory. *Commentationes Biologicae*, **16:** 1-117.
Semenova, L.M. 1979. [Key features in the biology of *Candona candida* (Ostracoda, Crustacea).] *Hydrobiol. J.*, **15:** 21-24. (In Russian).
Sohn, I.G. 1969. Non-marine ostracodes of Early Cretaceous age from Pine Valley Quadrangle, Nevada. *US Geol. Surv. Professional Paper*, **643:** B1-B9, 2 pls.
Sohn, I.G. 1979. Non-marine ostracodes in the Lakota Formation (Lower Cretaceous) from South Dakota and Wyoming. *US Geol. Surv. Professional Paper*, **1069:** I-IV, 1-41, 8 pls.
Sohn, I.G. 1987. The ubiquitous ostracode *Darwinula stevensoni* (Brady & Robertson, 1870), redescription of the species and lectotype designation. *Micropaleontology*, **33:** 150-163.
Staplin, F.L. 1963a. Pleistocene Ostracoda of Illinois. Part I. Subfamilies Candoninae, Cyprinae, general ecology, morphology. *J. Paleontol.*, **37:** 758-769.
Staplin, F.L. 1963b. Pleistocene Ostracoda of Illinois. Part II. Subfamilies Cyclocyprinae, Cypridopsinae, Ilyocyprinae; Families Darwinulidae and Cytheridae, stratigraphic ranges and assemblage patterns. *J. Paleontol.*, **37:** 1164-1203.
Stout, L.N. 1975. Polytypic species and hybridization in Quaternary freshwater ostracods of North America. *Palaeogeogr., Palaeoclimatol., Palaeoecol.*, **17:** 257-266.
Sylvester-Bradley, P.C. 1947. The shell structure and evolution of the Mesozoic ostracod *Cypridea. Quart. J. Geol. Soc. Lond.*, **103** (Proceedings)**:** VIII.
Sywula, T. & Pietrzeniuk, E. 1989. Type Arthropoda, Class Ostracoda Latreille, 1806. *in*: Rühle, E. (ed), *Geology of Poland, Volume III. Atlas of guide and characteristic fossils, Part 3b, Cainozoic, Quaternary* [English ed. (1994). Brochwicz-Lewiñsky, W. (trans.)], Publishing House Polish Ecological Agency: 75-90, 195-198, pls. 48-63.

Tambareau, Y., Hottinger, L., Rodriguez-Lazaro, J., Villatte, J., Babinot, J.-F., Colin, J.-P., Garcia-Zarraga, E., Rocchia, R. & Guerrero, N. 1997. Communautés fossiles benthiques aux alentours de la limite Crétacé/Tertiaire dans les Pyrénées. *Bull. géol. Soc. Franc.*, **168:** 795-804.

Taylor, D.J. & Hebert, P.D.N. 1993. Habitat-dependent hybrid parentage and differential introgression between neighbouring sympatric *Daphnia* species. *Proc. Natl. Acad. Sci. USA*, **90:** 7079-7083.

Turgeon, J. 1993. *Genetic diversity in sexual and asexual members of Cyprinotus and Cypricercus (Ostracoda: Crustacea).* MSc thesis, Univ. Guelph.

Turgeon, J. & Hebert, P.D.N. 1994. Evolutionary interactions between sexual and all-female taxa of *Cyprinotus* (Ostracoda: Cyprididae). *Evolution*, **48:** 1855-1865.

Vandel, A. 1928. La Parthénogénèse géographique. *Bull. Biol. Franc. Belg.*, **62:** 164-281, 3 pls.

Vrijenhoek, R.C. 1993. The origin and evolution of clones versus maintenance of sex in *Poeciliopsis. J. Heredity*, **84:** 388-395.

Weider, L.J. 1993. A test of the 'general-purpose' genotype hypothesis: differential tolerance to stress among *Daphnia* clones. *Evolution*, **47:** 965-969.

West, R.G. 1977. *Pleistocene geology and biology,* (2nd ed), Longman.

Whatley, R.C. 1992. The reproductive and dispersal strategies of Cretaceous non-marine Ostracoda. *in*: Mateer, M.J. & Chen, P.-J. (eds), *Aspects of non-marine Cretaceous geology,* China Ocean Press: 177-192.

Williams, D.D. 1987. *The ecology of temporary waters,* Croom Helm.

Willis, K.J. 1994. The vegetational history of the Balkans. *Quart. Sci. rev.,* **13:** 769-788.

Yin, Y. 1997. Contribution to the morphology and ecology of Ostracoda Limnocytheridae and Candonidae. Comparative studies between Austrian and Chinese Crustacea. PhD thesis, Univ. Vienna: 163 pp.

CHAPTER 7

CYTOPLASMIC BACTERIA IN *HETEROCYPRIS INCONGRUENS* (CRUSTACEA, OSTRACODA)

Tom T.M. Vandekerckhove

Abstract

Cytoplasmically (= transovarially) inherited intracellular microorganisms may manipulate their host's sexuality in order to increase the proportion of (infected) females available for transmission. They may feminize genetic males, kill male embryos and induce thelytokous (= female-to-female) parthenogenesis and cytoplasmic incompatibility. We discovered a cytoplasmic infection in *Heterocypris incongruens* (Crustacea, Ostracoda) using Transmission Electron Microscopy. The microorganisms are especially abundant in the ovaries, suggesting transovarial transmission, and in the mantle epithelium. Morphologically, they resemble none of the endosymbionts known so far, and their biological function is not at all clear. From their position and their transmission mode, an effect upon the host's reproduction cannot be ruled out, all the more because *H. incongruens* is composed of sexual as well as parthenogenetic and mixed populations, possibly reflecting different infection levels.

Introduction

Several arthropod species are known to harbour intracellular symbiotic or parasitic microorganisms. Obligate endocytic parasites, which often kill their host, cannot be transmitted unless they produce a resistant stage able to withstand the adverse environment once the host has died. However, a good alternative – especially for mutualists and weak parasites – is to invade the female gonadal tissue at least during the time of reproduction so as to be transmitted transovarially to the next generation; sperm cells, by contrast, are not readily accessible because the vast majority of cytoplasmic constituents, including mitochondria and bacteria, is discarded in the so-called waste-bags during spermiogenesis. Therefore, maternally inherited microorganisms are thought to evolve toward the ability of favouring female offspring in order to maintain themselves; as such, manipulation of the host's sexuality results in an adaptive advantage. Microorganisms may distort their host's offspring sex ratio in a number of ways.

First, by interfering with hormonal sex determination systems (which occur in many crustaceans), or more precisely, by preventing the development of the androgenic gland, they can transform genetic males into functional females during embryogenesis (Bulnheim 1978, Ginsburger-Vogel *et al.* 1980, Juchault & Legrand 1989, Dunn *et al.* 1993, Juchault *et al.* 1994). By contrast, if sexual development follows directly from the ratio of male determining to female determining loci (e.g.

Sex and parthenogenesis: evolutionary ecology of reproductive modes in non-marine ostracods, edited by K. Martens, pp. 119–125

in most insects), the system seems too rigid to be 'deceived', and infected male embryos may be killed – a dead end for both host and parasite – whereas their female counterparts develop normally (Lanier & Oliver 1966, Williamson & Poulson 1979, Skinner 1985, Werren *et al.* 1986, Chang *et al.* 1991, Hurst *et al.* 1992). Paradoxically, this partial microbial suicide will increase the proportion of infected individuals within the population. Anyway, in both situations the imminent lack of males to inseminate females would lead to a population crash (Hamilton 1967) unless the microbial influences are counterbalanced, e.g. by incomplete trans-ovarial transmission (Rigaud & Juchault 1992), autosomal masculinizing genes (Rigaud & Juchault 1993) or natural curing agents such as antibiotics in the environment or periodically elevated temperatures.

A third distorting mechanism, which, in addition, circumvents the reproductive problems of all-female sexual populations, is found in a number of parasitoid wasps (Zchori-Fein *et al.* 1992, Stouthamer & Werren 1993, Zchori-Fein *et al.* 1995, Pijls *et al.* 1996). When infected with *Wolbachia pipientis*, they switch to thelytokous parthenogenesis, whereby even virgin females produce 100% of daughters. The wasps have a haplodiploid constitution (i.e. males develop from unfertilized haploid eggs via arrhenotokous parthenogenesis, while fertilized diploid eggs give rise to females), implying that virgins should produce all-male broods. The bacteria render the haploid eggs diploid through gamete duplication (Stouthamer & Kazmer 1994), i.e. they inhibit chromosome segregation in the telophase of the first mitotic division of the haploid female pronucleus.

Lastly, a common phenomenon involving *Wolbachia pipientis* and known as cytoplasmic incompatibility usually leaves the sex ratio undisturbed. The bacteria are presumed to alter the chromosomal imprinting during spermatogenesis, so that insemination by infected males causes the embryos to abort unless the female bears the same bacterial strain, infected females being always compatible. Incompatible crosses abort because the paternal chromosome set is eliminated, rendering the egg inviable because of haploidy. Quite logically, in the haplodiploid wasp *Nasonia vitripennis* (Hymenoptera: Pteromalidae), cytoplasmic incompatibility has a masculinizing effect (Richardson *et al.* 1987).

O'Neill *et al.* (1997) provide an elaborate exposition on 'haunted' arthropods. From this it should be clear that aberrant sex ratios and parthenogenesis might be induced by cytoplasmically inherited microorganisms, especially when animal species comprise populations that differ fundamentally with regard to their sexuality: sexual vs. parthenogenetic, 'normal' vs. 'aberrant' sex ratio. Potential candidates in this respect are discussed in the review from Hurst (1993). Here we report the observation, by means of Transmission Electron Microscopy (TEM), of cytoplasmic bacteria within the ovaries and surrounding mantle tissue in *Heterocypris incongruens*, a cypridid ostracod reproducing parthenogenetically in our laboratory. The presence of the bacteria elsewhere in the Ostracoda is under investigation.

Material and methods

Animals

H. incongruens adults were collected from anostracan (*Streptocephalus torvicornis*; place of origin: Used, Zaragoza, Spain) stock cultures, where both stably coexisted and the former had established a parthenogenetic population. The bottom of the aquaria was covered with sand, detritus, exuvia and decomposing remains providing a good substrate for the Ostracoda to settle.

Tissue preparation

The animals were fixed at 4°C in 2% paraformaldehyde + 2% glutaraldehyde in a 0,06M cacodylate buffer, pH 7,2-7,4. In the fixative the valves (containing the ovaries) were manually removed from the ostracods for further use. Fixation time was 96h, after which the specimens were rinsed in cacodylate buffer and postfixed in a 1% OsO_4 solution in *aqua bidest.* for 4h. The specimens were dehydrated and embedded in Araldite CY212 resin following standard protocols (Raes *et al.* 1989). Thin sections (65nm) were made with the Reichert-Jung Ultracut E.

Staining

The specimens were stained *en bloc* with saturated uranyl acetate in 50% ethanol and with saturated lead acetate in water-free ethanol-acetone (1:1), 1h each after the corresponding dehydration step. Thin sections were automatically processed (using the Reichert-Jung Ultrostainer) in a saturated uranyl acetate solution in 50% ethanol (30' at 40°C) and in alkaline lead citrate (5' at 20°C), respectively.

Transmission Electron Microscopy

The specimens were viewed with a Zeiss EM 900.

Results

Ultrastructural features of the bacteria can be seen in Fig. 7.1. They appear pleomorphic, often with irregular borders, and measure 0,3 – 0,6 μm in diameter (cocci) or 0,3 – 0,4 μm × 0,6 – 1,3 μm (rods). Contrary to most endocytic microorganisms, they are not enclosed in vacuoles, but instead display a strikingly thick cell wall-like outer layer. The latter does not seem to be composed of peptidoglycan, a wall component of many Gram-negative and Gram-positive bacteria providing the cell with mechanical strength against turgor pressure inside. This phenomenon is rather common in intracellular microbes (especially the pleomorphic ones) because the rigidity of peptidoglycan is no longer needed or is even a disadvantage when living inside cells. Such bacteria typically have a trilaminar envelope, to wit (from inside to outside) a cytoplasmic membrane, an outer membrane of the Gram-negative cell wall and a host-derived vacuolar membrane. Neither has the outer envelope of the

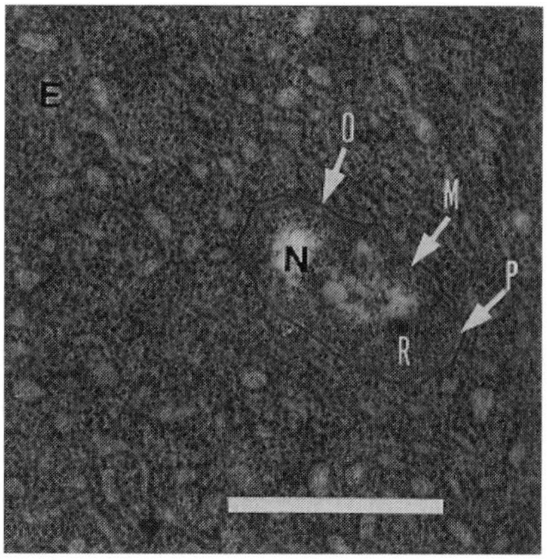

Fig. 7.1 Electron micrograph of a cytoplasmic bacterium in a gut diverticulum cell from *Heterocypris incongruens*; note the abundant rough endoplasmic reticulum (**E**) characteristic of a secretory cell. **M**: bacterial cytoplasmic membrane, **N**: nucleoid (with DNA filaments often visible), **O**: outer wall-like layer, **P**: periplasmic space, **R**: prokaryotic-type ribosomes (recognizable by their smaller size as compared with eukaryotic ribosomes from the RER). Bar: 600nm.

H. incongruens endosymbionts got a membranous structure typical of Gram-negative bacteria. However, a periplasmic space approximately 15 to 20 nm wide is apparent. The clear nucleoid part is sharply defined from the dense peripheral cytoplasm which consists mainly of tightly packed ribosomes of prokaryotic size. The microbes can be seen in great numbers and dividing in the gonadal cells (Fig. 7.2), suggesting that transovarial transmission may be an important spreading factor. Furthermore, they are also numerous in mantle epithelial cells and in macrophage-like cells residing in the vestibulum (not shown); they are much fewer in abdominal epithelium and gut diverticulum cells. A final morphological peculiarity of the *H. incongruens* microorganisms, which is still being studied, is their ability to develop stacks of closely apposed membranes invaginating from the cytoplasmic membrane (not shown).

Discussion

Our parthenogenetic all-female *H. incongruens* population is obviously infected with intracellular bacteria, but some questions as to their nature remain to be answered.

What bacterial species are we dealing with? At this early stage very little can be said but that they morphologically resemble none of the endosymbionts identified so far. Molecular characterization/identification is therefore planned. Hitherto, bacteria inducing parthenogenesis, feminization and cytoplasmic incompatibility have been consistently identified as *Wolbachia* (Werren *et al.* 1995), whereas male killers belong to several Proteobacteria subgroups (Gherna *et al.* 1991, Werren *et al.* 1994).

Secondly, is the association mutualistic, parasitic or neutral? This problem can be solved by means of antibiotic experiments (Stouthamer & Luck 1993) in conjunction with ecological study. Any mutualistic symbiosis of the type found in corals, where endocytic zooxanthelles favour deposition of calcareous skeleton

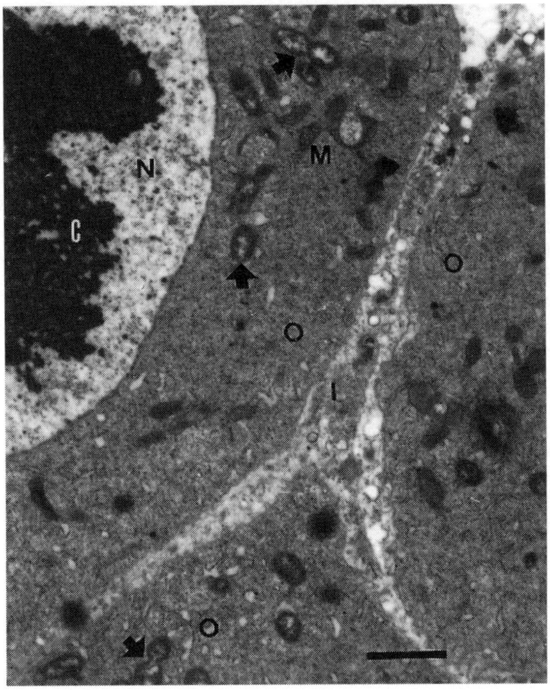

Fig. 7.2 Low-magnification electron micrograph of ovarial cells: 3 oocytes (**O**) and an interstitial (or nurse) cell (**I**). **C**: condensed chromatin from chromosomes, **M**: mitochondria, with dark matrix and numerous interspersed cristae, **N**: nucleoplasm of host cell. Arrows indicate bacteria (several of which are dividing), distinguishable from mitochondria by their clear nucleoid. Bar: 700nm.

($CaCO_3$) by absorbing CO_2 for photosynthesis, cannot be ruled out on the basis of the internal morphology of the microbes.

Do these bacteria occur in other (e.g. sexual) populations or species? This is presently under investigation (fluorescence microscopy and TEM). Parthenogenetic individuals of the related species *Cypridopsis vidua* from the same locality were found to be free of cytoplasmic bacteria.

Is there any sex ratio distortion? Again, antibiotics could reveal which type we are dealing with (Stouthamer *et al.* 1990).

Finally, are the microbes maternally (i.e. vertically, transovarially) inherited only, or do they spread horizontally (i.e. by contagion) as well? Water is an easy and relatively safe means of transport as compared with a terrestrial (micro) habitat. Moreover, from the place where the microbes are seen (the mantle and tissues enclosed herein), the outer environment is directly within range: many microbes would have to cross a one-membrane barrier only. Reinfection upon release from host cells may be the actual spreading strategy, indeed, without excluding vertical transmission as a supplementary means.

References

Bulnheim, H.P. 1978. Interaction between genetic, external and parasitic factors in sex determination of the crustacean amphipod *Gammarus duebeni*. *Helgol. Wiss. Meeresunters.*, **31:** 1-33.

Chang, K.S., Shiraishi, T., Nakasuji, F. & Morimoto, N. 1991. Abnormal sex-ratio condition in the walnut leaf beetle, *Gastrolina depressa* (Coleoptera, Chrysomelidae). *Appl. Entomol. Zool.*, **26:** 299-306.

Dunn, A.M., Adams, J. & Smith, J.E. 1993. Transovarial transmission and sex ratio distortion by a microsporidian parasite in a shrimp. *J. Invertebr. Pathol.*, **61:** 248-252.

Gherna, R.L., Werren, J.H., Weisburg, W., Cote, R., Woese, C.R., Mandelco, L. & Brenner, D.J. 1991. *Arsenophonus nasoniae* gen. nov., sp. nov., the causative agent of the son-killer trait in the parasitic wasp *Nasonia vitripennis*. *Int. J. Syst. Bacteriol.*, **41:** 563-565.

Ginsburger-Vogel, T., Carre-Lecuyer, M. & Fried-Montaufier, M.C. 1980. Transmission expérimentale de la thélygénie liée à l'intersexualité chez *Orchestia gammarellus* (Pallas). Analyse des phénotypes sexuels dans les descendances de femelles normales transformées en femelles thélygènes. *Arch. Zool. Exp. Gen.*, **122:** 261-270.

Hamilton, W.D. 1967. Extraordinary sex ratios. *Science*, **156:** 477-488.

Hurst, G.D.D., Majerus, M.E.N. & Walker, L.E. 1992. Cytoplasmic male killing elements in *Adalia bipunctata* (Linnaeus) (Coleoptera: Coccinellidae). *Heredity*, **69:** 84-91.

Hurst, L.D. 1993. The incidences, mechanisms and evolution of cytoplasmic sex ratio distorters in animals. *Biol. Rev.*, **68:** 121-193.

Juchault, P. & Legrand, J.J. 1989. Sex determination and monogeny in terrestrial isopods *Armadillidium vulgare* (Latreille, 1804) and *Armadillidium nasatum* (Budde-Lund, 1885). *Monit. Zool. Ital. (N.S.) Monogr.*, **4:** 359-375.

Juchault, P., Frelon, M., Bouchon, D. & Rigaud, T. 1994. New evidence for feminizing bacteria in terrestrial isopods: evolutionary implications. *C.R. Acad. Sci. Paris Sér. III*, **317:** 225-230.

Lanier, G.N. & Oliver, J.H., jr. 1966. 'Sex-ratio' condition: unusual mechanisms in bark beetles. *Science*, **153:** 208- 209.

O'Neill, S.L., Hoffmann, A.A. & Werren, J.H. (ed). 1997. *Influential passengers. Inherited microorganisms and arthropod reproduction*, Oxford Univ. Press: 214 pp..

Pijls, J.W.A.M., Van Steenbergen, H.J. & Van Alphen, J.J.M. 1996. Asexuality cured: the relations and differences between sexual and asexual *Apoanagyrus diversicornis*. *Heredity*, **76:** 506-513.

Raes, H., Bohyn, W., De Rycke, P.H. & Jacobs, F. 1989. Membrane-bound iron-rich granules in fat cells and midgut cells of the adult honeybee. *Apidologie*, **20:** 327-337.

Richardson, P.M., Holmes, W.P. & Saul, G.B. 1987. The effect of tetracycline on non-reciprocal cross incompatibility in *Mormoniella (=Nasonia) vitripennis*. *J. Invertebr. Pathol.*, **50:** 176-183.

Rigaud, T. & Juchault, P. 1992. Genetic control of the vertical transmission of a cytoplasmic sex factor in *Armadillidium vulgare* Latr. (Crustacea, Oniscidea). *Heredity*, **68:** 47-52.

Rigaud, T. & Juchault, P. 1993. Conflict between feminizing sex ratio distorters and an autosomal masculinizing gene in the terrestrial isopod *Armadillidium vulgare* Latr. *Genetics*, **133:** 247-252.

Skinner, S.W. 1985. Son-killer: a third extrachromosomal factor affecting the sex ratio in the parasitoid wasp, *Nasonia (=Mormoniella) vitripennis*. *Genetics*, **109:** 745-759.

Stouthamer, R. & Kazmer, D.J. 1994. Cytogenetics of microbe-associated parthenogenesis and its consequences for gene flow in *Trichogramma* wasps. *Heredity*, **73:** 317-327.

Stouthamer, R. & Luck, R.F. 1993. Influence of microbe-associated parthenogenesis on the fecundity of *Trichogramma deion* and *T. pretiosum*. *Entomol. Exp. Appl.*, **67:** 183-192.

Stouthamer, R. & Werren, J.H. 1993. Microbes associated with parthenogenesis in wasps of the genus *Trichogramma*. *J. Invertebr. Pathol.*, **61:** 6-9.

Stouthamer, R., Luck, R.F. & Hamilton, W.D. 1990. Antibiotics cause parthenogenetic *Trichogramma* (Hymenoptera/Trichogrammatidae) to revert to sex. *Proc. Natl. Acad. Sci. USA*, **87:** 2424-2427.

Werren, J.H., Hurst, G.D.D., Zhang, W., Breeuwer, J.A.J., Stouthamer, R. & Majerus, M.E.N. 1994. Rickettsial relative associated with male killing in the ladybird beetle (*Adalia bipunctata*). *J. Bacteriol.*, **176:** 388-394.

Werren, J.H., Skinner, S.W. & Huger, A.M. 1986. Male-killing bacteria in a parasitic wasp. *Science*, **231:** 990-992.

Werren, J.H., Windsor, D. & Guo, L. 1995. Distribution of *Wolbachia* among neotropical arthropods. *Proc. R. Soc. Lond. B.*, **262:** 197-204.

Williamson, D.L. & Poulson, D.F. 1979. Sex ratio organisms (Spiroplasmas) of *Drosophila*. *in*: Whitcomb, R.F. & Tully, J.G. (eds), *The Mycoplasmas*, Vol. III, Academic Press: 175-207.
Zchori-Fein, E., Roush, R.T. & Hunter, M. 1992. Male production induced by antibiotic treatment in *Encarsia formosa*, an asexual species. *Experientia*, **48:** 102-105.
Zchori-Fein, E., Faktor, O., Zeidan, M., Gottlieb, Y., Czosnek, H. & Rosen, D. 1995. Parthenogenesis-inducing microorganisms in *Aphytis* (Hymenoptera: Aphelinidae). *Insect Mol. Biol.*, **4:** 173-178.

CHAPTER 8

INTRASPECIFIC MORPHOLOGICAL VARIABILITY: MORPHOMETRY OF VALVE OUTLINES

A. Baltanás and W. Geiger

Abstract

Size and shape variability in ostracod carapace is addressed through the study of two taxa belonging to the major non-marine lineages in the group: *Eucypris virens* (Cypridoidea) and *Limnocythere inopinata* (Cytheroidea). Size is studied using 'traditional' meristic measurements (length/height relationships), whereas shape has been explored with 'new' morphometric tools, namely the elliptic Fourier analysis. Variability in both features has been explored at the intraspecific level including comparisons between populations (for both species) and between clones (for *L. inopinata*). Comparisons extend to reproductive mode, as sexual populations have been analysed, together with parthenogenetic ones, for both species. Quite unexpectedly, similar amounts of size and shape variability occur in both species at between-populations and between-clones levels. Genotypic and environmental factors and their interactions seem to play a different role in determining the valve shape in both lineages, an issue of great importance for taxonomic and palaeoenvironmental reconstruction studies.

Introduction

When interpreting the diversity of the organic world, human beings have practised a typological approach which takes advantage of the overwhelming variety of forms we are confronted with. Following that line of reasoning, technology helped us by providing powerful instruments to address the morphological variety of nature in a range beyond of what is 'visible' to us (this mainly refers to microscopy but also to all kinds of devices developed in order to make it possible for us to measure morphological features). We know at present that our perception of the world around us is one among many existing (see, for instance, the recent discovery of mate recognition using the ultraviolet part of the spectrum in blue tits; Andersson *et al.* 1998, Hunt *et al.* 1998). But still most of our knowledge is based on visual (=morphological) descriptions.

Taxonomy, the theory and the practice of classifying organisms (Mayr & Ashlock 1991), mainly relies on morphological features. Molecular techniques have recently been incorporated to the array of methods for taxonomic and systematic assessment, but their contribution (quantitatively considered) to taxonomy is, as yet, far from equaling that of classical morphology (see chapter 10 for a more complete discussion). 'Ostracodology', the study of ostracods (Crustacea, Ostracoda), is a

Sex and parthenogenesis: evolutionary ecology of reproductive modes in non-marine ostracods, edited by K. Martens, pp. 127–142

good representative of such a kind of approach. Some advances have been made in using molecular techniques (see chapter 13), but most attempts are still in the realm of morphological assessment. For recent ostracod species, although it is common to use carapace attributes as well, soft part anatomy is preferred as it provides a finer expression of phylogenetic relationships. There are indeed a larger number of characters available in the soft parts compared to the carapace; thus, information on the evolution of the group is larger and more readily available in the soft parts. However, when working with fossil (or sub-fossil) material, carapace features are the only available information for taxonomic identification. Fortunately, the carapace of the ostracod is not just a structural epiphenomenon, it is an important functioning part, a carefully engineered framework which allows its bearer to withstand mechanical forces and adds stability to his movements (Benson 1981). Thus, carapace shape, together with structural elements (inner lamella, selvage, lists, etc.) and ornamentation (nodes, ridges, spines, tubercles, etc.), becomes a unique feature for many species (Van Morkhoven 1962). Strikingly enough, and in spite of their remarkable importance, neither carapace nor soft parts morphological variability are properly known at the specific or population level.

In the following paragraphs we attempt to assess the amount of morphological variability at different intraspecific levels (between populations, within populations and within clones) and to identify causes (genotypic-environmental) for such variability. Because ostracods are highly diverse organisms in terms of taxa described, ecological requirements, life-histories, reproductive modes, etc., it is not easy to grasp general insights into morphological patterns. Thus, we have approached the problem by considering taxa from the two major ostracod lineages with representatives in the non-marine water bodies (the Cytheroidea and the Cypridoidea), relying primarily on recent data compiled in the framework of the EU HCM or its associated projects. The basic body plan of the different groups have been described in chapter 4, which should be consulted as an introduction to the present part.

Morphological variability in the carapace

Traditionally, ostracod carapace shape has been qualitatively described using drawings and, more recently, SEM photographs. Both techniques are of invaluable help to the ostracodologist as they provide useful insights of qualitative aspects of the valves like their general appearance, texture, etc. This information facilitates species recognition as long as differences between species are large enough to be appreciated by the naked (microscope-aided) eye. Thus, experience plays a main role and the whole process is laden with subjectivity (which is not necessarily bad, but frequently inconvenient).

In order to explore shape variability in a quantitative, reproducible way, we have used methods for shape description based on Elliptic Fourier outline analysis (for a more complete description see Kaesler & Waters 1972, Kaesler & Maddocks 1984, Schweitzer *et al.* 1986, Burke *et al.* 1987, Schweitzer & Lohmann 1990). Outline analysis is especially suited to poorly ornamented ostracods in which the lack of easily identifiable diagnostic homologous sites ('landmarks') prevents us from the use of other morphometric techniques (Kaesler & Foster 1987, Reyment *et al.* 1988,

Abe *et al.* 1988, Reyment & Bookstein, 1993, Reyment, 1995). The aim of the method is to describe the outline shape mathematically through a series of coefficients that can be afterwards reconstructed, compared and classified using standard multivariate techniques (e.g. Rohlf & Archie 1984) (Fig. 8.1).

Background information on carapace morphology of selected species

The two species in the analysis differ considerably with regard to size, shape and ornamental structures of the shells. In *Eucypris virens*, size and shape have been often recognised as highly variable. Females' (most of the populations studied are parthenogenetic) size range is remarkably large: 1.5-2.4 mm (Stephanides 1948, Nüchterlein 1969, Crossetti & Margaritora 1982, Smith & Martens 1996).

Concerning the shape, several morphological varieties have been recognised by several authors (Müller 1900, Wolf 1919, Bronshtein 1947), although their meanings have been largely debated (Alm 1915, Sars 1928, Bronshtein 1947). Tétart (1982) concluded that morphological variation of shape in this species is discontinuous, arguing that the continuous pattern generated by syngamic lineages is split into fragments and frozen when reproductive mode shifts to parthenogenesis. The valves are smooth and lack tubercles and reticulation, but show a distinct dimorphism between left and right valve. Left valve is larger and overlaps the right one.

Limnocythere inopinata is considerably smaller in size (0.58-0.64 mm) with some populations showing a distinct nodation pattern whereas others are smooth (Fuhrmann & Pietrzeniuk 1991). There is no clear dimorphism between left and right valves in terms of shape, but the left valve usually has small but distinct teeth on the postero-ventral margin. Left valve embraces the right one at the rear but not in front. The Chinese material we worked on was previously described as different species (Huang 1984) but soft part analysis showed that they fit well within the reported range of morphological variability of *L. inopinata* (Yin 1997).

Both species reproduce throughout Europe mainly parthenogenetically.

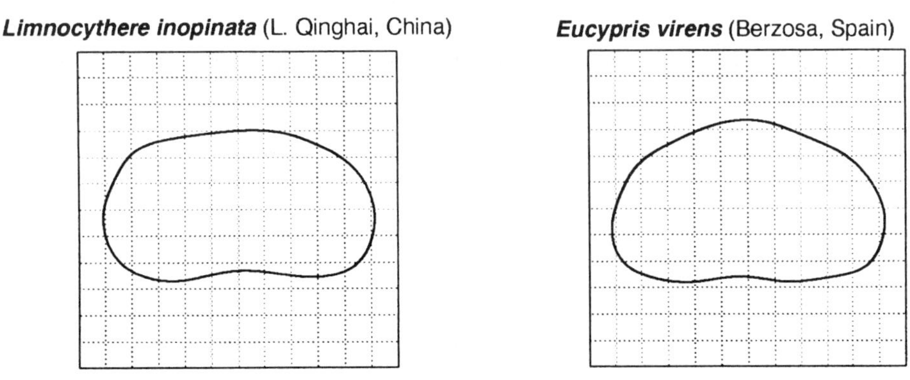

Fig. 8.1 Outline reconstruction of the mean shape of two populations of the species considered using harmonic coefficients obtained from elliptic Fourier analysis.

Material

For our comparative approach we investigated a total of 25 populations (11 of *L. inopinata* and 14 of *E. virens*) sampled throughout Europe, including also one site in Tibet (PR China). A total of 2108 single valves, both left and right ones, were analysed (Table 8.1). Most of the populations studied reproduce by apomictic parthenogenesis but L-7 and the fossil L-10 (for *L. inopinata*) and E-14 (for *E. virens*). From the bisexual *L. inopinata* populations, only females were sampled and their outlines recorded, but male and female valves have been used from the E-14 (*E. virens*). For most of the populations the genetic diversity has been determined by DNA analysis and/or electrophoretic methods (see chapter 15, 16). Furthermore, during sampling and from laboratory studies on selected populations, additional ecological information has been gathered. Only adult shells were included in the analysis (for an assessment of shape changes during ontogenetic development see Baltanás *et al.* in prep.).

In addition to field populations, six clones (C1 to C6) were isolated from L-5 (East Austria) and cultured in the lab under controlled environmental conditions (12:12 light cycle, temp. 20°C) (Yin 1997). Two further clones – a supposed generalist (C7) and a specialist (C8) – were raised at two different temperatures (15° and 25°C) (Yin loc. cit.). Both clones show important ecological differences: C7 is the most abundant clone in Neusiedlersee area (which includes L-3, L-4 and L-5) throughout the year, whereas C8 is restricted to the summer when salinity and temperature rise significantly in the lake. Finally, two other clones (C9a and C9b) were isolated from a temporary lake in Central Spain (different to L-6) and raised in the lab in the same conditions than C1-C6.

Population and clonal size variability

We first measured 'traditional' morphometric variables, namely valve length and height. To do so, valves are oriented with their ventral margins resting on a base line and length is measured as the longest horizontal axis cutting the valve outline at two points (Farkas 1974). Height is the orthogonal segment which crosses the valve at its highest point. Although both measurements are reasonable surrogates for ostracod size, valve area seems to be a better option as long as it is not dependent on how the 'base-line', for the supposedly horizontal position, is defined. Indeed, what we are measuring is not the valve area, but the area occupied by the projection of the carapace side in lateral view.

Cytheroidea: *Limnocythere inopinata*

Limnocythere inopinata clones raised in the lab under controlled conditions tend to be smaller than those sampled in the field (Fig. 8.2). As long as lab temperatures (20°C) were higher than those in the lakes where the natural populations are living, or at least much more stable, those results agree with the expectations because low temperatures increase the inter-moult period, resulting in larger ostracod adult size (Herman & Heip 1982, Martens 1985, Latifa 1987). It seems, however, that temperature cannot explain all variation in size and that reaction norms are different for

Table 8.1. Sampling localities, sampling site co-ordinates and number of shells analysed for the two species studied. (RV=right valves; LV=left valves)

Codes	Locality		Co-ordinates	Altitude (m)	RV	LV
L-1	Salzburg (AUSTRIA)	Freshwater Pond	47°48'N 13°03'E	550	53	45
L-2	Mondsee (AUSTRIA)	Freshwater Lake	47°50'N 13°23'E	550	57	51
L-3	Neusiedlersee (AUSTRIA)	Alkaline Lake	47°56'N 16°48'E	273	50	51
L-4	Unterer Stinkersee (AUSTRIA)	Alkaline Pond	47°47'N 16°48'E	273	36	36
L-5	Oberer Stinkersee (AUSTRIA)	Alkaline Pond	47°48'N 16°48'E	273	50	52
L-6	Las Madres (SPAIN)	Oligohaline Grave-pit lake	40°18'N 3°31'W	600	53	47
L-7	Lake Qinghai (PR China)	Alkaline Lake	37°28'N 97°53'E	3193	51	50
L-8	ZPBG Bay-Lake Qinghai (PR China)	with freshwater influence	37°28'N 97°53'E	3193	50	49
L-9	Lake Erhai (PR China)	Oligohaline-Alkaline lake	37°28'N 97°53'E	3193	48	49
L-10	Neumark-Nord (GERMANY) (Bisexual)	Eemian fossils (Freshwater)			21	29
L-11	Neumark-Nord (GERMANY) (Parthenogenetic)	Eemian fossils (Saline)			24	26
E-1	Doñana (SPAIN)	Freshwater Pond	37°0'N 6°29'W	sea level	40	35
E-2	Tagarnillares-1 (SPAIN)	Freshwater Pond	38°30'N 6°28'W	~700	6	3
E-3	Tagarnillares-2 (SPAIN)	Freshwater Pond	38°30'N 6°28'W	~700	11	6
E-4	Cabeza de Buey (SPAIN)	Freshwater Pond	38°44'N 5°13'W	~550	5	6
E-5	Cáceres (SPAIN)	Freshwater Pond	39°31'N 6°15'W	~400	6	5
E-6	Talaván (SPAIN)	Freshwater Pond	39°39'N 6°17'W	~390	6	6
E-7	Serradilla (SPAIN)	Freshwater Pond	39°45'N 6°10'W	~400	10	9
E-8	Torrejón (SPAIN)	Freshwater Pond	39°46'N 6°1'W	~290	8	7
E-9	Canto del Pico (SPAIN)	Freshwater Pond	40°37'N 3°56'W	~850	48	44
E-10	Berzosa (SPAIN)	Freshwater Pond	40°37'N 3°56'W	~850	8	9
E-11	Morcuera (SPAIN)	Freshwater Pond	40°48'N 3°45'W	~1700	46	38
E-12	Haantjen (BELGIUM)	Freshwater Ditch	51°6'N 3°38'E	sea level	11	16
E-13	Bramhope (U.K.)	Freshwater Pond	53°50'N 1°35'W	~125	9	10
E-14a	Fitou (bisexual/females) (FRANCE)	Freshwater Pond	42°53'N 2°58'E	sea level	166	171
E-14b	Fitou (bisexual/males) (FRANCE)	Freshwater Pond	42°53'N 2°58'E	sea level	166	169

different genotypes. For instance, clone C8 grows larger at 15°C than at 25°C, as expected, but clone C7 shows no statistically significant differences in size for both culture temperatures.

At the population level there are also remarkable differences in valve size for *L. inopinata*. Within a restricted geographic area (Austria), populations from less fluctuating freshwater habitats (L-1) are larger than those inhabiting the oligo-mesohaline – totally dissolved solids (TDS) ≈ 1-16‰ – more fluctuating water bodies (L-3 to L-5). Chinese populations are generally larger than the European ones and within them, those inhabiting the large lake Qinghai (L-7, TDS ≈ 14‰) are larger than those living in areas irregularly subjected to inputs of freshwater (L-8 and L-9, TDS 1-20‰). The reverse trend in the correlation between size and salinity in China and

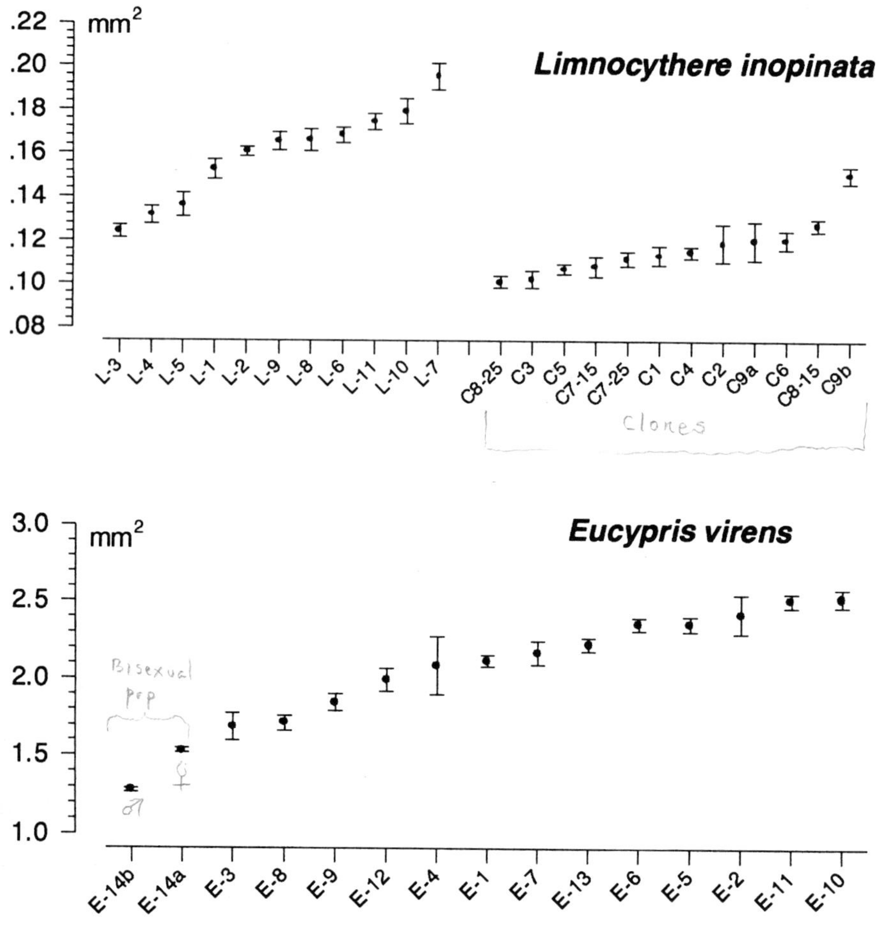

Fig. 8.2 Mean outline area (±95% confident limits of the mean) for each population and clone studied (only right valves).

Europe could be explained in terms of habitat stability. *L. inopinata* populations occur in habitats which differ in environmental stability (lakes and ponds), covering a large geographic range. In large lakes with predictable seasonal changes, fast development is not as important as in water bodies where optimal conditions change rapidly (be they salinity and/or temperature). It is not salinity as such but the life history strategies associated with a life in fluctuating environments which lead to small adult sizes. It is worth noting here the large size of the specimens from L-6, an oligohaline gravel-pit lake (TDS ≈ 2-3‰) of modest dimensions (surface=7.83 ha, max. depth=19 m) but highly stable from a physical point of view (Álvarez Cobelas *et al.* 1993).

Cypridoidea: *Eucypris virens*

Concerning *E. virens*, no such clear patterns, ecological or geographical, exist. The only remarkable element is that individuals in the bisexual population (E-14) have a smaller size, with males being significantly smaller than females (sexual or parthenogenetic) in all other populations (Fig. 8.2). Throughout a large geographic range, *Eucypris virens* occurs in similar types of water bodies (mainly astatic freshwater ponds) which differ from each other in a stochastic (local) rather than a systematic (regional) way. Thus, despite the fact that genetic resemblance should increase with geographical vicinity, developmental 'noise' produced by local environmental conditions is likely to diffuse spatial patterns in size change.

Comparative aspects

Length-height and length-area regression models yield almost identical coefficients between the two species besides their dissimilarity in shape (Table 8.2). Valve design in both species is quite constant throughout the populations, at least at the coarse-grained scale defined by these meristic variables as indicated by the high predictive power (r^2) of the regression models.

Variability in length, height and outline area has been evaluated by computing the coefficient of variation for each population and for both left and right valves. Results are depicted in Fig. 8.3 showing, somewhat unexpectedly, that similar levels of variation are found within populations and clones in both species.

Given that lab-cultured clones are sets of genotypically identical individuals which have been raised under the same environmental conditions, low variability was expected when compared to animals belonging to natural populations. We would also expect higher variability in bisexual populations than in parthenogenetic ones due to their higher genetic diversity. And, similarly, genetically diverse parthenogenetic populations (i.e. multi-clonal populations) should be also more variable in terms of shell morphology than the monoclonal ones. However, the data from our analysis reveal no such relationship. Not only does variability of the few bisexual populations available lie well within the variability range for parthenogens, but, in addition, parthenogenetic populations which show the lowest genetic diversity, rank high with regard to carapace size variability (e.g. E-11 for *E. virens* and L-1 for *L. inopinata*). Besides the restrictions set by the method which reduces complex shapes to meristic measurements, this might indicate some upper limit to carapace variation imposed by mechanic constraints of the carapace architecture. We can speculate on that by proposing that although GxA (genotype x environment) sets the mean values for morphological variables, physiological and biochemical random effects on the individual are fixed for the species (or even at a supraspecific level). It would be interesting to know whether species from other groups react in a similar way.

Table 8.2. Results from regression analysis between size related variables. Right and left valves have been grouped together.[1] Length-area relationship is allometric (area = a.(length)^b)

	Length (mm) vs.	Slope	Intercept	r^2	N
Eucypris virens	Height (mm)	0.561 ($p<0.001$)	0.067 ($p<0.001$)	0.936	1080
	Area[1] (mm^2)	1.996 ($p<0.001$	0.473 ($p<0.001$)	0.982	
Limnocythere inopinata	Height (mm)	0.544 ($p<0.001$)	0.009 (n.s.)	0.925	1028
	Area[1] (mm^2)	1.993 ($p<0.001$)	0.466 ($p<0.001$)	0.982	

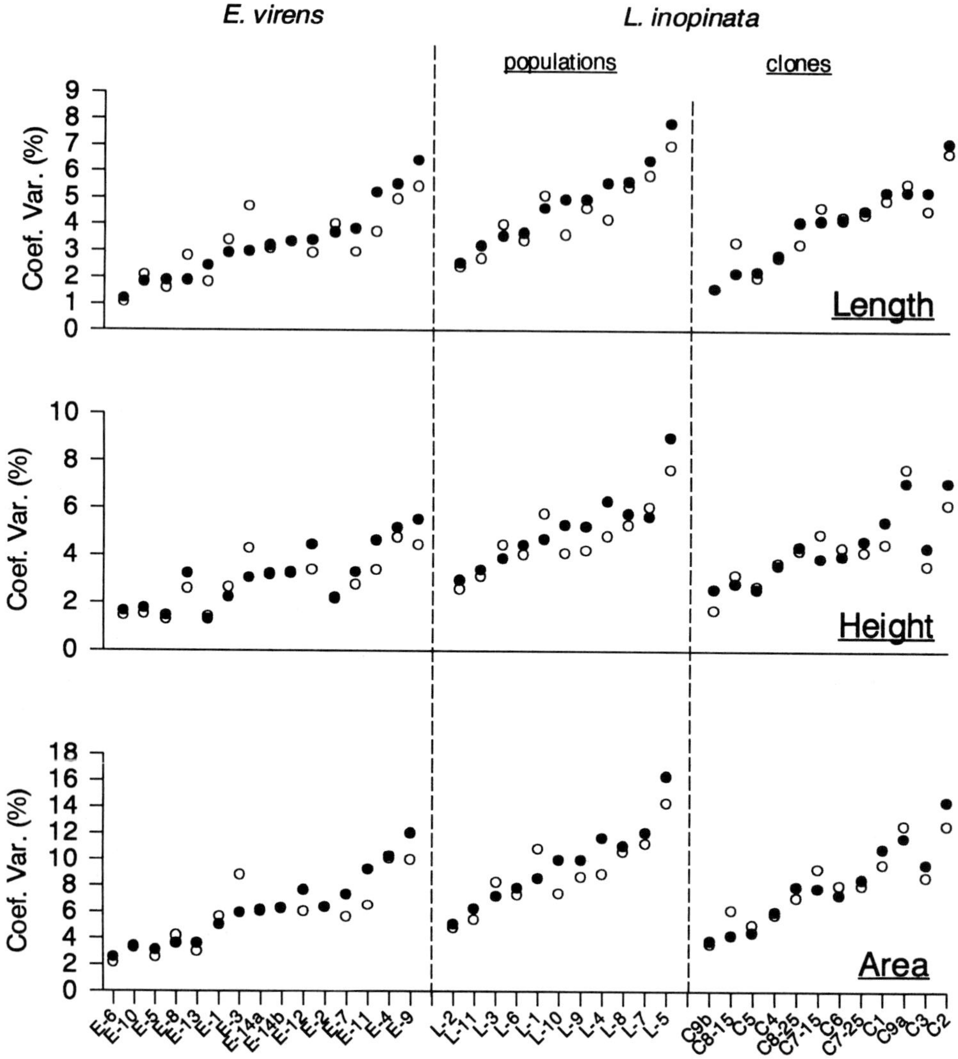

Fig. 8.3 Coefficients of variation of length, height and outline area for each of the populations and clones examined of both species. Open symbols are for right valves and closed symbols for the left ones.

Population shape variability

Shape descriptors rendered by outline analysis have been used to explore shape resemblance among all the different populations considered. First of all, elliptic Fourier coefficients were mathematically normalised to be invariant to size, rotation and starting position of the outline trace (Ferson *et al.* 1985). In such way, differences in size were removed from the analysis, making comparisons based on shape alone feasible. Because within a population shape variability is supposed to reflect differences at the individual level, mean shapes for each group were computed by averaging harmonic coefficients elementwise (Ferson *et al.* 1985). These mean shapes are assumed to represent the characteristic shape of each population.

To avoid random effects due to small sample size, some groups have been discarded (e.g. E-4, E-5 and E-6). Between-population shape similarity was measured using Mahalanobis distance between average shapes in order to adequately account for correlation among the variables (all the Fourier coefficients). A cluster analysis (UPGMA) was finally performed on the resulting distance matrix (Fig. 8.4).

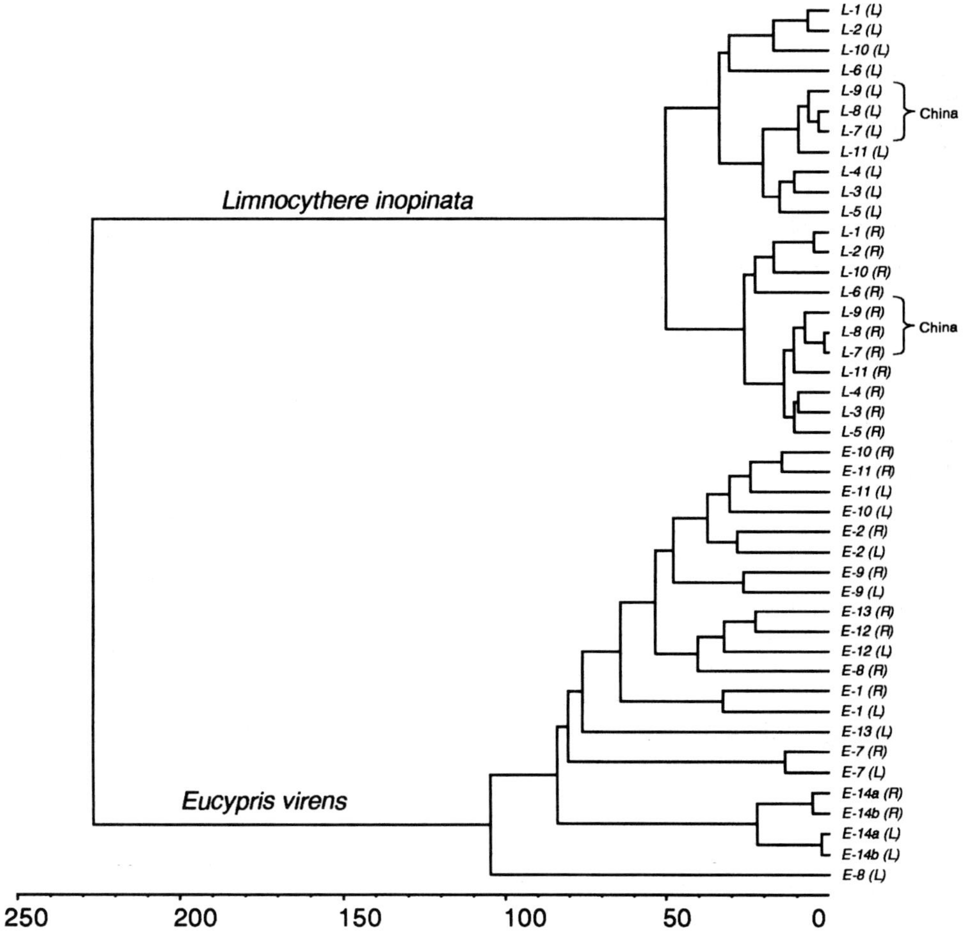

Fig. 8.4 Cluster analysis (UPGMA) of populations performed on a distance (Mahalanobis) matrix. (R)-Right valve; (L)-Left valve

As it was expected, both species are clearly discriminated into two large groups, but overall similarity among *Limnocythere inopinata* populations is higher (linkage distances are lower) than among populations of *E. virens* (Fig. 8.4). This shows that shape is a much more diverse feature in the latter species.

The analysis of the patterns observed will be conducted now for each species in turn:

Cytheroidea: *Limnocythere inopinata*

The same clustering pattern is reproduced for the left and the right valves (Fig. 8.4). This points to the existence of subtle but consistent differences in valve outlines at the population level which results in a coherent pattern of relationships among groups regardless of the valve (right or left) considered. But the pattern observed (any mean right valve is more similar to any other right valve than to its corresponding left valve) also suggests that steady differences between left and right valves exist, independently of the population they belong to.

The Chinese populations cluster together and also join the Austrian oligo-mesohaline lakes (L-3 to L-5). All Chinese samples come from the Lake Qinghai area: one from the Lake Qinghai itself, another from a bay in the lake (Zhang Peng Bin Guan bay) which remains isolated from the lake for several months each year and which is influenced by an incoming river, and the last one from lake Erhai, a lagoon nearby the Qinghai lake (Yin 1997). Salinity decreases as we move from the lake to the surrounding lagoons (TDS ≈ from 14 to 1‰) but pH (7.5-9.1) is high in all sampling locations (Yin 1997). The Austrian 'saline' lakes also have high pH (9-10) and alkalinity (9-13 meq l^{-1}) values, with salinity fluctuating from 1 to 16.5‰ (Yin 1997).

The other cluster is a combination of the Austrian freshwater sites (L-1 and L-2), the bisexual Eemian population (L-10) and the population from Central Spain (L-6). This cluster is linked at a relatively large distance with the other group, making it tempting to associate each one with a different type of water chemistry (freshwater *vs.* saline). L-6, in spite of being oligohaline (see above), clearly differs from its Austrian and Chinese 'saline' counterparts in terms of water chemistry. It is not an alkaline lake (pH is always close to 7.0 and not beyond 8.0; alkalinity is close to 3-4 meq l^{-1}) but rich in sulphates, with lower and more stable conductivity than the Austrian lakes. More interesting, the fossil (Eemian) bisexual population, known to occur in a freshwater environment (Fuhrmann & Pietrzeniuk 1991), is placed close to the other 'freshwater' populations and not close to the other Eemian fossils, which indeed belong to a parthenogentic population living in a saline environment (Fuhrmann & Pietrzeniuk 1991).

Cypridoidea: *Eucypris virens*

Populations of this species are not arranged in two groups according to left and right valves as occurs for *L. inopinata*. Instead, left and right valve from a single population are clustered together before the data are joined with other populations similar in shape (Fig. 8.4). It seems that differences in shape between different populations are greater than differences between left and right valves within one population. No clear latitudinal pattern appears but still some geographic trace can be detected (the

Northernmost populations –E-12 and E-13, some Central Spanish sites –E-10 and E-11). Only two 'aberrant' samples are not clustered with their corresponding pair, E-13 (L) and E-8(L). Both samples are small in size (10 and 7 individuals, respectively) so random effects could be responsible for their lack of congruence.

The bisexual population differs in shape from all the others and all subsets in it (males and females have been kept apart) not only cluster together but also according to valve side (right valves with right valves and left with left) regardless of the gender. This means that males and females are indeed quite similar in shape and, without their clear size dimorphism, it would be extremely difficult to distinguish them by outline. This has some important implications for the study of fossil assemblages as the presence of males could be overlooked, if no size dimorphism exists at the adult stage.

But relationships among populations with regard to their outline shape do not inform about the variability within each site. This was addressed by computing the coefficient of variation of the Mahalanobis distances from each individual to its group centroid (Fig. 8.5). As concerning meristic measurements, levels of variation are not too different in both species although they seem to be more evenly distributed among *Limnocythere inopinata* populations. *Eucypris virens* bisexual groups from E-14 (males-females, right-left valve) show the highest coefficients of variation (Fig. 8.5). This agrees with the prediction of a larger amount of variability associated with higher genetic diversity. However, the opposite occurs for *Limnocythere inopinata* (Fig. 8.5) in which the fossil bisexual population (L-10) shows the lowest shape variability. The other bisexual population in the *L. inopinata* data set (L-7) has a variation level similar to the asexually reproducing populations.

When discussing polyploidy or the coexistence of bisexual and parthenogenetic populations, morphological similarity between parthenogenetic and bisexual females, together with the genetic mechanisms underlying apomixis, is highly important. Differences in shape could hamper the likelihood of successful copulation leaving the two reproductive modes effectively isolated. Otherwise, isolation would be much weaker and coexistence models would require to include potential interbreeding. Indeed, the existence of polyploid clones of *E. virens* (Tétart 1978) suggests that successful interbreeding between both sexual and parthenogenetic individuals occurs in natural populations. At least in the laboratory, when males had no choice, some sort of copulation between males and parthenogenetic females occurred in both species, indicating that males did not perceive any difference throughout the complex copulatory act (chapter 10).

In *E. virens*, the bisexual females from E-14 cluster apart from parthenogenetic females suggesting shape differences between the two reproductive modes. This could also be due to geographic (or environmental) factors which are difficult to disentangle at this step. Without studying more bisexual populations (specially needed are samples from sites where both modes of reproduction co-occur), it is difficult to reject either of the two hypotheses. In the case of *L. inopinata*, the effect of reproductive mode on shape is less clear as there is an environment factor (water chemistry) that seems to play a main role in determining carapace shape. The two bisexual populations studied (L-10 from Germany and L-7 from PR China) are from different environments: L-10 is a freshwater habitat, whereas L-7 is saline.

Or to prove!

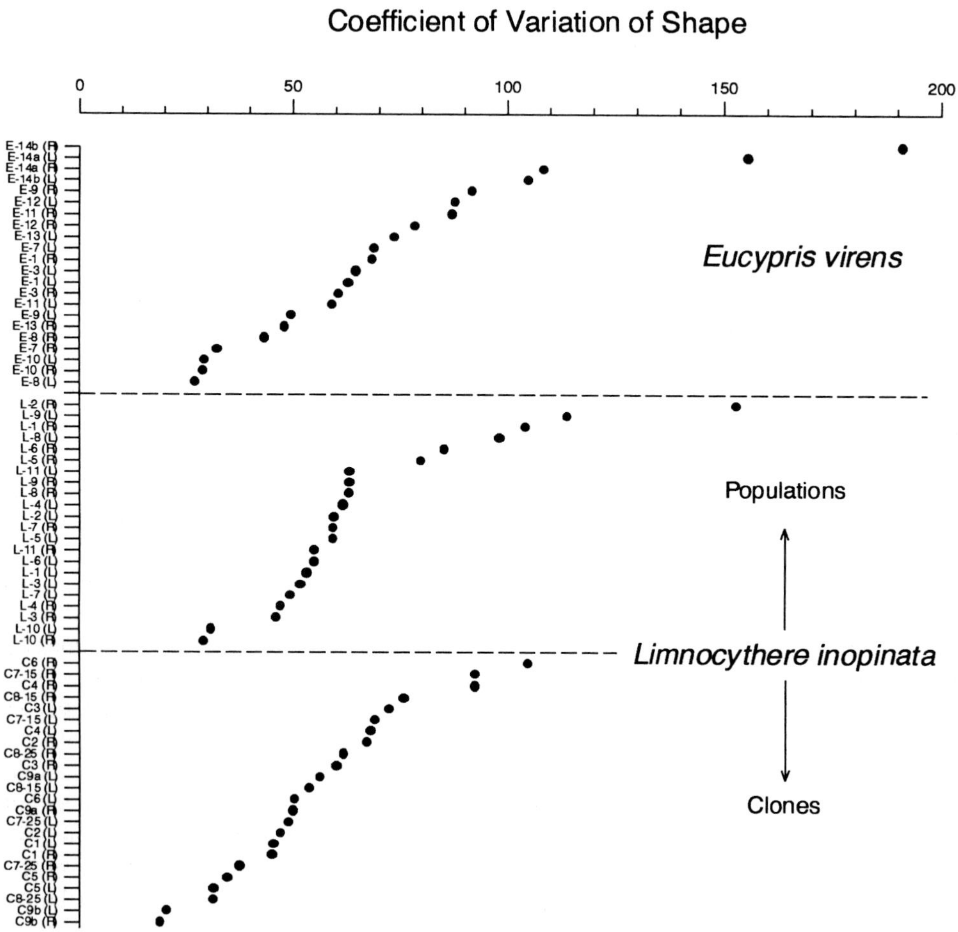

Fig. 8.5 Shape variability within population groups evaluated as the coefficient of variation of squared Mahalanobis distances from each case to its mean group shape.

Following the environmental affinity rather than the reproductive mode, L-10 clusters with the other freshwater populations instead of joining with females from L-11, which are from the same locality but from a saline environment. On the contrary, the Chinese bisexual females cluster first together with the freshwater Chinese and last with the saline European populations.

Most of the information on parthenogenetic populations comes from the analysis of sex ratios. Whilst for living specimens the identification of genders is an easy task as the sexual structures are in most cases easily recognisable, it is quite different when dealing with subfossil or fossil material and carapace shape does not display sexual dimorphism. Whereas in *L. inopinata*, a distinctive shape dimorphism between male and female shells exists (Petkovski 1959, Fuhrmann & Pietrzeniuk 1991, Yin 1997), sex assignment of carapaces in *E. virens* relies mainly on size differences. Males and females from the French bisexual population (Fig. 8.4) are very similar in shape even if males tend to be much smaller. In palaeo-assemblages males might be misidentified as last-stage female juveniles and therefore the number of bisexual populations underestimated. A careful analysis of juvenile stages might reveal additional information to solve this problem.

Clonal shape variability

Analysis of size and shape at the population level provides information about the magnitude and nature of within and between population variability and tools for enhancing the interpretative power of subfossil and fossil assemblages for palaeoecological reconstruction. To better elucidate the relative importance of genotype and environment, as well as their interactions in determining the shape of the ostracod carapace, clones were raised under controlled environmental conditions. In contrast with our expectations, clones showed an amount of variation in size (length, height, outline area) similar to the one found at the population level (Fig. 8.2). Concerning the shape, clones display a wide range of morphological variability despite being raised under constantly controlled environmental conditions (Fig. 8.6). This high within-clone variability explains why electrophoretically monoclonal populations are not less variable than genetically diverse ones (see above).

The effect of environmental factors on ostracod size and shape was explored by culturing two clones, C7 (a supposedly wide-niche clone) and C8 (a narrow-niche clone) under two different temperature regimes (15 and 25°C). Size changes are as expected in the 'specialist' (C8), lower temperature results in larger size, but the generalist (C7) does not vary significantly (Fig. 8.2). And, compared to size, shape is fairly conservative and does not change a lot with temperature (Fig. 8.6). As long as the variability observed in animals cultured under constant conditions is not lower than the variability observed in natural populations, neither for size (Fig. 8.3) nor for shape (Fig. 8.5), we should assume that environmental factors can affect mean levels of morphological features (mainly size, less shape), but do not act upon the variability of such features, a trait that seems to be determined by individual random variability on physiological and/or biochemical processes. At first glance, these results might seem to contradict what has been said above, i.e. that we could observe a clear separation between *L. inopinata* freshwater and saline populations. Indeed, we were suggesting a large influence of environmental factors on shape. However, freshwater and saline populations are likely to be genotypically different and, therefore, the separation might be explained by differences in genetic composition as well. Environment might play a role in favouring some genotypes but not acting directly on shape.

Possible applications

Our results demonstrate that the method we used detects very small differences in carapace shape and provides data suitable for statistical treatment. Shape as such bears much more information than length/height measurements and could therefore be of high potential value in the analysis of subfossil and fossil ostracod assemblages. This, together with the low costs of the equipment and the relative simplicity of the mathematical background, should favour its application amongst palaeontologists.

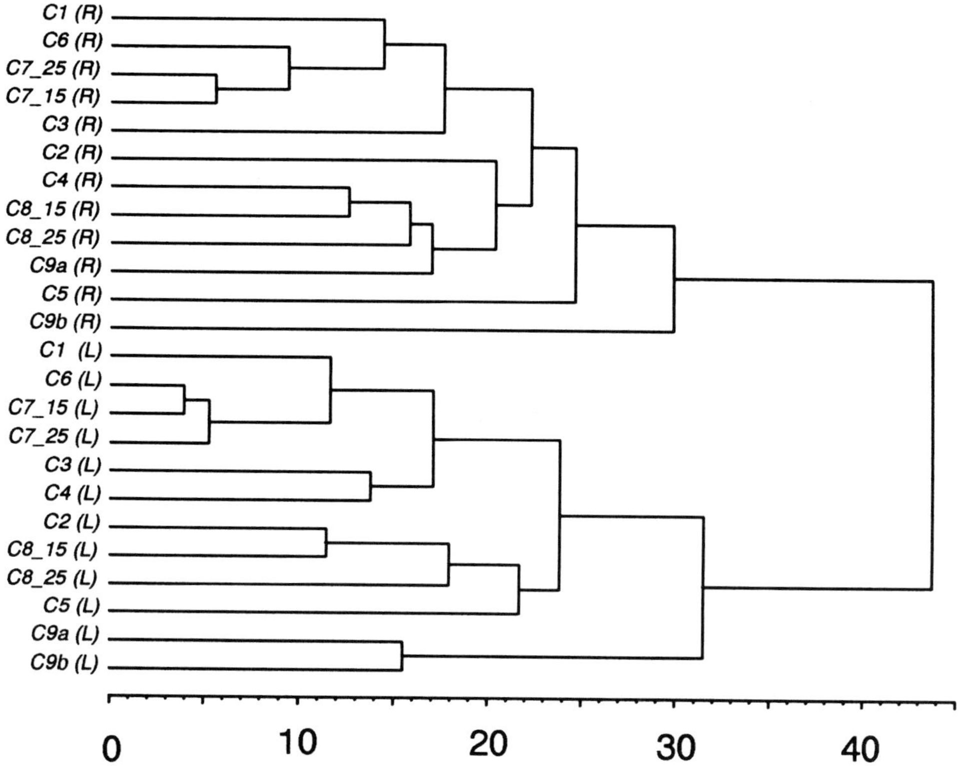

Fig. 8.6 Cluster analysis (UPGMA) of clones performed on a distance (Mahalanobis) matrix. (R)-Right valve; (L)-Left valve.

Although, at least in the two freshwater lineages investigated, shape is the result of complex interactions between genotype and environment, the application of shape analysis to taxonomic and phylogenetic purposes seems to be more straightforward than its use for palaeoenvironment reconstruction. However, cases might exist like in *Limnocythere inopinata*, where genetic similarity coincides with environmental preferences. In our test case, specimens from freshwater palaeoenvironments (Eemian-Interglacial) cluster together with recent European freshwater populations whereas those from fossil saline environments are more similar to the recent saline populations studied. But, for a more precise interpretation, we would need a more complete mapping of the relationships among extant and fossil populations on the ostracod morphospace. The presented study is just the first step in that direction.

References

Abe, K., Reyment, R. A., Bookstein, F. L., Honigstein, A., Almogi-Labins, A., Rosenfeld, A. & Hermelin, O. 1988. Microevolution in two species of ostracods from the Santonian (Cretaceous) of Israel. *Hist. Biol.*, **1:** 303-322.

Alm, G. 1915. Monographie der schwedischen Süsswasser-Ostracoden. *Zool. Bidr., Uppsala,* **4:** 1-248.

Alvarez Cobelas, M., Baltanás A., Velasco, J.L., Valladolid, M., Izquierdo, M. & Martín, E. 1993. Slow overturn in a gravel pit-lake. *Verh. Internat. Verein. Limnol*.**,** **25:** 83-87.

Andersson, S., Örnborg, J. & Andersson, M. 1998. Ultraviolet sexual dimorphism and assortative mating in blue tits. *Proc. R. Soc. Lond. B,* **265:** 445–450.

Benson, R. H. 1981. Form, function, and achitecture of ostracode shells. *Ann. Rev. Earth Planet. Sci.*, **9:** 59-80.

Bronshtein, Z.S. 1947. *Freshwater Ostracoda. Fauna of the USSR. Crustaceans. Volume II, Number I.* [English ed. (1988). Nayar, I. (trans.).], Amerind Publishing. for US Department of the Interior/National Science Foundation.

Burke, C. D., Full, W. E. & Gernant, R. E. 1987. Recognition of fossil fresh water ostracodes: Fourier shape analysis. *Lethaia*, **20:** 307-314.

Crosetti, D. & Margaritora, F. G. 1982. Osservazioni su popolazioni anfigoniche di *Eucypris virens* (Jur.) (Crustacea, Ostracoda) rinvenute in Sicilia. *Animalia*, **9:** 123-129.

Farkas, H. 1974. Morphological analysis of Ostracods (Crustacea). I. The problem of *Cypridopsis vidua* and *C. obesa*. *Acta Zool. Academ. Scient. Hung.*, **1-2:** 33-46.

Ferson, S., Rohlf, F. J. & Koehn, R. K. 1985. Measuring shape variation of two-dimensional outlines. *System. Zool.*, **34:** 59-68.

Fuhrmann, R., & Pietrzeniuk, E. 1991. Die Ostrakodenfauna des Interglazials von Gröbern (Kreis Gräfenhainichen). *Altenbg. nat. wiss. Forsch. Altenburg,* **5:** 168-193.

Herman, P. M. J. & Heip, C. 1982. Growth and respiration of *Cyprideis torosa* Jones 1850 (Crustacea, Ostracoda). *Oecologia*, **54:** 300-303.

Huang, B. 1984. Quaternary Ostracoda from Qinghai Lake region. *Mem. Nanjing Inst. Geol. Paleont., Acad. Sinica,* **7:** 188-232.

Hunt, S., Bennett, A. T. D., Cuthill, I. C. & Griffiths, R. 1998. Blue tits are ultraviolet tits. *Proc. R. Soc. Lond. B,* **265:** 451–455.

Kaesler, R.L. & Foster, D.W. 1987. Ontogeny of *Bradleya normani* (Brady): Shape analysis of landmarks. *in:* Hanai, T., Ikeya, N. & Ishizaki, K. (eds*)*, *Proceedings of the Ninth International Symposium on Ostracoda*.

Kaesler, R. L. & Maddocks, R. F. 1984. Preliminary harmonic analysis of outlines of recent Macrocypridid Ostracoda. *VII International Symposium on Ostracoda*, **2:** 169-174.

Kaesler, R.L. & Waters, J.A. 1972. Fourier analysis of the ostracode margin. *Geol. Soc. Am. Bull.*, **83:** 1169-1178.

Latifa, G.A. 1987. Effect of temperature on the development of embryonic and postembryonic stages of *Heterocypris incongruens* (Ramdohr, 1808) (Ostracoda). *Crustaceana,* **52:** 95-100.

Martens, K. 1985. Effects of temperature and salinity on postembryonic growth in *Mytilocypris henricae* (Chapman)(Crustacea, Ostracoda). *J. crust. Biol.*, **5:** 258-272

Mayr, E. & Ashlock, P.D. 1991. *Principles of Systematic Zoology (2nd Edition)*. McGraw-Hill.

Müller, G.W. 1900. Deutschlands Süsswasser-Ostracoden. *Zoologica, Original-Abh. Ges.gebiet Zool.,* **12:** 1-112.

Nüchterlein, H. 1969. Freshwater ostracods from Franconia. A contribution to the knowledge of systematic and ecology of Ostracoda. *Int. Revue ges. Hydrobiol.*, **54:** 223-287.

Petkovski, T. K. 1959. Beitrag zur Kenntnis der Ostracoden-Fauna Jugoslawiens (V), **4:** 158-185.

Reyment, R. A. 1995. On multivariate morphometrics applied to Ostracoda. *in:* Riha, J. (ed), *Ostracoda and Biostratigraphy,* Balkema: 43-48.

Reyment, R. A. & Bookstein, F. L. 1993. Infraspecific variability in shape in Neobuntonia airella: an exposition of geometric morphometry. *in:* McKenzie, K. G. & Jones, P. J. (eds), *Ostracoda in the Earth and Life Sciences,* Balkema: 291-314.

Reyment, R. A., Bookstein, F. L., McKenzie, K. G. & Majoran, S. 1988. Ecophenotypic variation in *Mitilus pumilus* (Ostracoda) from Australia, studied by canonical variate analysis and tensor biometrics. *J. micropalaeontol.*, **7:** 11-20.

Rohlf, F. J. & Archie, J. W. 1984. A comparison of Fourier methods for the description of wing shape in mosquitoes (Diptera: Culicidae). *Syst. Zool.*, **33:** 302-317.

Sars, G.O. 1928. An account of Crustacea of Norway. *Bergen Museum*, **9:** 1-277.

Schweitzer, P. N., Kaesler, R. L. & Lohmann, G. P. 1986. Ontogeny and heterochrony in the ostracode *Cavellina* Coryell from Lower Permian rocks in Kansas. *Paleobiology*, **12:** 290-301.

Schweitzer, P. N. & Lohmann, G. P. 1990. Life-history and the evolution of ontogeny in the ostracode genus Cyprideis. *Paleobiology*, **16**: 107-125

Smith, R. & Martens, K. 1996. On *Eucypris virens* (Jurine). *Stereo-Atlas of Ostracod Shells* **23:** 61-68.
Stephanides, T. 1948. A survey of the freshwater biology of Corfú and of certain other regions of Greece. *Praktika Helenica*, **2:** 1-263.
Tétart, J. 1978. Les garnitures chromosomiques des Ostracodes d'eau douce. *Trav. Lab. Hydrobiol. Univ. Grenoble*, **69-70:** 113-140.
Tétart, J. 1982. Etude de la variation morphologique de la carapace chez *Eucypris virens* (Ostracode Cypridide). *Arch. Zool. Exp. Gén.*, **122:** 341-351.
Van Morkhoven, F.P.C.M. 1962. *Post-Paleozoic Ostracoda* (Vol I & II), Elsevier.
Wolf, J. P. 1919. Die Ostracoden der Umgebung von Basel. *Arch. Naturgesch.,* **85:** 1-100.
Yin, Y. 1997. *Contributions to the Morphology and Ecology of Ostracoda Limnocytheridae and Candonidae. Comparative Studies between Austrian and Chinese Crustacea.* Ph.D. Thesis, Univers. Vienna. 163 pp.

CHAPTER 9

INTRASPECIFIC MORPHOLOGICAL VARIABILITY OF LIMBS

K. Martens, G. Rossetti and W. Geiger

Abstract

The extant knowledge on causality of morphological variability in (non-sexual) non-marine ostracod limbs is reviewed. Significant effects of genome and environment on lengths and ratios of setae and segments are experimentally demonstrated using three clonal lineages of *Limnocythere inopinata*. Different setae show different responses to treatments (clone, temperature, environment) and this could be related to either different ontogenetic sequences or direct underlying genetic causes. The former is refuted by data on natatory setae in *Herpetocypris*. There are indications, though no proof, that morphological variability in limbs is comparable in sexual and asexual populations in a cypridinid ostracod (thus corroborating the data on the valve morphometry), but species with mixed reproduction appear to be more variable than species without sexual populations. The morphological uniformity in darwinulids matches the low genetic variability.

Introduction

Evolution works through a combination of two, seemingly opposing, forces: the creation of (largely random) variation through mutation (and in sexually reproducing organisms also recombination), on which natural selection can act (Mayr 1970). For evolution to work, both forces must be active. Evolution is assumed to go faster in groups with sexual reproduction than in asexually reproducing lineages (Bell 1982) and this is usually linked to the fact that the former reproductive mode produces variation at higher rates (Maynard Smith 1998). If this is the case, then variation (both genetic and phenotypic) present in a sexual population must be at all times greater than in an asexually reproducing population of comparable size, since we can assume that selection prunes in comparable ways in both types of gene pools.

Morphological variability, as observed in micro-invertebrates such as ostracods, can have different origins. Structural variation in clades (number of setae on a segment, position of setae and segments, etc.) nearly invariably has underlying genetic causes and as such is mostly translated into taxonomically discrete units (species). Intraspecific variability mostly consists of variation in shapes, size and length ratio of claws, setae and segments. Such observed phenotypic variability can be due to artefacts (i.e. position of limbs in microscopic preparations) or be genuine; in the latter case it can reflect underlying genotypic variability or be environmentally induced (ecophenotypic) or be a result of interactive factors between both genotype and the

Sex and parthenogenesis: evolutionary ecology of reproductive modes in non-marine ostracods, edited by K. Martens, pp. 143–155

environment. Maddocks (1990) claims that most so-called intraspecific variability is introduced by "... dissection damage, pressure of the cover glass, artistic limitations and the author's ignorance...", rather than from genuine within-population sources. In the following paragraphs, we attempt to assess the effect of environment and of genotype on genuine phenotypic (intraspecific) variability in ostracod morphologies and wish to determine if reproductive mode has an effect on these ranges of variability.

Because most ostracodologists were traditionally palaeontologists, ostracod taxonomy was for a long time based on structure and shape of ostracod valves. However, it is also true that the soft parts of ostracods *are* the animal and an increasing effort has been devoted to the comparative description of ostracod appendages during the past decades. However, most of these data deal with differences relevant for taxonomy, i.e. interspecific differences. In most cases, qualitative differences are illustrated on which taxonomic decisions (descriptions of new taxa, synonymies, etc.) are then based. Maddocks (1976, 1988) used quantitative data on ostracod limbs to derive phyletic cladograms of podocopid, resp. macrocypridinid ostracods, but no intraspecific variability was used. Data on intraspecific variability in limb chaetotaxy are indeed rare in the literature, as compared to the available database on variability of valve features (see above).

Ostracod appendages are generally considered to be very conservative, for two main reasons. Firstly, as in ostracods the entire body is enveloped by the two valves, ostracod limbs in general are reduced in number as well as in individual complexity of the chaetotaxy, compared to other Crustacea (chapter 4). Both types of structurally-induced reductions limit the total potential variability. Secondly, as limbs are (again because of their position between two valves) less in contact with the natural environment, selection can act only in a limited way upon them. Both mechanisms can explain the extremely conserved morphology of the limbs in ostracods as a group (i.e. at the interspecific level), compared to other benthic microcrustaceans such as harpacticoid copepods, but might also act at the intraspecific level. Writing from a taxonomic point of view in the impressive revision of the marine group Macrocyprididae, Maddocks (1990: 22) stated: "The soft-part anatomy is very consistent within species of Macrocyprididae. Individual variation does exist, but for the most part it is noticeable only in vestigial characters that are not likely to have taxonomic value." Broodbakker (1982, 1983a-d, 1984a, b) is one of the few to have presented ranges of variability for different lengths and ratio of setae and segments in populations of several cypridinid species, but does not analyse the results. We will refer to some of his data in the present chapter.

We report on morphological assessments of taxa from the three major ostracod lineages with representatives in non-marine water bodies, relying primarily on recent data, compiled in the framework of the EU HCM or its associated projects. The basic body plan of the different groups has been described in chapter 4, which should be consulted as an introduction to the present part. Note that only limbs with no known function in mate recognition are used here, as this could strongly bias the comparison of variability between the two sexes. The importance of intrapopulational variability in copulatory appendages is discussed in chapter 10.

The present chapter reviews extant knowledge on morphological variability in non-marine ostracod limbs. For the Cytheroidea, both quantitative experimental results on clones and field data at population levels for the species *Limnocythere inopinata* are reviewed (Yin *et al.* in press). For the Darwinuloidea, a qualitative comparison of morphological variability between 4 populations of *Darwinula stevensoni* was effected (Rossetti & Martens 1996). Of these two species, only all-female populations were assessed. Only scattered information on the Cypridoidea is available and the discussion of this group will be limited.

Cytheroidea: *Limnocythere inopinata*

Yin *et al.* (in press) cultured three clones of *L. inopinata*, derived from Austrian waterbodies, in the laboratory under various temperature and salinity conditions, after which morphological traits for soft parts (11 limb setae) were observed. Field populations from Chinese salt lake populations and from Austrian lakes were also compared. Details of laboratory procedures and field collections are given in Yin (1997) and Yin *et al.* (in press). Preliminary experiments had shown that clone M from Mondsee did not survive in conductivities higher than 800 S cm^{-1} for more than 2 days. Similarly, clones N and S, isolated from Neusiedlersee and Oberer Stinkersee, respectively, do not tolerate conductivities below 1200 S cm^{-1} for more than 3 days. Therefore, specimens of clone M were cultured at three temperatures (15, 20 and 25°C) but at the lowest salinity (expressed as E.C.: 400 S cm^{-1}) only, whereas those of clones N and S were reared at three temperatures (15, 20 and 25°C) and in three conductivities (2000, 5000 and 8000 S cm^{-1}) for each experimental temperature.

Setae and segments of limbs measured for the present analyses are illustrated in Fig. 9.1. Both absolute size, as well as ratios of lengths of setae and of setae and segment lengths were used. A three-way analysis of variance (ANOVA) on these sizes and ratios (Yin *et al.* in press) showed that, in most cases, both absolute and relative lengths of setae and limb segments were highly significantly affected by the genetic component (clone), by temperature and by salinity (environment), and that there were significant interactions amongst these major effects. Such a correlation was never before unequivocally and experimentally demonstrated in non-marine ostracods.

The taxonomic consequences of these results become clear from further analyses. A hierarchical clustering of 11 limb ratios and 15 experimental conditions (3 clones and three temperatures for M and, in the case of clones N and S, also two different salinities - Fig. 9.2) shows that all high-salinity experiments form one group (cluster 1), whilst the freshwater clones cluster together with the high temperature / low salinity experiments of clones N and S (cluster 2). A third cluster is formed by the low salinity / low temperature experiments (cluster 3). The clusters can be characterised by the following differences: limb ratios of cluster 1 are all significantly higher than those of cluster 2; cluster 1 has larger limb ratios at limbs 1 and 2 than cluster 3. Although the three temperature treatments of clone M cluster together, they take a position amongst the various treatments of clones N and S. The three clones thus form an inclusive genetic cluster and this supports the view which considers all three clones conspecific.

Although this cluster analysis already indicates that environment has a considerable effect on the limb ratios of the different clones, the genetic component of the limb-variability does not become clear from this analysis. It is possible to break down the overall variability, which differs considerably, into its components by a

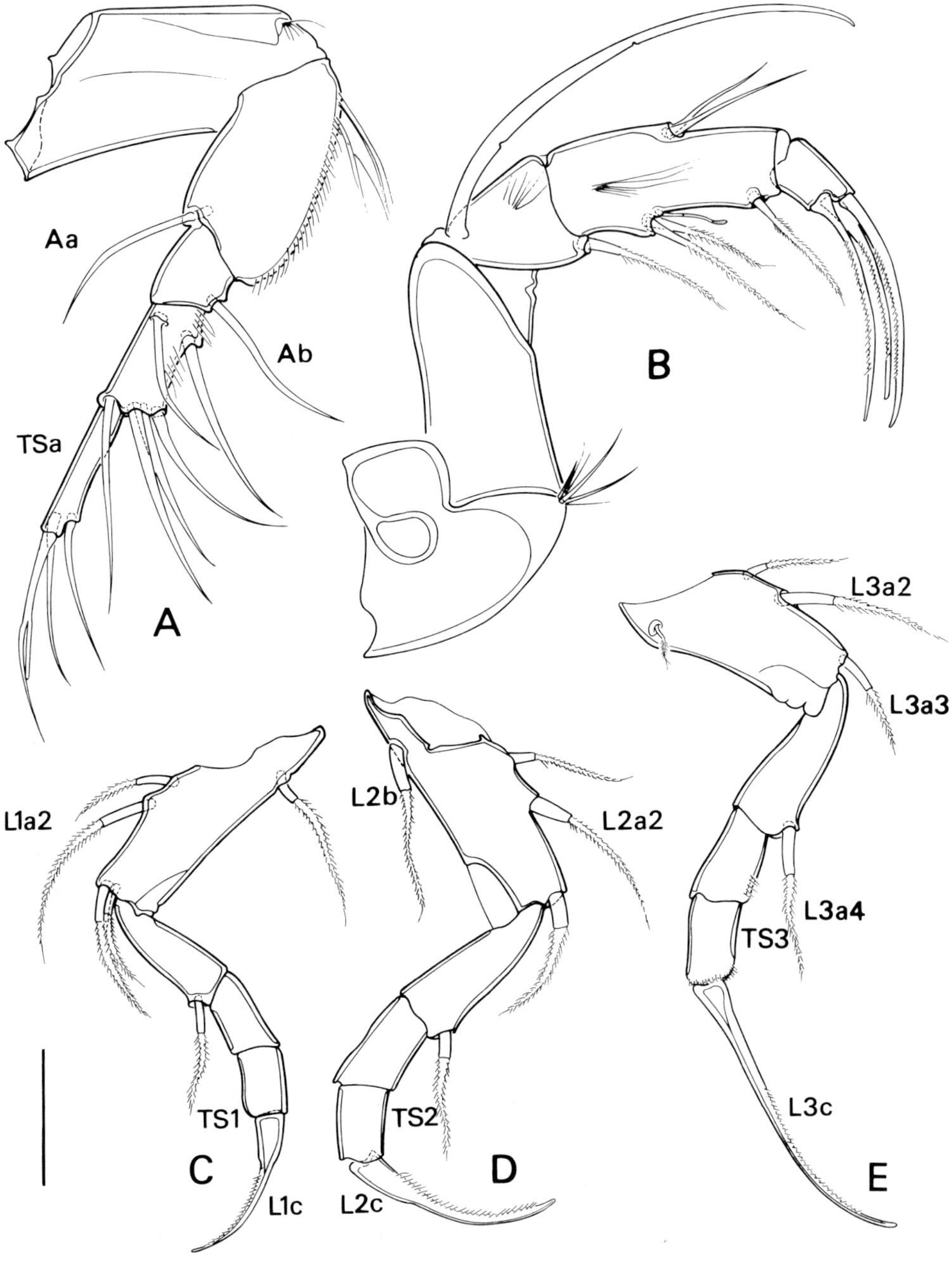

Fig. 9.1 Setae and segments measured for the analysis of effects of genetic and environmental components on limb variability in *Limnocythere inopinata*. (redrawn after Yin *et al.* in press).

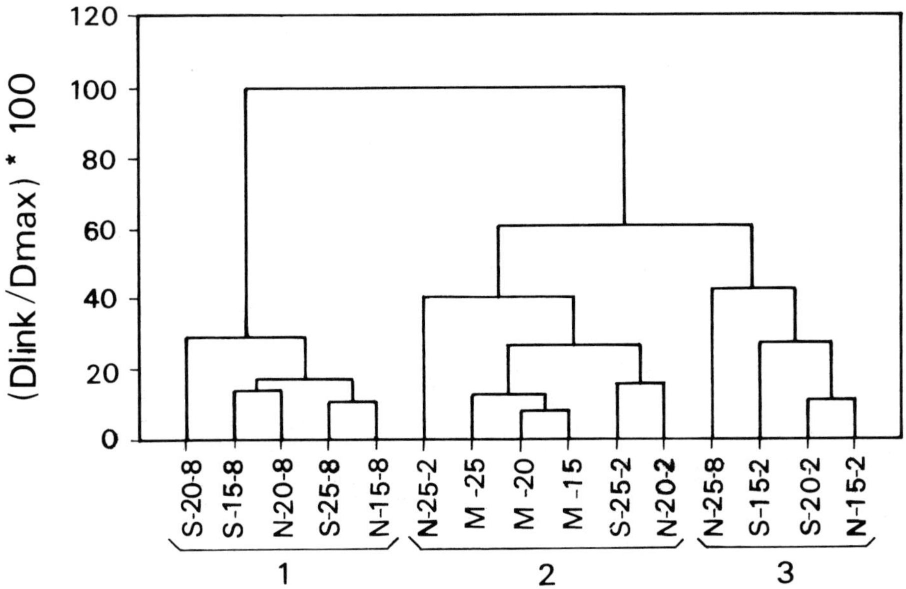

Fig. 9.2 Limnocythere inopinata, cluster analysis on setae/limb segment ratios and experimental situations. Weighted pair-group averaging, squared euclidean distance. (redrawn after Yin *et al.* in press).

MANOVA. This type of analysis shows the individual sums of squares, a measure of the variability due to clone, temperature, salinity, and various interactions between these factors and individual, within-experiment, variability. These components, as part of the overall variability for clones N and S, are shown in Fig. 9.3. Variability due to genotype (clone) in this analysis is generally low and never exceeds 15%. This is not surprising, as the saline clones N and S are indeed genetically very similar (genetic similarity of 99% – Rossi *et al.* 1995, chapter 15). Most of the variability in the comparison between these two clones is due to environmental factors. Interactions between these three factors (clone, temperature and salinity) are an important component in all limb ratios except for the antennal setae. A similar analysis on the effects of temperature and genotype (not of salinity, as clone M was not subjected to different salinities) on all three clones shows a much larger effect of genotype, clearly due to the inclusion of clone M (genetic similarity 96% with clones N and S – Rossi *et al.*, *loc.cit.*, chapter 15) in this analysis (Fig. 9.4). Note, however, that the factor 'clone' in this analysis does not purely reflect the genotype, as clone M on the one hand and N and S on the other have been reared in different salinities and the results of the latter analysis thus have value relatively to that of the previous one.

Both analyses show a large difference in total variation between different setae. For example, the measured setae on the A1 show a low variability, while the end claws on the three thoracal limbs are most variable. Further differences are discussed in detail in Yin *et al.* (in press), but no apparent reasons for these differences could be invoked. Yin *et al.* (in press) postulated that groups of setae and claws

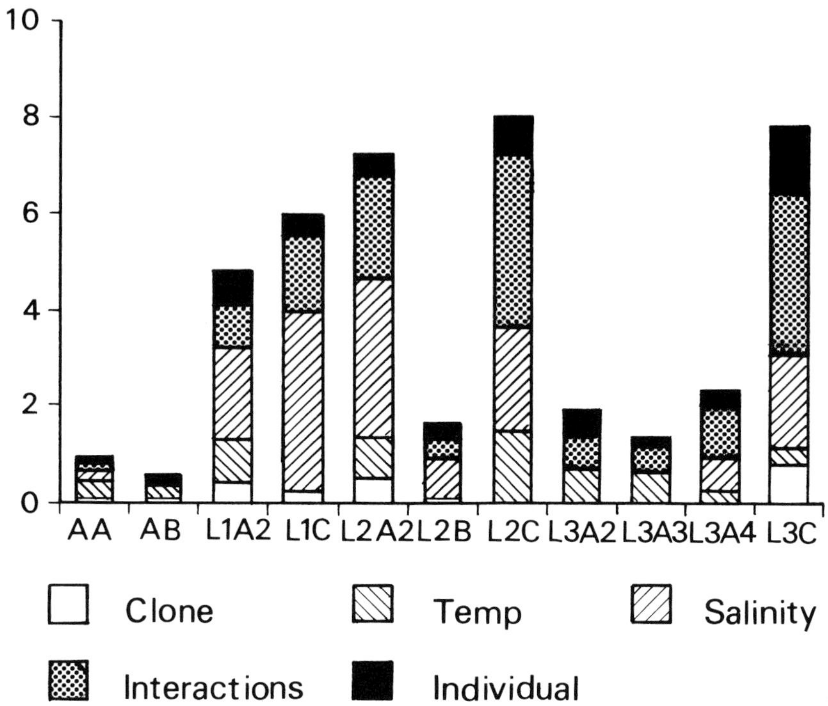

Fig. 9.3 Limnocythere inopinata, individual (phenotypic or within – experiment) variability of clones N and S expressed as sums of squares. Bars represent the means of all setae ratios. (redrawn after Yin *et al.* in press).

showing different reactions in the above analysis will also have different ontogenetic histories. Different limbs and setae indeed appear in different postembryonic juvenile stages (as shown for *Eucypris virens* by Smith & Martens, in press), but this causality remains to be demonstrated, as also underlying genetic differences could be the direct cause.

Darwinuloidea: *Darwinula stevensoni*

The morphology of soft parts for specimens of four populations of *Darwinula stevensoni* (Finland, Belgium, Italy and South Africa) was investigated qualitatively with both Scanning Electron Microscopy (SEM) and light microscopy (Rossetti & Martens 1996). As an example, the Antenna (A2) of one specimen from the four populations is illustrated in Fig. 9.5. The same populations were also screened for molecular variability, with allozyme markers (Rossi *et al.* in prep.) and for both mitochondrial (COI) and nuclear (ITS) DNA (Schön *et al.* 1998), and were found to be either monoclonal or have a markedly low genetic variability.

The most significant differences between the populations investigated, apart from the carapace size which is not discussed here, are in absolute sizes of append-

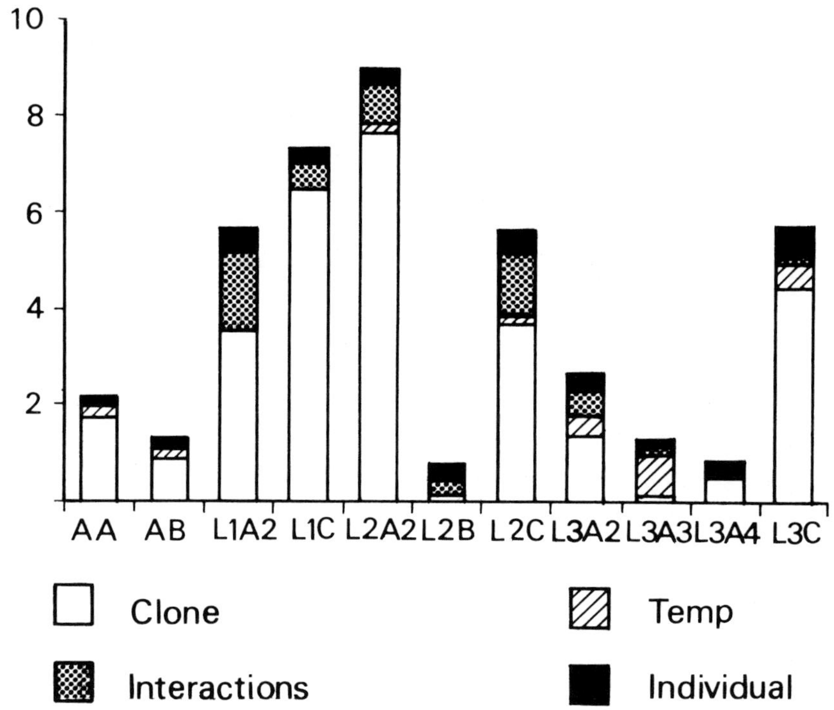

Fig. 9.4 Limnocythere inopinata, individual (phenotypic or within – experiment) variability of clones M, N and S expressed as sums of squares. Bars represent the means of all setae ratios (redrawn after Yin *et al.* in press).

ages. As preliminary analyses did not show any evident variations in size ratios between podomeres or in length ratios in the chaetotaxy (although no full morphometric analysis of sizes and ratios as for *L. inopinata* were carried out), Rossetti & Martens (1996) concentrated on comparisons of small details in micro-morphology of limbs amongst specimens of the different populations.

The number of aesthetascs in the clump on the second podomere of the A2 varied slightly: the most common pattern is a group of 4 aesthetascs, but 5 were occasionally observed. Number of setae along the distal margin of the T_1 branchial plate varies between 24 and 27 and a similar variation (20-26) was also reported by Sohn (1987). As both aesthetascs and branchial rays are small and weakly sclerified, it is difficult to ascertain whether this represents a bias in the observed variation (Maddocks 1990), or whether it reflects genuine variability. No differences were found in the number of respiratory rays of Md and T_1 branchial plates, which showed a constant number of 8 and 11, respectively. Small details (such as degree of sclerotisation, presence of setulae, barbulate setae, spines, hirsute parts, etc.) were mostly at the limit of observation with light microscopy, and are not discussed here. Overall, the morphological variability amongst specimens from four different populations, geographically widely apart, is minimal.

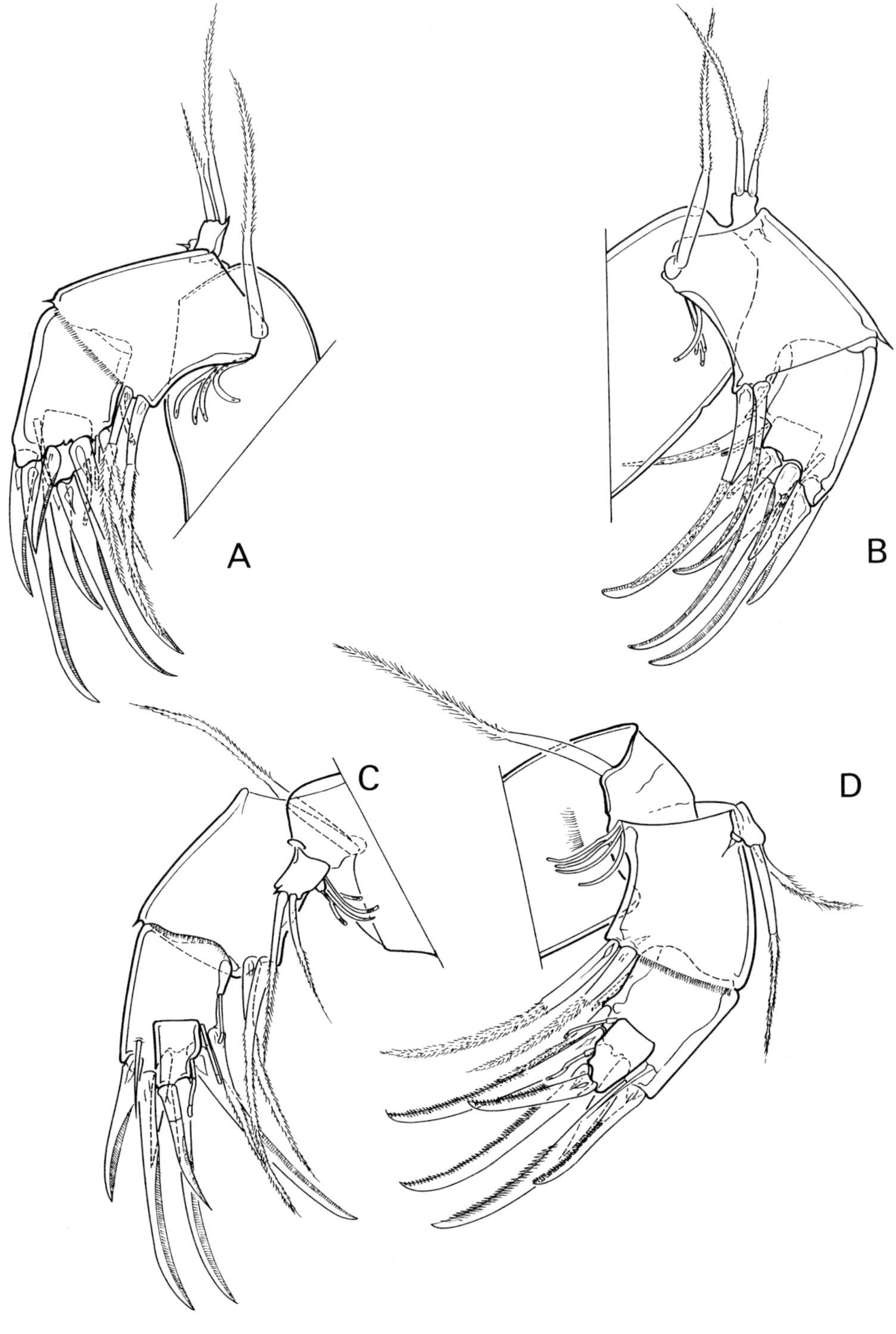

Fig. 9.5 Darwinula stevensoni, drawings of the Antenna (A2) of specimens from 4 different localities: A. Hollandersgatkreek, Belgium. B. Canale Corniano, Italy. C. Lake Pääjärvi, Finland. D. Lake Sibaya, South Africa. (redrawn after Rossetti & Martens 1996).

A feature worth noting is the incidence of aberrant individuals in darwinulids, an example of which is provided in the study of Rossetti & Martens (1996). Fifteen gravid females of *D. stevensoni* from Lake Montorfano were dissected. One of these presented anomalous T_1, with asymmetrical limbs, both with only one lateral seta on the second endopodial segment (t1 and t2, respectively) and with d2 on the third endopodal segment claw-like and of the same length as d3; d1 was furthermore absent in one of the limbs (Fig. 9.6, arrow). All other appendages, as well as the carapace morphology, were typical of *D. stevensoni*. This observed asymmetry is furthermore situated in the T_1, a feature not common in females, which indicates that this female most likely had a deficient development. This specimen was therefore considered an isolated teratology rather than a stable (clonal) morphology by Rossetti & Martens (1996).

Cypridoidea-Cyprididae

As already cited above, morphological limb-variability in species of the most common group of non-marine ostracods, the Cypridoidea, has thus far not yet been investigated in a systematic way and what little information is available is scattered in a few isolated papers. Broodbakker (1982) provided measurements and ratios of various setae and segment lengths in three different species of *Heterocypris* from West Indian islands. Two important deductions can be made from his tables: variation (as indicated by Standard Deviation) was found to be different between different limbs and setae, but appeared similar between males and females. The first observation corroborates the results from the experiments with *Limnocythere inopinata* (see above). The second is unfortunately not informative, as no distinction was made between parthenogenetic and sexual females, assuming that Broodbakker's sexual populations actually contained both sexual and parthenogenetic females.

In their revision of European *Herpetocypris* species, Gonzalez Mozo *et al.* (1996) illustrated the morphological variability in Antennal natatory setae, length of setae on the first thoracopodal palp, and variability in size and shape of spines on the caudal rami. Some conclusions from these qualitative assessments are useful here. The length of the natatory setae on A2 in *H. brevicaudata* did not appear to show different variability in populations from Belgium, Spain and Algeria (qualitatively assessed only – Gonzalez Mozo *et al.* 1996). This is significant, as the Algerian population was sexual (or at least mixed) while the other two were all-female populations, so that both modes of reproduction within one species do not appear to show different morphological variability, at least as assessed from this one feature. It should be clear that the variability of the sexual lineage is compared to the variability of all clonal lineages present in the population, not to one clonal lineage. However, a larger variability was found in *H. brevicaudata*, the only species of which sexual populations are known in this genus, than in the other species. Gonzalez Mozo *et al.* *(loc.cit.)* speculated that this was at least partly due to the presence of the sexual root, an assumption taken further in chapter 17. Finally, the difference in variability between different setae (see above for *L. inopinata*) is further corroborated for *Herpetocypris*, as within the natatory setae of the A2 it was found that each seta

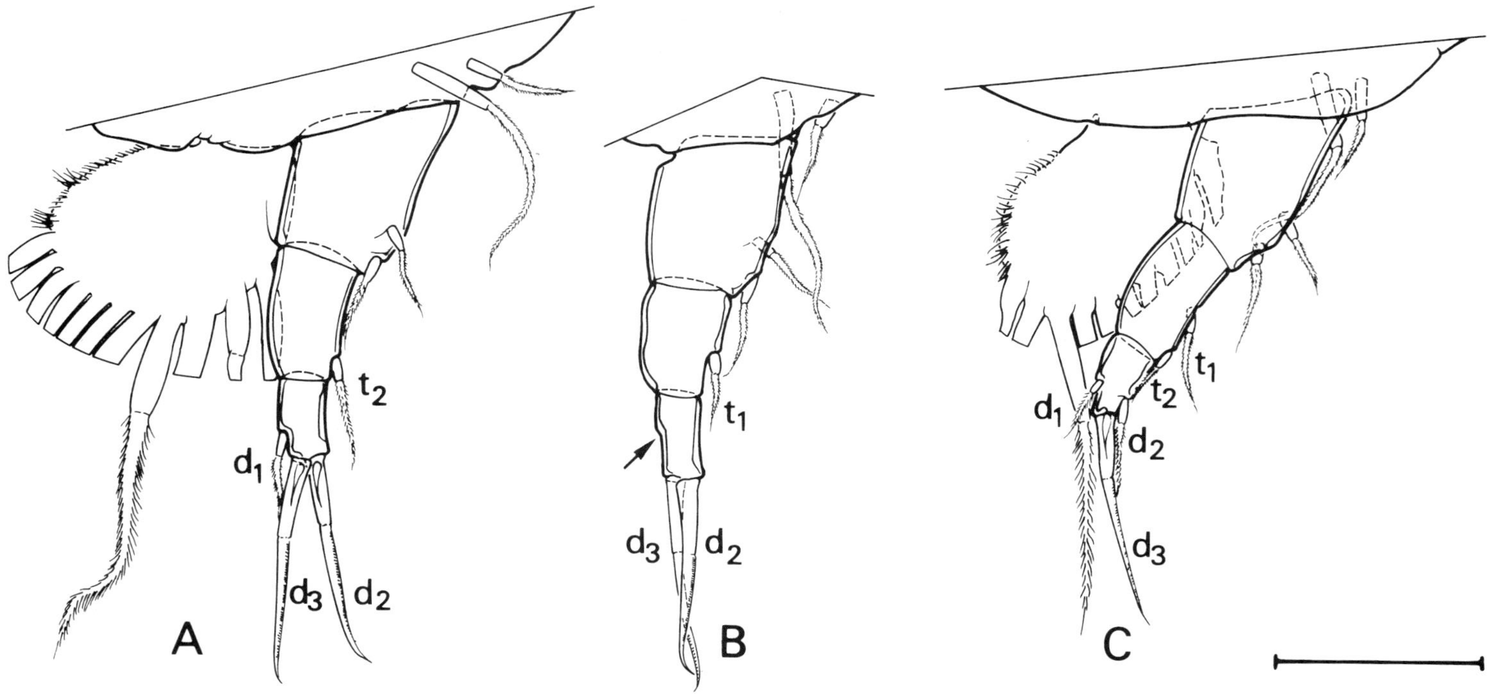

Fig. 9.6 Darwinula stevensoni, Lake Montorfano, Italy. A,B. T_1 (OC.1841): aberrant form. C. T_1 (OC.1842): normal form. (redrawn after Rossetti & Martens 1996).

has to be assessed separately and different natatory setae showed a strikingly different variability, for example in *H. reptans,* seta 1 was nearly always the longest (always longer than 5), 6 invariably the shortest (always shorter than 5). In *H. brevicaudata*, in spite of its large variability, seta 5 was always longest, with either 1 or 6 the shortest. These patterns are so consistent that they can even be used to identify these species. The difference in the behaviour of these different setae in these two species does not, however, support the hypothesis that such differences in variability are related to ontogeny, as in both *H. reptans* and *H. brevicaudata* the ontogenetic sequence of appearance of these natatory setae is identical.

Discussion

The available data on causality of morphological variability in appendages of non-marine ostracods is very limited and the present summary is preliminary at best. Detailed conclusions, like those on valve shape in the previous chapter, are unfortunately at present impossible. However, a number of initial observations can be made and integrated in the matrix of morphological, palaeontological, ecological and genetic data, discussed in the present volume. Last but not least, some of these observations suggest exciting directions for future research.

For the first time in non-marine ostracods, environmental (temperature and salinity), genetic and interactive components of limb-variability have been demonstrated experimentally, albeit thus far for one species, *Limnocythere inopinata*, only (Yin *et al.,* in press). Results from different clones and different treatments constitute an integrated cluster, supporting the idea that clonal lineages can be part of an inclusive genetic and morphological cluster (Mallet 1995, chapter 11). Different setae and segments show different variability and different reactions towards environmental treatments, and this was hypothetically linked to a different ontogenetic appearance of these structures. However, this cannot be corroborated by the (preliminary) results on cypridinids, as two species of *Herpetocypris* show clear interspecific differences in different setae (Gonzalez Mozo *et al.* 1996), which nevertheless have identical ontogenetic sequences. Most likely, the underlying causes for these differences are genetic, with different genes coding for different limbs and setae, and this opens possible avenues for future genetic research on linkage of genes. Further in *Herpetocypris*, variability does not seem to differ between sexual and asexual lineages of the same species (although this needs confirmation by quantitative analyses), but it seems firmly established that the only species with known sexual (or mixed) populations (*H. brevicaudata*) also appears to be the most variable species (Gonzalez Mozo *et al.* 1996).

At least one species in the Darwinulocopina, *Darwinula stevensoni,* shows little or no morphological variation between geographically and ecologically significantly different populations: the latitudinal gradient and the related ecological differences (temperature, salinity) are only reflected in differences in size (valves and limbs), as almost no variability in shape exists (Rossetti & Martens 1996). When this is compared to differences in shape of valves in, for example, *Eucypris virens* (Tétart 1982, chapter 8) or in differences in natatory setae on A2 within and between populations in *Herpetocypris brevicaudata* (Gonzalez Mozo *et al.* 1996), the morphological uniformity in *D. stevensoni* is at first glance astonishing, but of course relates to the genetic variability within and amongst populations of this species (chapters 15, 16). Rossetti & Martens (1998) extrapolate this morphological uniformity to all Darwinulidae, with interspecific differences in this group being small (low interspecific variability) but constant (low intraspecific variability).

The incidence of aberrant individuals in darwinulid populations in general and in *D. stevensoni* in particular, however, is worth mentioning. The aberrant T_1 in the teratological female from Lake Montorfano was here deemed epigenetic, rather than genetically induced, for example as a consequence of problems experienced during one of the 9 moults in this particular specimen (see above). However, other aberrations include the valve reversal of a specimen of *Penthesilenula brasiliensis* from southern France (Martens *et al.* 1997), and the double caudal ramus in specimens of the same species found in both South America (and there described as a new species) and in South Africa, the latter amongst an abundant number of species with normal furca (Rossetti & Martens 1998). These two examples are almost certainly genetically induced.

It is possible that aberrant individuals are more conspicuous in darwinulids than in cytherids and cypridinids, exactly because of the generally low morphological variability in the former as compared to the latter two lineages. However, if this incidence of such aberrant individuals amongst darwinulids is indeed higher, then again this is a direction for future research worth following, given the genetic particularities in this group (Schön *et al.* 1998, chapters 15, 16).

References

Bell, G. 1982. *The Masterpiece of Evolution, the Evolution and Genetics of Sexuality.* Croom Helm: 635 pp.

Broodbakker, N.W. 1982. The genus *Heterocypris* (Crustacea, Ostracoda) in the West Indies. Part I. Taxonomic characters. *Bijdr. Dierk.*, **52:** 207-227.

Broodbakker, N.W. 1983a. The genus *Heterocypris* (Crustacea, Ostracoda) in the West Indies. Part II. Carapace length, ecology and zoogeography. *Bijdr. Dierk.*, **53:** 115-134.

Broodbakker, N.W. 1983b. The genus *Hemicypris* (Crustacea, Ostracoda) in the West Indies. *Bijdr. Dierk.*, **53:** 135-157.

Broodbakker, N.W. 1983c. The subfamily Candoninae (Crustacea, Ostracoda) in the West Indies. *Bijdr. Dierk.*, **53:** 287-326.

Broodbakker, N.W. 1983d. The genus *Strandesia* and other Cypricercini (Crustacea, Ostracoda) in the West Indies. Part I. Taxonomy. *Bijdr. Dierk.*, **53:** 327-368.

Broodbakker, N.W. 1984a. The genus *Strandesia* and other Cypricercini (Crustacea, Ostracoda) in the West Indies. Part II. Carapace length, ecology and distribution of two *Strandesia* species. *Bijdr. Dierk.*, **54:** 1-14.

Broodbakker, N.W. 1984b. The genus *Tanycypris* (Crustacea, Ostracoda) in the West Indies. *Bijdr. Dierk.*, **54:** 15-24.

Gonzalez Mozo, M., Martens, K. & Baltanas, A. 1996. A taxonomic revision of European *Herpetocypris* Brady and Norman, 1889 (Crustacea, Ostracoda). *Bull. k. belg. Inst. Natuurw. Biol.*, **66:** 93-132

Maddocks, R.F. 1976. Quest for the ancestral podocopid: numerical cladistic analysis of ostracode appendages, a preliminary report. *Abh. Verh. naturwiss. Ver. Hamburg.* (N/F) **18/19** (Suppl.): 39-53.

Maddocks, R.F. 1988. Multivariate analysis of leg morphology of Macrocyprididae. *in*: Hanai, T., Ikeya, N & Ishizaki, K. (eds), *Evolutionary biology of Ostracoda.* Elsevier: 219-234.

Maddocks, R.F. 1990. Living and Fossil Macrocyprididae (Ostracoda). *Univ. Kansas paleontol. Contr. Mongr.*, **2:** 410 pp.

Mallet, J. 1995. A species definition for the modern synthesis. *Trends Ecol. Evol.*, **10:** 294-299.

Martens, K., Rossetti, G. & Fuhrmann, R. 1997. Pleistocene and recent species of the family Darwinulidae Brady & Norman, 1889 (Crustacea, Ostracoda) in Europe. *Hydrobiologia,* **357:** 99-116.

Maynard-Smith, J. 1998. *Evolutionary genetics,* Oxford Univ. Press: 330 pp.

Mayr, E. 1970. *Populations, species and evolution,* Belknap: 453 pp.

Rossetti, G. & Martens, K. 1996. Redescription and morphological variability of *Darwinula stevensoni* (Brady & Robertson, 1870) (Crustacea, Ostracoda). *Bull. k. belg. Inst. Natuurw. Biol.,* **66:** 73-92.

Rossetti, G. & Martens, K. 1998. Taxonomic revision of the recent and holocene representatives of the family Darwinulidae (Crustacea, Ostracoda), with a description of three new genera. *Bull. k. belg. inst. Natuurw. Biol.,* **68**.

Rossi, V., Pospisil, P., Geiger W. & Menozzi P. 1995. Genetic structure of Austrian and Italian populations of *Limnocythere inopinata. in: Proceedings of a European Scientific Network meeting – Evolutionary Ecology of Reproductive Modes in Non-marine Ostracods*. Unpubl. Doc., Madrid: 2-6.

Schön, I., Butlin, R.K., Griffiths, H.I. & Martens, K. 1998. Slow evolution in an ancient asexual ostracod. *Proc. R. Soc. Lond. B,* **265:** 235-242.

Smith, R. & Martens, K. in press. The ontogeny of the cypridid ostracod *Eucypris virens* (Jurine, 1820) (Crustacea, Ostracoda). *in*: Horne, D.J. & Martens, K. (eds), Proceedings of the XIII International Symposium for Ostracodologists. *Hydrobiologia.*

Sohn, I.G., 1987. The ubiquitous ostracode *Darwinula stevensoni* (Brady & Robertson, 1870), redescription of the species and lectotype designation. *Micropaleontology*, **33:** 150-163.

Tétart, J. 1982. Etude de la variation morphologique de la carapace chez *Eucypris virens* (Ostracoda, Cyprididés). *Arch. Zool. exp. gén.,* **122:** 341-351.

Yin, Y. 1997. Contribution to the morphology and ecology of ostracoda Limnocytheridae and Candonidae: Comparative studies between Austrian and Chinese Crustacea. PhD Thesis, Uni. Vienna, 163 pp.

Yin, Y., Geiger, W. & Martens, K. (in press). Effects of genotype and environment on phenotypic variability in *Limnocythere inopinata* (Crustacea, Ostracoda). *Hydrobiologia.*

CHAPTER 10
REPRODUCTIVE BEHAVIOUR

D.J. Horne, D.L. Danielopol and K. Martens

Abstract

Ostracods are remarkable for their complex and diverse male copulatory organs, sometimes also giant spermatozoa, as well as for the occurrence of female brood care in certain lineages. We review the sexual behaviour of ostracods, mainly of the benthic podocopids. It is shown that the best way to understand the evolution of morphology of these ostracods, related to sexual activities, is to approach it through the concept of functionally morphological modules. The main stages of the mating process are described, i.e. the mate recognition, the copulation followed by the insemination phases and finally the post-mating stage. Explanatory hypotheses for the diversity of the male dimorphic traits are provided. It is shown that the paradigmatic *lock and key model* does not fit the ostracod data while *the sexual selection model* and especially the *cryptic choice model* of Eberhard (1996) has a wide application to the morphological data set offered by this crustacean group. Other processes, like natural selection (*sensu stricto*), or organismal developmental constraints, apparently played a more restricted role in the origination of the diverse morphological traits of the limbs involved in sexual activities of the ostracods.

Seven different forms of brood care are recognized; these evolved independently in various lineages. We review the development of brood care in both marine and non-marine taxa and discuss its evolutionary significance. The ability of the female ostracods to brood eggs and/or juveniles within the carapace may serve to protect them from adverse environmental conditions and predators. Brood care may be advantageous for the dispersal of ostracods such as Timiriaseviinae, which might not have desiccation-resistant eggs. Finally, it is shown that the theoretical possibility of brood selection might occur in an ancient asexual.

Preamble

One of the original aspects of ostracods, as compared to other crustacean groups, e.g. the copepods, is the presence of extremely large and structurally complex sexual organs. It is frequent to find, in both marine and non-marine environments, ostracod species with male copulatory appendages (the paired hemipenes) representing about one third of the whole body length and volume, while some non-marine species have spermatozoa extending up to five or even ten times the total length of the animal. Moreover, male morphological traits used during the mating process can be strongly dimorphic and cover a wide range of shapes. Many male appendages are used during sexual activities, which involve several successive steps culminating with the copulation and insemination of the female (Figs. 10.1-2, 10.7).

Ostracoda have a rich fossil record, hence they offer unique opportunities to reconstruct some of the functional solutions used for reproductive activities, includ-

Sex and parthenogenesis: evolutionary ecology of reproductive modes in non-marine ostracods, edited by K. Martens, pp. 157–195

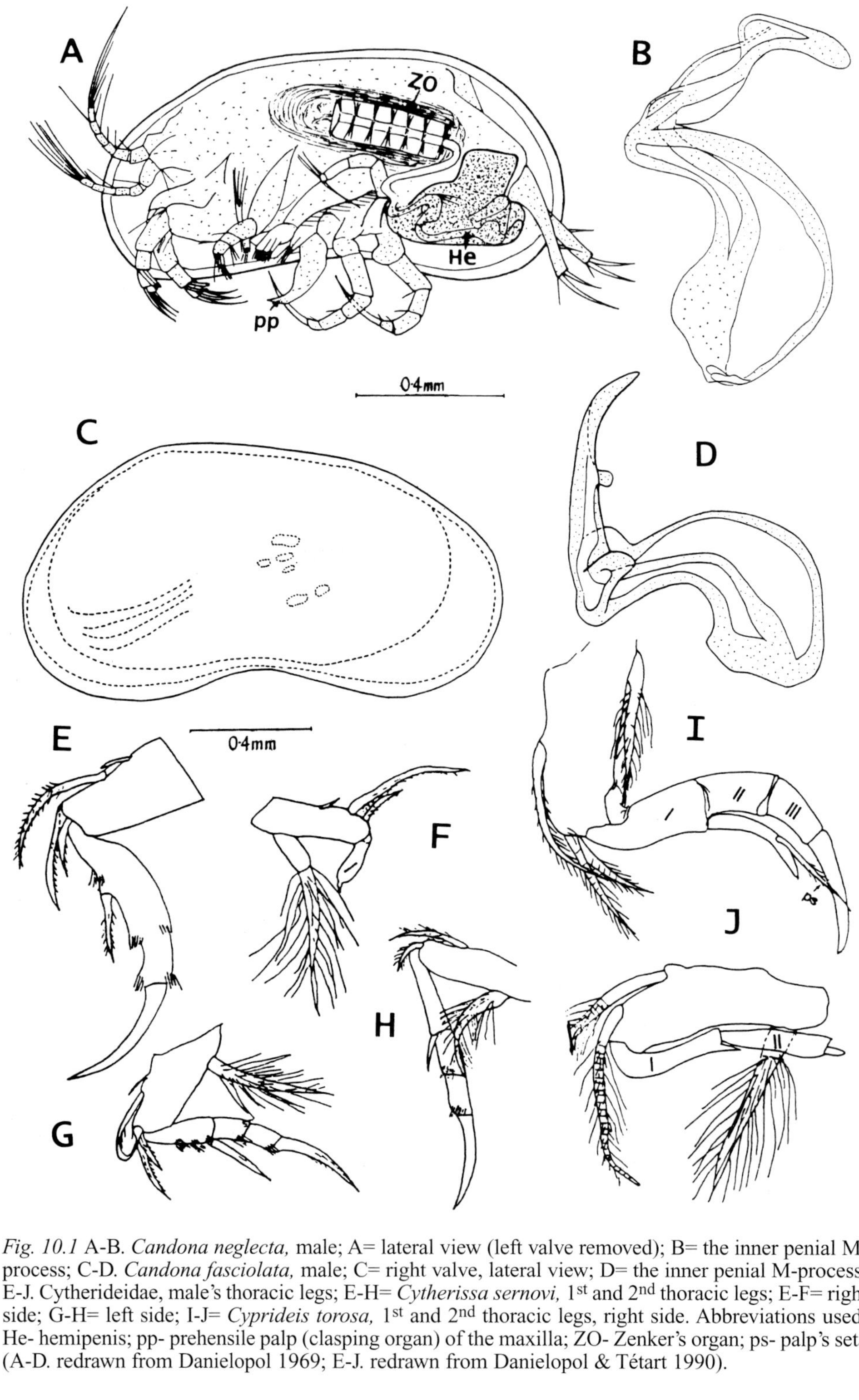

Fig. 10.1 A-B. *Candona neglecta,* male; A= lateral view (left valve removed); B= the inner penial M-process; C-D. *Candona fasciolata,* male; C= right valve, lateral view; D= the inner penial M-process; E-J. Cytherideidae, male's thoracic legs; E-H= *Cytherissa sernovi,* 1st and 2nd thoracic legs; E-F= right side; G-H= left side; I-J= *Cyprideis torosa,* 1st and 2nd thoracic legs, right side. Abbreviations used: He- hemipenis; pp- prehensile palp (clasping organ) of the maxilla; ZO- Zenker's organ; ps- palp's seta (A-D. redrawn from Danielopol 1969; E-J. redrawn from Danielopol & Tétart 1990).

ing brood care, by species known from various geological periods, from the Palaeozoic onwards. Several ostracod groups are now extinct, and functional interpretations related to those groups are difficult (chapter 6). However, the combined study of living and fossil forms of groups with both Recent and Fossil representatives, allows to decipher functional aspects of peculiar morphological traits, suspected to have been developed for reproductive purposes. The present chapter deals with two aspects of reproductive ethology, the sexual behaviour and the brood care, reviewed within an evolutionary perspective.

Sexual behaviour

Introduction

During a keynote address at a recent Symposium on Ostracoda (Chatham, 1997), Roger Butlin & Paolo Menozzi stressed that the evaluation of the role of sexual selection in the elaboration of the complex and diverse structures of the ostracod genitalia and their large spermatozoa is one of the potentially most exciting topics for evolutionary biologists (Butlin & Menozzi, in press). To answer at least part of this challenging question, we present a brief description of the various steps of the mating process observed, mainly in non-marine podocopid ostracods, with special emphasis on mate recognition, followed by an overview of the diversity of the male's dimorphic morphological traits, which are potentially involved during the sexual union. Explanatory hypotheses for the origin of this type of morphological diversity are further examined in order to evaluate which one best fits the extant data on ostracod morphology and behaviour.

Morphological traits involved in the mating process: copulatory modules

The morphology of an organism can be viewed through a functional analysis of the various character states. We can classify the morphological characters of an ostracod in several functional groups of traits or functional morphological *modules* (Danielopol *et al.* 1990). For example, most of the morphological characters of the carapace cooperate to one main function: the protection of the soft body structures. The main part of the second antenna (A2) and of the walking limbs cooperate as one module to the locomotory function. The sensorial setae of the antennae (A_1 and A_2) and parts of the thoracic legs (especially the third walking limb), as well as the caudal ramus and the copulatory appendages, participate in the mating process. For the feeding function, we have a module represented by the antennae (parts of their structure), the mandible, the maxillula and, partly, the first walking limb. Parts of some of the limbs, for example the distal endopodal segments of the antennae, are involved in several functional modules at once: in this case the locomotion, the feeding and the mating.

The copulatory or mating module is used by the male to stimulate the female into accepting him for copulation, hence reproduction. We can thus expect specific, sexually dimorphic structures in the male morphology, not only in the hemipenis, but in all limbs and parts of the carapace involved in (pre-) mating contact between males and females to have been evolved as adaptations for sexual reproduction. For the above reason, the ruling theory on sexual selection by the female choice (see below, Eberhard 1985, 1996, Andersson 1994), predicts that more specific morphologies can be expected in males than in females.

We start our review presenting general aspects of the mating process elaborated by ostracods (for a general introduction to the morphology of this crustacean group, see chapter 4).

The mating process

Methodological problems

Our information on the diversity of the mating patterns found in ostracods derive from both direct observation and indirect evidence obtained through comparative anatomical studies. Direct observation of the mating process in both limnic and marine ostracods is difficult to obtain mainly because of two reasons: (1) the mating act proceeds generally during a restricted period of the adult's life cycle, which for most species is difficult to foresee; (2) the successive steps of the mating process are completed, generally, within a few minutes (Fig. 10.3). Hence, for many ostracod species only episodic information has thus far been presented. McGregor & Kesling (1969) and Cohen & Morin (1990) summarized the major part of these data.

Important progress for the detailed documentation of the copulatory behaviour of ostracods was achieved using systematic microscopic observations during longer periods of time under controlled laboratory conditions (Danielopol 1980a, Kamiya 1988). More recently, video recording techniques (e.g. Abe & Vannier 1991, Danielopol & Horne 1996, Parker 1997, Horne unpubl. and below) were added to these procedures.

Comparative (functional) anatomical studies (Kesling 1965, Hart & Hart 1969, 1974, Danielopol 1969, 1978, Martens 1990a) are helpful for the reconstruction of evolutionary aspects of the mating behaviour (see below). By combining information from both approaches, i.e. direct observation of living animals and morphological investigation on preserved material, we can obtain robust general ideas on mating behaviour of living ostracods, which can then be extrapolated to the past using data from palaeontological studies.

The main mating stages

The main successive mating stages are as follows: (1) the *mate recognition* phase, which occurs either from a distance or in direct contact, or both, (2) the *copulation* stage, represented by the mate formation (or copula) and the intromission of the copulatory appendages within the female's carapace space, mostly followed by (3) the *insemination* stage (the sperm is moved into the female genital organ), and finally by (4) the *post-mating* phase when the two individuals separate, but during which selection still acts on the inserted sperm. These four stages are now discussed in some detail.

Mate recognition

Recognition at a distance

Sensorial communication during mating in ostracods is achieved either by distant (chemical, and to a lesser extent, optical) signalling or by direct (mechanical and/or chemical) contact. Each sex has the capacity to send signals that can be recognized as attractive by the opposite sex of any given species. There is indirect evidence for distant pheromonal signalling, derived from comparative morphological studies on planktonic marine halocyprids, as especially males of some species of the genera *Paraconchoecia* and *Spinoecia* display unusually strongly developed antennal aesthetascs (Poulsen 1973). The ultrastructure of the antennular sensory setae of the planktonic halocyprid *Conchoecia spinirostris* (Heimann 1979) confirms their chemo-sensorial function, and these organs could thus be useful to detect (conspecific) members of the opposite sex, even while swimming in oceanic waters.

Visual signalling with light patterns (bioluminescence) in some myodocopid marine ostracod lineages (e.g. the speciose genus *Vargula*) has been documented extensively (review in Cohen & Morin 1990, Morin & Cohen 1991). Males of different myodocopid species have specific flashing patterns which attract females. This behaviour is more complex and involves further adaptations. For example, in certain species conspecific non-flashing males accompany a flashing male, awaiting the attracted females. The lateral eyes of the myodocopid males are better developed than those of the female. Males of such myodocopid species, which have compound lateral eyes, are able to perceive the blue light of the iridescent structures of the females at short distances (Morin & Cohen 1991, Parker 1995, 1997).

In the non-marine groups, no such signalling has thus far been documented. However, females in species of the non-marine hyponeustic *Notodromas* (belonging in one of the four families of the Cypridoidea, the Notodromadidae – see chapter 3) are strongly pigmented, while males are able to recognize females thanks to their excellent optic devices, as the naupliar eye in this group is ramified in three ocelli, while strongly developed biconvex lenses on the valves correspond to the two lateral ones (Fox 1966, Bonaduce & Danielopol 1988). Females also have such lenses, but they are smaller than those of the males. Optical mate recognition in this group is thus a distinct possibility.

Direct contact

However, most evidence indicates that benthic ostracods recognize their mate through direct contact of their limbs. In many podocopid ostracods, mating is initiated by the fact that the male will grab the female and will start handling her. Males of many podocopid ostracods will generally spend time grabbing whatever resembles an ostracod, specimens from other species, sand grains, etc. When contact is made, the male will attempt to stimulate the female into accepting him for copulation and reproduction. To this end, three possible types of signalling are used: behavioural, morphological and chemical. The first two types have been established by direct observation (e.g. Danielopol 1978, 1980a, Kamiya 1988, Abe & Vannier 1991, Danielopol & Horne 1996); the third one is inferred from data on the use and position of the aesthetascs during mating and from comparative morphology of these organs. The combination of the morphology (valves, limbs) and the behaviour of the male (position of the valves, touching of the female parts by the male antennae, hemipenes, etc.) can be used by the female to determine whether the potential mate is indeed conspecific (*recognition*) and whether or not within the variation of the species this mate is preferable (*selection*). During this recognition phase, it is almost certain that chemical recognition is also active. For instance, in subterranean candonids, the male recipient organs (e.g. the aesthetasc Y of the antennae) are sometimes larger than those of the female, even within the same species, pointing towards a greater importance of this type of signalling for the exclusively subterranean dwelling species (Rogulj *et al.* 1993). Other setae presumably play both mechano- and chemo-sensorial roles. For example, the antennal setae t_2 and t_3 of males in the candonids, have a distally inserted membranous button, and come in direct contact with the female during the initial phase of mating when the male tries to fix the female. Also the endopodites of the male maxilla, which are transformed in clasping organs, terminate with a membranous tip, the latter most likely with sensorial functions (Kesling 1965, Danielopol 1978).

In the case of *Eucypris virens* (Fig. 10.2) and *Metacypris cordata* (Fig. 10.3), males start to touch the female, using the chaetotaxy of the 2nd antenna with a preference for the lateral and posterior sides of the female's carapace. For the marine cytherids *Bicornucythere bisanensis*, *Loxoconcha japonica* and *L. uranouchiensis,* it was observed that the males touch the lateral side of the female's carapace with both the antennae and the first two pairs of thoracic legs, turning them in a lateral position which later allows the intromission of the hemipenes inside the carapace space (Abe & Vannier 1991, Kamiya 1988). Note, furthermore, that the right male thoracic legs in various cytherid groups are much reduced as compared to the left side, while the distal part of their endopodite is transformed either in a clasping organ or is provided with well-developed sensorial structures. The shape and structure of the thoracic legs of the males belonging to the cytherideids *Cytherissa sernovi* and *Cyprideis torosa* (Fig. 10.1 E-J) are good examples of this. Such morphological asymmetry could be related to both the recognition and this lateral copulatory position (see below).

This tactile activity of the male will provoke a decision from the female: if she recognizes it as a courtship signal and accepts it, she will slow down her movement or even completely stop, which will allow mate pair formation. By fixing the female, other limbs from both the female and the male will come in contact. Males of cypridoids like *Eucypris virens* and *Herpetocypris brevicaudata* touch the female with a strong unipectinate comb-like claw inserted on the distal endopodal segment of the antenna (Fig. 10.4 A). Similar dimorphic claws evolved parallel in other ostracod groups (Fig. 10.4, B, C), e.g. in the cytheroid Timiriaseviinae (Limnocytheridae) and Sphaeromicolinae (Entocytheridae). From comparative morphological studies (Danielopol *et al.* 1990, Martens 1991), one can infer that in the case of the limnocytherines of the genus *Leucocythere,* the contact of the female by the male proceeds with a strongly modified endopodite of the third pair of thoracic legs and with the caudal ramus (Fig. 10.4, D-F). In the case of the cytheroid *Bicornucythere bisanensis,* Abe & Vannier (1991) noted that males generally have a poor ability to recognize the correct female for mating, and therefore 'false' copula between two males also occur.

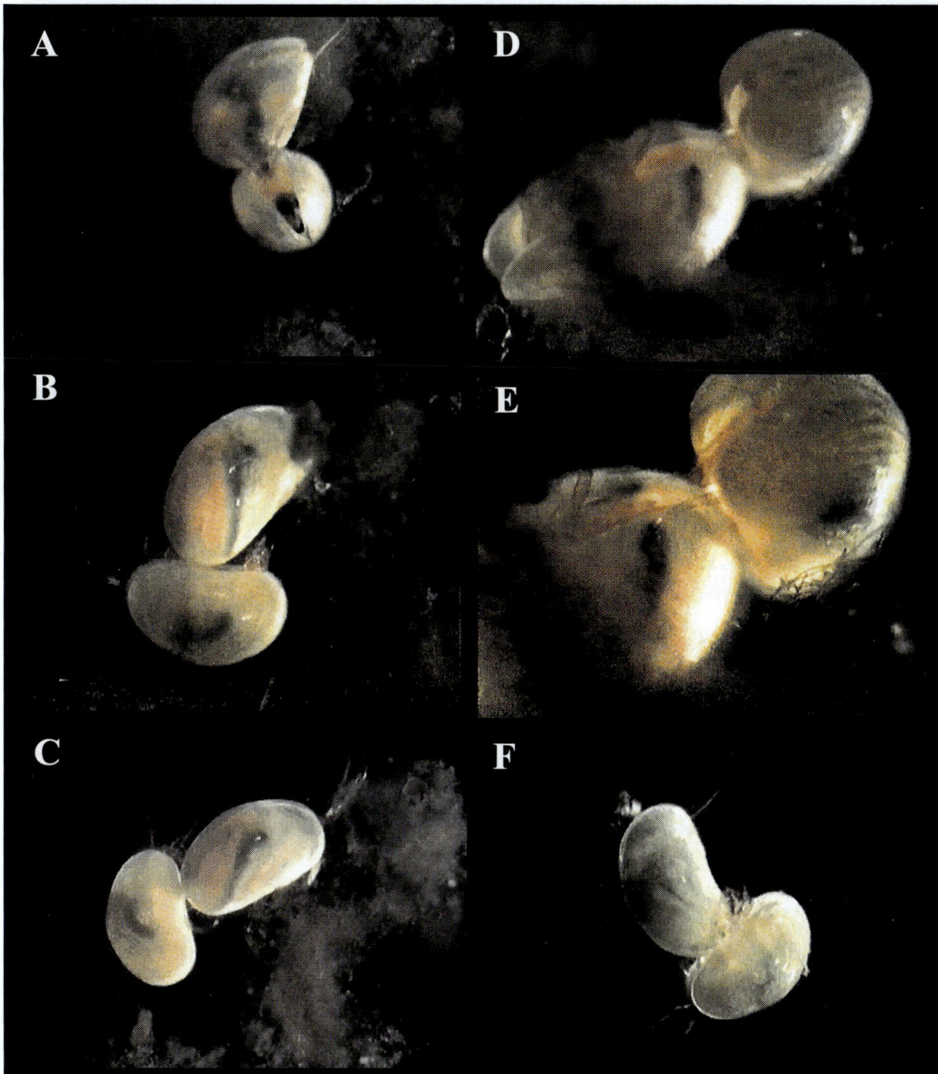

Fig. 10.2 Copulatory behaviour in *Eucypris virens* (Cyprididae), from Spanish sexual populations (from original high-resolution videorecordings by P. Alcorlo); A= the male (below) makes contact with the female (above); B= the male (below) is in position for copulation, gripping the posterior margin of the female's valves with his prehensile palp (clasping organ), the copulatory appendages (hemipenes) are extended ready for penetration into the female's carapace; C= the male (below) has just withdrawn the hemipenes, but is still in the normal copulatory position; D= postero-ventral view of a male (right) about to copulate with a female; E= the same view as (D), but a few seconds later, the hemipenes have been inserted postero-ventrally between the female's valves; F= a different copula, the hemipenes are clearly visible (below) as they are extended immediately prior to penetration of the female (above).

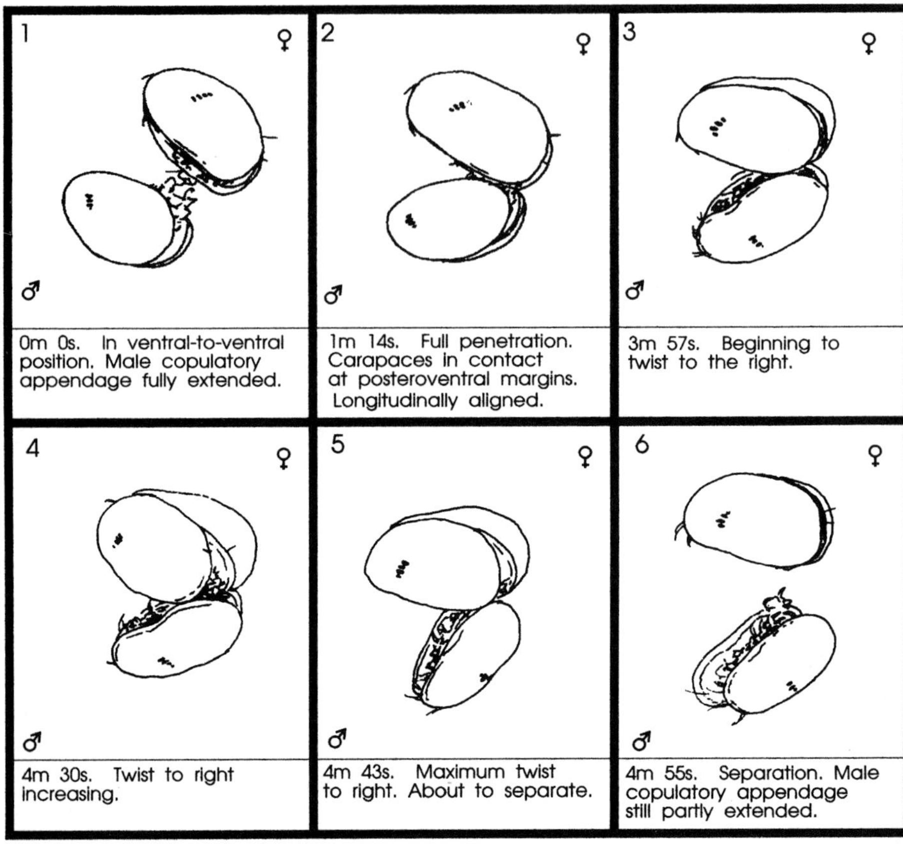

Fig. 10.3 Copulatory behaviour of *Metacypris cordata.* Sketches based on video recording (from Danielopol & Horne 1996).

In conclusion, copula in most non-marine ostracod lineages are established, because a male will grab a female and will use morphological and behavioural, and possibly chemical, signals to stimulate the female into recognizing him as a potential mate. The female signals her acceptance by slowing down her movements, allowing the male to more firmly fix the copula, during which further (tactile) contacts between other limbs are established and the hemipenes are introduced within the female's carapace space.

Copulation

WITH SEXUAL FEMALES

Six different mating positions have thus far been recognized in ostracods (Fig. 10.6). These are: ventral/dorsal, ventral/postero-dorsal, ventral/ventral, gaped ventral/ventral, reverse ventral/ventral and skewed ventral/ventral (McGregor & Kesling 1969, Cohen & Morin 1990). Compared with other crustacean groups, like the Copepoda (Boxshall 1998) and/or the malacostracan Isopoda, Amphipoda, etc. (review in Schram 1986), this diversity is remarkable.

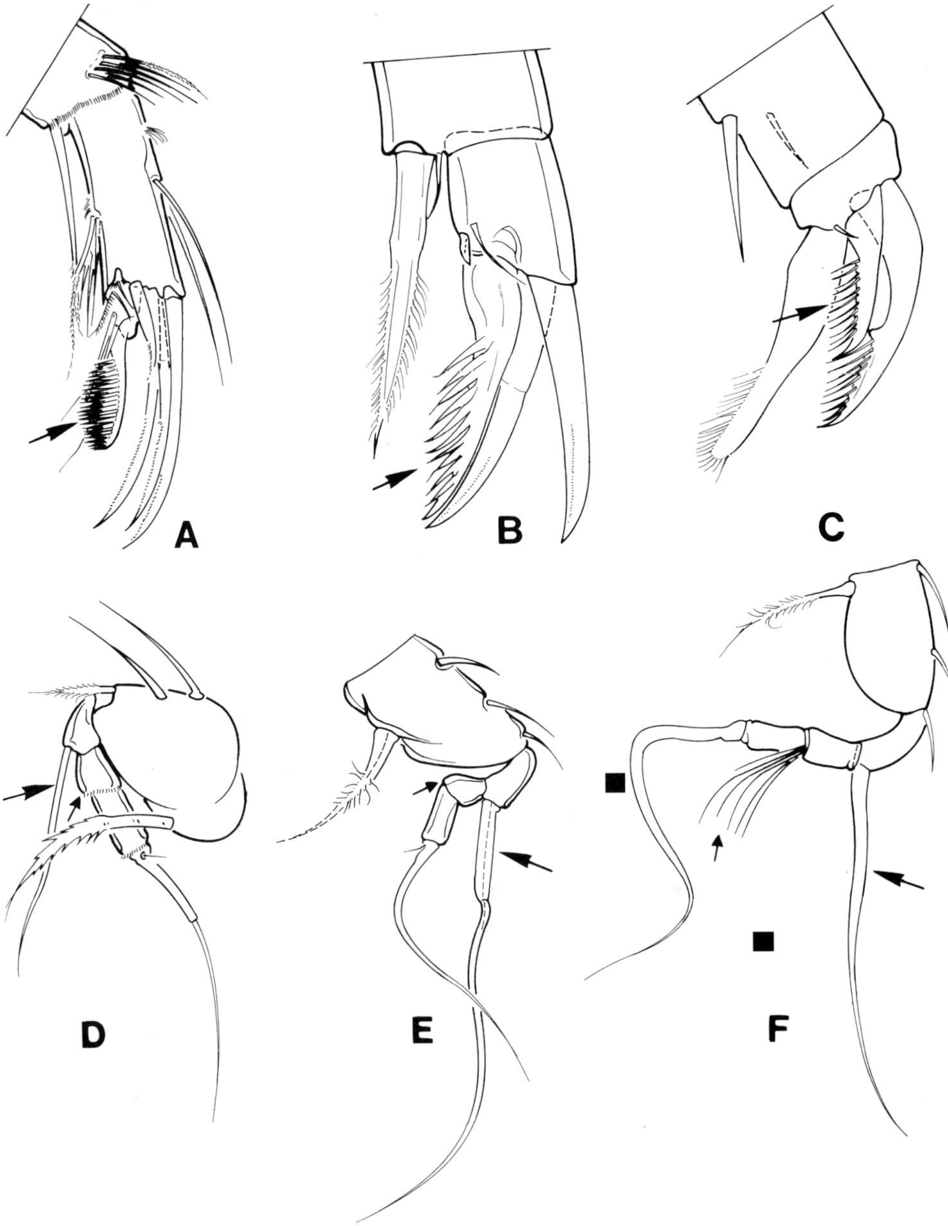

Fig. 10.4 A-C. Male's dimorphic distal antennal claws (arrow); A= *Herpetocypris brevicaudata*; B= *Elpidium inaequivalvis*; C= *Sphaeromicola cebennica juberthiei;*
D-F. Male's dimorphic traits (arrows) on the 3rd thoracic leg of *Leucocythere*; D= *L. helenae*; E= *L. algeriensis;* F= *L. mirabilis* (A redrawn from Gonzalez Mozo *et al*. 1996; B from Danielopol 1975; C from Danielopol 1977; D-F from Martens, in press).

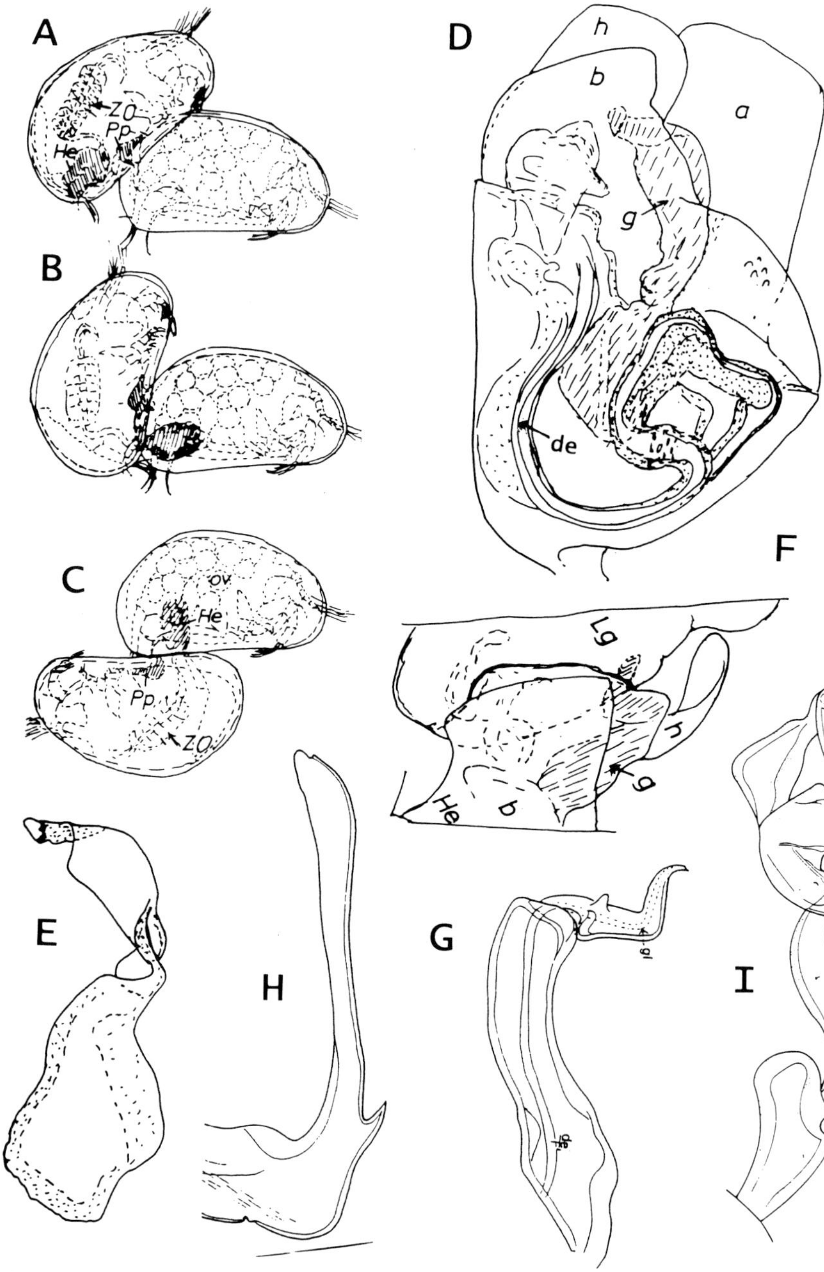

Fig. 10.5 A-F. *Pseudocandona pratensis*; A-C= Successive copulatory phases (see also text); D= hemipenis, medial view; E= penial M-process; F= detail of the outer and inner contacts of the hemipenis with the female's genital organ; G-H. *Limnocythere inopinata*, male's copulatory organ, details; G= copulatory process; H= the hook-like process; I. *Leucocythere mirabilis,* male copulatory organ, detail of copulatory process.
Abbreviations used*:* a, b, h- distal lobes of the hemipenis; g- penial M-process; de- ductus ejaculatorius; gl- glans; Lg- female's genital lobe; ov- ovocytes, He, ZO, pp- see explanations in Fig.1 (A-F redrawn after Danielopol 1980a; G-H after Martens 1990; I after Danielopol *et al.* 1990).

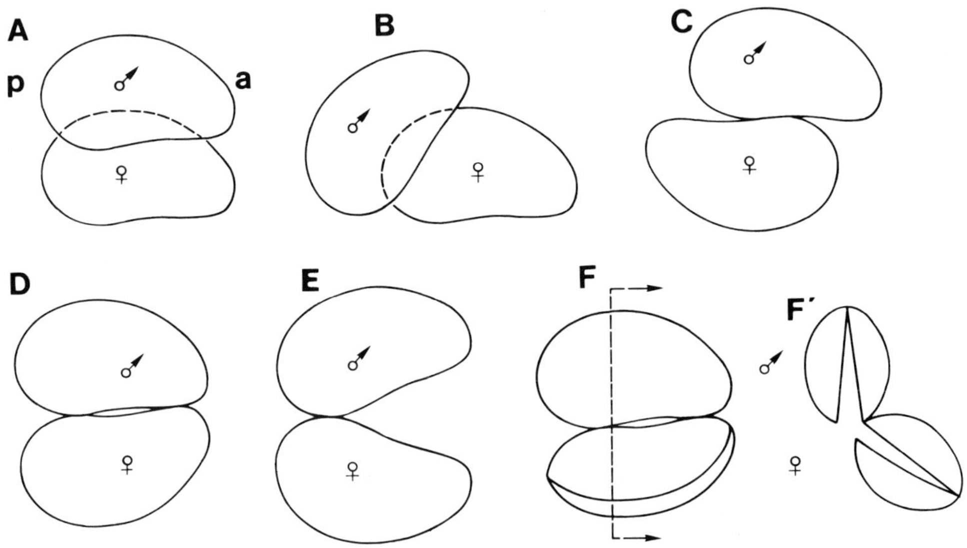

Fig. 10.6 Male – female copulatory positions: A= ventral / dorsal; B= ventral / postero-dorsal; C= reverse ventral / ventral; D= ventral / ventral; E= gaped-ventral / ventral; F= skewed ventral / ventral, dashed line showing the oblique position in F´; F´= oblique position of F (redrawn after Cohen & Morin 1990). Abbreviations used: a= anterior, p= posterior.

Copulation lasts from a few seconds to about 30 minutes or longer in most podocopids. Before the formation of the copula, the male protracts the posterior extremity of the body towards the ventro-posterior part of its own carapace. The hemipenes are in this way extruded (Figs. 10.2-3). Parallel to this, a semi-erection of the hemipenis occurs. In many podocopids, the hemipenes rotate 180° clockwise, either already in the male carapace (e.g. in the case of *Physocypria* species observed by McGregor & Kesling 1969) or outside, before the insertion into the female (most podocopids). Cytheroids developed a complex sclerified system to allow movement of the copulatory appendages outside of the carapace. Its functional structure was thoroughly described by Hart & Hart (1969) for the Entocytheridae and by Martens (1990a) for the Limnocytheridae. The muscles which support the hemipenes protrusion outside of the carapace are inserted in the postero-distal sclerified system. Note that bairdioids, like *Pussella,* do not have such a sclerified system (Danielopol 1976), but still the body is able to extend to about one third of its total length (see fig.1 in Danielopol 1976).

Detailed observations on a number of candonid species, e.g. *Fabaeformiscandona suburbana* (McGregor & Kesling 1969), *Candona sanociensis* (syn. C. *dancaui*), *Pseudocandona pratensis*, *P. rouchi* (Danielopol 1980a) can be summarized as follows. The species of the genera *Candona* and *Fabaeformiscandona* copulate in postero-dorsal / ventral position, while those of the genera *Pseudocandona* and *Limnocythere* in reverse ventral to ventral position. The male of *P. pratensis* starts to copulate in the ventral / postero-dorsal position, moving the penes inside the female carapace (Fig. 10.5, A-C - note the position of the clasping palps of the male's maxilla fixing posteriorly the female in B), followed by the final reverse ventral / ventral position. Through a double rotation with a 180° shift of the hemipenes, the right side of the male copulatory organ will fix the reversed female's genital organ and the posteriorly distal side of the hemipenis becomes anteriorly oriented (Danielopol 1980 a). The distal part of the inner M-process of the erected hemipenis (Fig. 10.5 E) will fix on the inner side of the vagina, while the *bursa copulatrix* (see chapter 4) and the copulatory tube are deeply inserted into the vaginal space (Fig. 10.5 F). Other penial lobes (b and h in Fig. 10.5F) touch the female's outer genital surface only marginally.

The copulation position adopted by *Eucypris virens* (studied by P. Alcorlo in Chatham) is postero / dorsal (Fig. 10.2), and this appears to be the most common position in cypridoids, e.g. in *Cypria* (McGregor & Kesling 1969). A rare gaped ventral / ventral copulatory position without rotation of the penes is shown by Elofson (1941) for *Leptocythere castanea*. Differences in copulation position between two related species of *Loxoconcha* were described by Kamiya (1988): *L. uranouchiensis* (living on sand) copulates in a ventral / ventral position while the carapaces lay on their lateral side; *L. japonica* (a phytophilous species) uses a skewed ventral / ventral position, while the female remains fixed with her strong distal claws of the thoracopods (especially the 3rd pair) on the algal substrate. *Bicornucythere bisanensis* copulates in the ventral / ventral position (Abe & Vannier 1991), while *Metacypris cordata* copulates either using the same position or the gaped position (Danielopol & Horne 1996, Fig. 10.3 here). Males of *B. bisanensis* were observed (Abe & Vannier 1991) to copulate unsuccessfully with other conspecific males and juveniles in crowded conditions. Copulation of myodocopids of the genus *Skosbergia* is in ventral / ventral position and is maintained by the interlocked position of the setae and claws of the antennae, the mandibular palps and the caudal rami of both sexes (Parker 1997).

The Entocytheridae copulate in ventral / dorsal position, with adult males guarding last instar females (Hart & Hart 1974, Danielopol 1977, Fig. 10.7). In this case, throughout the guarding phase, males fix females anteriorly with the second antennae and posteriorly through insertion of the hemipenes inside the female's carapace and through attachment of that organ to an inner postero-dorsal membranous lamella. After moulting, the male most probably copulates very quickly and inseminates the adult *triunguis* female.

With parthenogenetic females

The capacity of males to choose and/or to discriminate between females reproducing sexually and those reproducing parthenogenetically were investigated, albeit preliminarily, through video-recording of mating couples of *Eucypris virens* and *Limnocythere inopinata*. In *E. virens*, males from a sexual Spanish population were brought together with females from a parthenogenetic English population (studied by P. Alcorlo in Chatham). Copulation success in this latter case was more limited than between males and females originating from the syngamic Spanish population. A number of successful copulations could be registered, with the male hemipenes inserted within the parthenogenetic female for an extended period of time (several minutes). Although it could not be observed that in those particular cases the parthenogenetic female was inseminated (because of the opacity and pigmentation of the carapace), the attempted copulations add circumstantial evidence to the origin

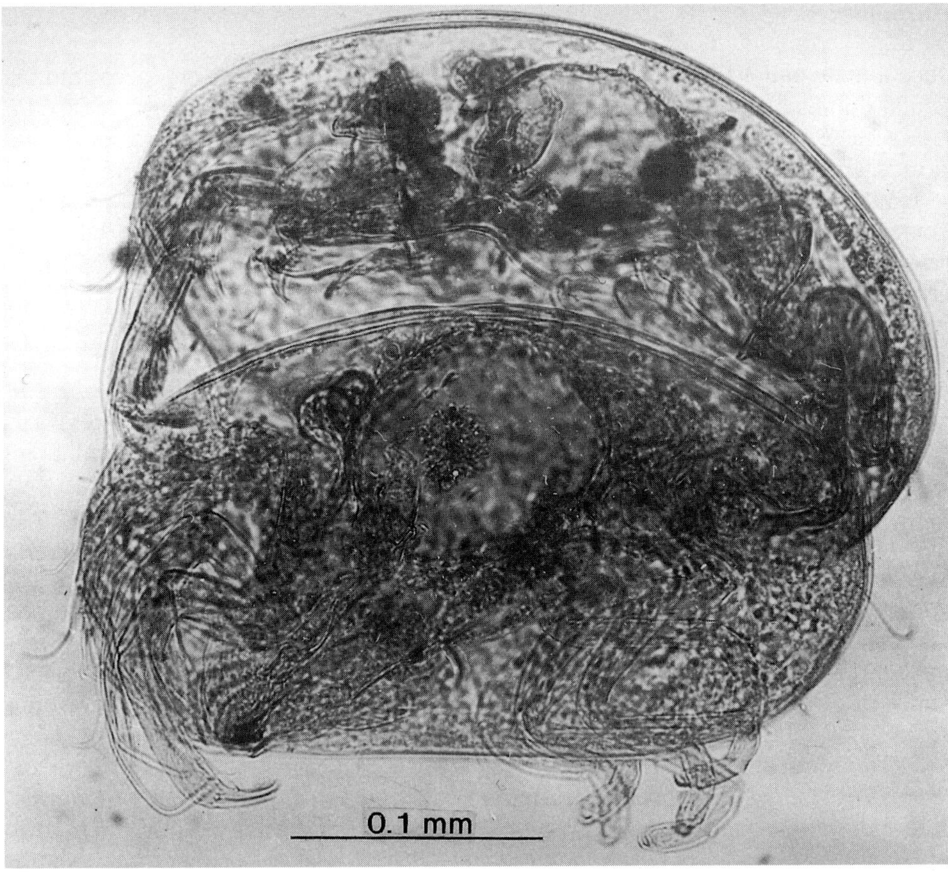

Fig. 10.7 Sphaeromicola cebennica juberthiei (Entocytheridae) in copulatory position. (original).

of polyploid clones through intraspecific hybridisation between males belonging to bisexual populations and exclusively parthenogenetic females. In *L. inopinata,* attempted copulations between single males (produced by clonal females) and parthenogenetic females were registered on videotapes (studied by Y. Yin and W. Geiger in Mondsee) and were found to be far less successful. Only a few times did the male manage to introduce the hemipenis within the female carapace, but for a few seconds only. It is unknown how long a 'normal' copulation between male and sexual female in this species lasts, but judging from the complexity of the glans on the copulatory process in the species of this genus (Martens 1990a), it appears unlikely that a few seconds would be sufficient to introduce sperm in the female.

Although these observations are qualitative and preliminary at best, they already indicate that intraspecific hybridisation is a possible mechanism for (polyploid) clonal origin (Butlin *et al.* in press), but that it is not necessarily of equal importance in all lineages. Further observations of such attempted hybridisations, with subsequent genetic analysis of the offspring, is necessary to determine the incidence of this phenomenon in species with mixed reproduction (chapter 17).

Proven for Heterocypris

Insemination

Once a male and a female are well fixed, the transmission of spermatozoa into the genital organs of the female is initiated. Danielopol (1980a) noted that the insemination in *Pseudocandona pratensis* started two minutes after the copula was formed, and lasted for three minutes. The position of the Zenker's organ moves from horizontal to more vertical (Hartmann 1975, McGregor & Kesling 1969). The fast contractions and extensions of the Zenker's pump can be observed through the transparent candonid carapaces. Abe & Vannier (1991) noted that during the copulation of *Bicornucythere bisanensis* the male's carapace vibrates, which could be interpreted as movements of the body assisting the hemipenes trabeculi which press on the *vesicula ejaculatoria.* This latter effectively pumps out the sperm of cytheroids, as described by Martens (1990a). For myodocopids of the genus *Skogsbergia,* Parker (1997) mentions that the insemination lasted a few seconds only. Note, finally, that all insemination is internal in Ostracoda; no cases of external insemination are thus far known.

Post-mating behaviour

Once copulation and/or insemination is finished, the mate pair will separate and no post-mate guarding has thus far been described in Ostracoda. During the post-mating stage, both females and males either remain active as in the case of most podocopids or a resting pause follows, as was observed by Parker (1997) for a myodocopid species.

In the podocopid *Metacypris cordata* (Fig. 10.3), the male starts to twist relative to the female and removes the copulatory appendages from the female carapace. Both individuals continue to move on the substrate, but during the initial period of this stage, the hemipenes remain partly erect. Only after a few seconds are the limbs rearranged in the non-erected position, i.e. the frontal plate which during copulation moves to the lateral side by more than 90° returns to its longitudinal position and retracts within the peniferal shield. The whole complex of the hemipenes and the posterior body part of the male are retracted back into the carapace. The same was observed in *Pseudocandona pratensis* and *P. rouchi* (Danielopol 1980a). Senescent male ostracods, e.g. males of *Limnocythere* (*Limnocytherina*) *sanctipatricii* and *Candona neglecta*, do not succeed in resetting completely and immediately the erected hemipenis after copulation (Danielopol pers. observ.).

Ostracods are able to re-mate quickly. Under controlled conditions, Danielopol (1980a) observed that a male of *Candona sanociensis* (syn. *Candona dancaui)* mated six times during its life time. An additional two females were presented to a male of this species, which had already successfully copulated with (and inseminated) a female, and three copulations succeeded within the next 30 minutes (Danielopol 1980a). Myodocopid females, once inseminated, lose their natatory setae so that they are confined to the benthic environment and can no longer be attracted to another flashing male (Cohen & Morin 1990, Morin & Cohen 1991).

During the post-mating stage, the females which received the spermatozoa probably act on the movement of spermatozoa toward the ovocytes in the uterus. This assumption is based on the observation that only part of the spermatozoa which accumulate into the seminal receptacle move back through the long spiral spermiduct which has a reduced diameter (for additional anatomical details see Kesling 1965, Hartmann 1975). Their movement seems to be induced by the female's activity or by her gametes. It is in the uterus that syngamy occurs, forming the ova which will be subsequently released outside the body. Eggs will either be stored within the carapace space or be deposited separately or in clutches in the surrounding environment.

Explanatory hypotheses for the diversity of the male dimorphic traits

The paradigmatic explanation: the lock and key model

During the first half of this century, taxonomists were amazed by the incredible morphological variability in male reproductive organs. Especially in insects, the actual copulatory organs could build complex and elaborate structures. It was proposed (review in Eberhard 1985) that mate recognition (by which conspecific males and females identify one another as potential mates) was primarily a mechanical process, whereby non-conspecifics were excluded, because male intromittent organs did not fit into female genitalia, i.e. both organs had to fit like a key in a lock. Eberhard (1985) extensively documented several strong counter arguments which also apply to the ostracods:

(1) Although the *keys* can build very diverse structures, female genitalia do not show the matching correspondence in the *locks*. Broodbakker (1982) showed that the male copulatory organs of three related species of the cypridid genus *Heterocypris* (*H. antillensis, H. margaritae* and *H. punctata)* display various types of latches on the inner side of the peniferum. The sclerotized latches come in contact during the mate process with the female genital lobe and vagina and represent excellent criteria to recognize the species. The shape of the genital lobe and vagina of these species do not show any major interspecific differences which correlate with the peculiar shapes of the latches.

(2) Many secondary sexual traits in males are not related to mechanical functions during copulation but to chemical or sensorial ones. This is the case of the male sensorial setae on the antennae in ostracods. In males of the Candoninae, the setae t_2 and t_3 of the second antenna have their extremities transformed in button-like structures (Danielopol 1978) while the homologue female setae are of the normal mechano-receptor type (Broodbakker & Danielopol 1982). Therefore, other processes leading to the development of the complex and diverse male's dimorphic traits must exist. The commonality of intersexual communication brings us to the examination of the arguments mainly related to Darwinian evolutionary explanations.

Sexual selection

When Darwin (1859) developed his theory on speciation by natural selection, he was left with the problem as to why males and females of several species are so different in appearance and, even more puzzling, to what extravagant proportions male traits can evolve, for example, the bright colours, large feathers and fins or conspicuous vocal signalling, none of which can improve survival, rather the opposite. He, therefore, introduced the concept of *sexual selection* (1871). Whereas natural selection implies competition over natural resources with the aim of body growth and survival for individuals, sexual selection refers to competition over mates and their gametes, aiming towards successful insemination and/or reproduction (Bell 1997). If an individual is unfit under natural selection, it will die. An individual with a low sexual fitness will fail to reproduce.

Sexual selection can act at several moments during the reproductive process and in various ways. Most commonly, it occurs before the actual mating, during courtship, either by *contests* (males fight over possession of females, e.g. in deer) or by *female choice,* when the male attempts to stimulate the female into accepting him for copulation (Eberhard 1985). It has been argued that *choice* is a very human-contrived term, and that *preference* or *selection* would be more appropriate for this prospect (Paterson 1985). A combination of both processes is found in *leks*, where several males attempt to demonstrate their fitness in a so-called *arena*, while females observe and eventually select a mate.

Paterson (1985) developed the concept of Specific-Mate Recognition System (SMRS) for the "subset of adaptations, which are involved in signalling between mating partners or their cells". Using this concept, Paterson (1985) defined species as the *most inclusive population of individual biparental organisms which share a common fertilisation system*. Several models attempt to explain the origin of such, sometimes elaborate, recognition systems by which males attempt to stimulate females into accepting them for reproduction. Two well-known ones are the *Fisherian runaway process* and the *Condition-dependence hypothesis*. Other mechanisms of competition over mates more recently recognized are *scrambles* (early search and swift location of mates), *endurance rivalry* (ability to remain reproductively active during a large part of the season) and *sperm competition*, including mate guarding, frequent copulation, etc., but also the ability to remove rival sperm. All of these are reviewed by Andersson (1994).

Sexual selection in ostracods, leading to the diversification of various male dimorphic traits, proceeds in our opinion mainly through two processes: (1) pre-copulatory sensory communication and (2) through male courtship within internal (or cryptic) parts of the female genitalia during the copulation and the insemination phases.

Pre-copulatory sensory communication between mates

Ostracods which are involved in external pre-copulatory communication show less diversity in sexual morphological traits than those which are involved in internal courtship during the copulation and insemination phases (see below). The aesthetasc Y of the 2nd antenna in a group of Candoninae with exclusively hypogean dwelling species, the genus *Mixtacandona,* displays a significant dimorphism, with males having a significantly longer chemo-sensorial seta than females (Rogulj *et al.* 1993). There are also interspecific differences in the length of the aesthetascs Y within this genus, e.g. *M. pietrosanii* has a longer aesthetasc with a stronger sexual dimorphism than those of the closely-related species *M. lattingerae*. Certain marine planktonic halocyprids, e.g. *Paraconchoecia caudata* or *Spinoecia vibekensis,* display similar sexual dimorphisms (Poulsen 1973). The chemo-sensorial setae of the 1st antenna are differently developed in both sexes. There are four short chemo-sensorial antennular setae in the female, while the male has two short and two long aesthetascs. Interestingly, the frontal organ (especially the distal part) is larger in the female than in the male.

Morphological traits of non-marine ostracods involved during direct mechano-sensorial contact of individuals of the opposite sex are developed differently on the male's antennae, the thoracopods, caudal rami, etc. These sensorial limbs come in contact with the female during the first phase of intersexual contact, when the male tries to fix the female forming the mating couple, and also for a short period of time during the mating phase. The 2nd antenna of the male ostracods displays convergent shapes of the distal claws in various phylogenetically unrelated groups (Fig. 10.4, A-C) i.e. one of the normally bipectinate claws (i.e., set with a double row of spines) is transformed into a strong, unipectinate, comb-like claw (Danielopol 1978, Martens 1987). Such setae are visible in cypridoids like *Herpetocypris* (Gonzalez Mozo *et al.* 1996), cytheroids like the limnocytherid *Elpidium* (Danielopol 1975) and the ento-cytherid *Sphaeromicola* (Danielopol 1977). Other tactile setae used by the male during the first contact with the female are especially the antennal setae t_2 and t_3 of cypridoids and/or the distal extremities of the thoracopods of the cytheroids (Danielopol 1978, Danielopol & Tétart 1990, Gonzalez Mozo *et al.* 1996, Martens 1990a, in press). The endopodal structures of the cytheroid thoracopods as sensorial signal transmitter display other peculiar morphologies as compared to those of Cypridoidea. Note, for example, the curious shape and structure of the right two thoracopods of *Cytherissa sernovi* and *Cyprideis torosa* (Fig. 10.1 E, F, K, J), as compared to the normally developed ones (Fig. 10.1 G, H). The same applies for the strongly dimorphic endopodite of the 3rd thoracopod of the *Leucocythere* species (Fig. 10.4 D-F).

CRYPTIC FEMALE CHOICE DURING MATING AND POST-MATING PHASES

Sexual selection is still largely defined in terms of competition over mates, but the definition could be refined to competition over gametes (Eberhard 1985, 1996), as the process by no means ends with successful mating. Males, as we showed above, exert their courtship on females, not only through external communication, but also through internal signalling with their genitalia, especially during the intromission phase. This implies that females, by the so-called *cryptic choice* (Eberhard 1985, 1996) have much greater opportunities to exert influence over events in the post-intromission phase. Especially in prezygotic (but post-intromission) conditions, females can largely control whether successful copulation will eventually also result in offspring, but they even have some degree of postzygotic control, i.e. by induced abortion. These aspects of *cryptic female choice* (i.e. which cannot be measured from successful mating), potentially of large evolutionary importance, are extensively reviewed by Eberhard (1996), according to whom (p. 418): "... the concept of courtship needs to be expanded. All male behaviour prior to, during, and following copulation which serves to induce the female to perform both pre- and postintromission responses that increase the chances that his sperm, rather than that of another male, will fertilize her eggs, can reasonably be considered as courtship". Therefore, the main currency of selection in the Eberhard's model is the gamete fertilization, not energy optimization. Males of non-marine ostracods display a high diversity of genitalic morphological traits involved in the *internal female courtship*. These can be divided into two groups: those which penetrate deep into the internal space of the female carapace during the mating and touch partly the external side of the genital lobe (type 1), partly the vagina, and secondly, those which penetrate deep inside the vagina (type 2). The diversity of the clasping organs on the hemipenes of *Limnocythere* species (Martens 1990a) or those of the Entocytheridae (reviewed by Hart & Hart 1974) illustrate this, as in both cases these structures fix on the external side of the genital area of the female. Martens (1990a) showed that the highest inter- and intraspecific diversity of morphological structures displayed by the lower ramus of the hemipenis' clasping organ of *Limnocythere* occur either on two extremity areas, the hook-like process and the lateral process (type 1) or on the *glans* (type 2) (Fig. 10.5 G, I). Similar situations, i.e. a high shape diversity of the structural parts of the hemipenis which touch and/or fix the female (type 1) can be found in various groups of the Candoninae *Candona, Pseudocandona* and *Fabaeformiscandona* (Kesling 1965, Danielopol 1969, 1978, 1980a).

One of the peculiarities of the Ostracoda is the large size of their hemipenes. Cytheroidea developed an especially powerful muscular system which has to pump out the spermatozoa during the insemination stage (for details see Hartmann 1968, Martens 1990a, and chapter 4). In Cypridoidea, beside the well-developed inner structures in the hemipenis, e.g. the spermiduct and a complicated sclerified system which opens the peniferal shield during erection (details in McGregor & Kesling 1969, Danielopol 1969, 1978), also a strong spermatozoan pump, the Zenker's organ which lies outside the hemipenes, has developed (for a recent review on this morphological complex, see Matzke-Karasz 1997). We consider that the unusually large male sexual organs which repeatedly evolved within various phylogenetically unrelated groups of major ostracod groups, like the Cytheroidea, the Cypridoidea and the Bairdioidea (for this latter group see description of the copulatory complex in Danielopol 1972, 1976), represent good arguments for the active role played by sexual selection during the long evolution of these phylogenetic lineages, from the Palaeozoic to the Recent.

Sperm competition

The idea that various morphological peculiarities of the ostracod reproductive organs of both sexes play a potential role in sperm competition is attractive. Some of the Podocopa ostracod groups like the Cypridoidea, both in marine and nonmarine species, evolved giant spermatozoa, i.e. with long tails (see chapter 2) and strong acrosomal structures like crests, perforatorium, etc. (Wingstrand 1988, Hartmann & Guillaume 1996). Peculiar to the podocopid ostracods is the capacity of the spermatozoön to penetrate and to coil inside the ovocyte. Because of its large size, only the first most active spermatozoön will achieve fertilization of a female gamete. It is, hence, very likely that competition between spermatozoa occurs during the sperm transmission within the male copulatory organs, i.e. the strongest and the most motile gametes which succeed first in penetrating into an ovocyte will outcompete the others.

Sperm competition can occur either along the male's *ductus ejaculatorius*, which in many cases is thin, very long, like in the Cypricercini (Broodbakker 1983) and sometimes has large chambers (like in the labyrinthal area of the Herpetocypridinae, Gonzales Mozo *et al.* 1996) or within the female's *receptaculum seminis* and the coiled and thin *Befruchtungskanal* or seminal tube (Hartmann 1968). Especially these latter structures are interesting, because, following Eberhard's (1996) model, the female could realize her cryptic choice of the sperm through the constraints she imposes on the spermatozoa to move through her long genital tracts.

Martens (in press) postulated that another form of sperm competition could be produced through the activity of the copulatory process, which in some cases is corkscrew-shaped (e.g. in species of *Leucocythere*, see Fig. 10.5, I and Danielopol *et al.* 1990, Martens 1990a). The corkscrew-shaped copulatory process is in principle able to eliminate alien sperm when retracting out from the female genital organ and by a new intromission it could introduce its own sperm. Hartmann (1975) mentioned that bairdiid ostracods use this mechanism: with the retraction of the copulatory process from the vagina, the male eliminates a part of the sperm from previous copulations. One of the remaining spermatozoa will harden, becoming a spermatophore which plugs the female genital opening. In some cases of marine ostracods (Bairdioidea and Myodocopa), morphological and behavioural adaptations evolved presumably to reduce sperm competition. The spermatophore which plugs the female's vagina (Hartmann 1968, Cohen & Morin 1990) potentially prevents subsequent copulations. Butlin *et al.* (in press) link skewed sex ratios (with female excess) to lower sperm competition, which then enables the males to invest in few large spermatozoa rather than in many smaller ones. The relevance of sperm competition in sexual selection in the different ostracod groups remains unsolved, but forms exciting avenues for future research.

Natural selection

Danielopol & Horne (1996) noticed that the females of *Metacypris cordata* which rapidly accept a male to mate, slow down their movement, thus economising energy for both sexes. The opposite situation is represented by females which do not accept males for copulation and where both sexes increase their motility, with the female trying to escape from the male, the male trying to fix the female and insert the copulatory complex (paired hemipenes) inside the valves. The latter type of activity implies *de facto* an increase in energy expenditure and here energy economy is the classic selection currency (*sensu* Diamond 1993).

Martens (in press) analysed the variability of the copulatory modules in the Limnocytherinae (Table 10.1): the five lineages within this subfamily have developed very different morphological signalling systems, involving valves, antennula, exopodite of the antenna, third thoracopod and various aspects of the copulatory complex of the hemipenis. None of the potential mechanisms responsible for the origin of this seemingly high diversity have thus far been demonstrated directly, but the resulting patterns provide indirect evidence that three different mechanisms contributed. Stochastic processes (chance) are involved in the initial *choice* of modified structures. Subsequent sexual selection on traits of the recognition systems causes radiative speciation within lineages. Natural selection constrains the development of structures and systems and thus directs, or at least influences, the development and the diversity of morphological structures. For example, natural selection restricts the size and complexity of the copulatory complexes on the hemipenis, as these structures must always be able to fit within the closed valves. Although male valves in some of these lineages indeed are larger than female valves, with

Table 10.1 Specific sexually dimorphic morphologies in copulatory modules of Limnocytherinae (Ostracoda). A1= antennula, A2= antenna, T3= third thoracopod, Cp= copulatory process, CO= clasping organ, LR= lower ramus, UR= upper ramus, Fu= furca, gr = group (from Martens, in press).

	Valves	A1	A2	T3	Hemipenis		
					Cp	CO	Fu
Limnocythere-gr.	V	–	–	–	Cp	LR	–
Korannacythere	–	–	A2	–	–	–	–
Paralimnocythere-gr.	V	–	–	–	Cp	–	Fu
Neolimnocythere-gr.	V	–	–	–	–	–	Fu
Leucocythere-gr.	V	A1	–	T3	Cp	UR	Fu

ventral expansions allowing more space for the copulatory structures, there are severe structural constraints limiting the extent of such valve enlargements. It is assumed (Martens 1996a) that in *Korannacythere*, a small limnocytherid genus from South Africa of which all three species live in ion-poor temporary rock pools, the difficulty to calcify valves has caused carapaces to remain small, with almost undifferentiated hemipenis structures as a consequence. In such cases, the currency of natural selection is carapace space and not energy optimisation. Another example of dimorphic carapace shape, possibly originated through natural selection, is in *Candona* species. The male valves of *Candona neglecta* and *C. fasciolata* (Fig. 10.1, A, B) are widely arched in the posterior half, which allows optimal accomodation of the large hemipenes, while the females have more rectangular carapace shapes (Kesling 1965, McGregor & Kesling 1969, Danielopol 1969).

One aspect of the ostracod diversification is represented by the evolution of whole copulatory modules (Martens in press). Danielopol (1980b) compared the size of the copulatory appendages and the size of the inner carapace space of 21 species belonging to the genera *Mixtacandona, Candonopsis, Pseudocandona, Fabaeformiscandona* and *Candona*. The representatives of the genus *Mixtacandona,* which are known by a limited number of hypogean-dwelling species in the Palaearctic only (Danielopol 1985), have small hemipenes, without an inner M-process, poorly sclerified; the male prehensile palps display a slight asymmetry and are very flat; the Zenker's organs have relatively short lateral spikes (for additional details see Danielopol & Cvetkov 1979). The opposite is the case of the species belonging to the genera *Candona* and *Fabaeformiscandona,* two species-rich groups, widely spread in the epigean non-marine habitats of the Palaearctic area (Löffler & Danielopol 1978). The copulatory appendages of the species belonging to these genera are strongly sclerified, the Zenker's organs display very powerful rosettes of spikes, the clasping palps of the male maxilla have complex volumetric shapes and a strong asymmetry, while, finally, the inner carapace shape is much larger than those of *Mixtacandona* (Danielopol 1978, Danielopol 1980b). These data suggest that different copulatory modules of Candoninae evolved by the process of (both natural and sexual) selection, acting at a supra-specific level, following the ideas of Gerhart & Kirschner (1997), who consider that evolvability of morphology of whole groups of traits could occur through selection at a supra-specific level, that of the clade.

Alternative scenarios

Not always can the morphological structures of an organism be conveniently explained as an adaptation produced by natural or sexual selection processes (Gould & Lewontin 1979). There are morphological traits for which we do not see a functional role, their origins are apparently due to stochastic processes or to pleiotropic effects. For instance, *Cytherissa sernovi* from Lake Baikal has a *ductus ejaculatorius* of the hemipenis which is long and coiled; within its middle section, the lumen has a slight internal inflation to which corresponds a stronger wall sclerotization on the external side (Fig. 10.8, A, B). *Cytherissa tuberculata,* which occurs together with *C. sernovi,* displays a simple tubular *ductus ejaculatorius* (Fig. 10.8, C, D), as in most of the Cytherideidae (Danielopol & Tétart 1990). There is no obvious evidence that the peculiar structure of the *ductus ejaculatorius* in *C. sernovi* could have a functional role during the mating process, in order to improve the copulation efficiency. However, if this should be the case, one could consider the peculiar trait of *C. sernovi's* hemipenis an adaptation and it should be interesting to reconstruct its selective history.

One can notice other minute morphological traits which are most likely not functional in signalling or mating, but which presumably occurred by stochastic and/or pleiotropic mechanisms in various limnocytherid and entocytherid species. In the former group, Martens (1990a, 1994) and Yin (1997) showed that the outer margin of the hook-like process of the lower ramus of the hemipenis of *Limnocythere inopinata* (Fig. 10.5, H), which does not come in contact with the female's body during copulation, presents a minute tubercule which displays an intra-specific variability. Similar examples are documented for various ostracod groups, e.g. the Entocytheridae (Hart & Hart 1974).

Morphological traits of an organism are in some cases primarily the product of internal developmental processes in which, at a later time, natural selection can play an active role (Raff 1996, Wagner 1998). The diversity of the lateral peniferal shields of various candonids can be seen as a result of changes in the morphogenetic program responsible for the development of the peniferum. For instance, most of the *Candona neglecta* species-groups have a lateral peniferal shield with a rectangular lobe 'a' (Figs. 10.1 A, 10.10, A), which is furthermore dorsally oriented (Danielopol 1969). This lobe is generally oriented postero-distally in the case of *Pseudocandona* and *Mixtacandona* species (Klie 1938, Danielopol 1969, Danielopol & Cvetkov 1979). While *Mixtacandona* species (Fig. 10.10 B) have the lateral peniferal lobe straight and elongate, the homologue lobe of *Pseudocandona* is widely rounded (Figs. 10.5 D, 10.10 C). Because, during mating, this lobe is not directly involved in the process of the fixation of the hemipenis to the female genital structures (Danielopol 1980a), its morphological diversity within one genus, e.g. *Candona*, is much reduced as compared to those traits having a direct adaptive role during copulation. For this latter case we take as example the high morphological variability in the distal part of the inner M-process of the *Candona* and *Pseudocandona* species (Figs. 10.1 B, D, 10.5 E). This copulatory limb is well sclerified and penetrates in the female's vagina, fixing the hemipenis tightly to the female genital area (Fig. 10.5 D, F and Danielopol 1969, 1980a).

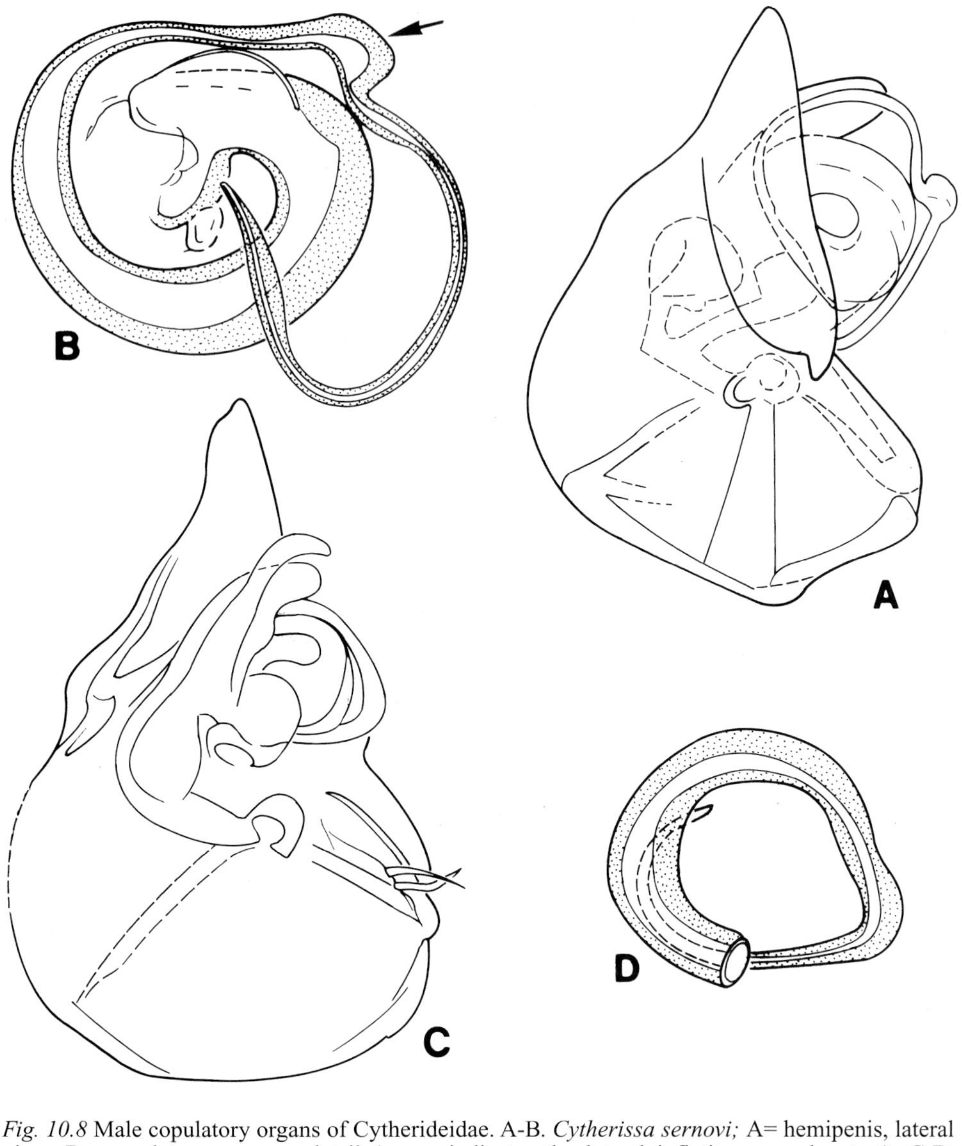

Fig. 10.8 Male copulatory organs of Cytherideidae. A-B. *Cytherissa sernovi;* A= hemipenis, lateral view; B= copulatory process, detail (arrow indicates the lateral inflation, see also text). C-D. *Cytherissa tuberculata;* C= hemipenis, lateral view; D= copulatory process, detail (redrawn after Danielopol & Tétart 1990).

Similar morphological differences as those mentioned here for the distal lobe of the Candonidae hemipenis are also present, with various degrees of interspecific variability, within the Cyprididae. Martens (1986, chapter 4) amply documented that in some lineages (*Sclerocypris*), the distal lobes of these peniferal sheats have distinct specific morphologies, while in others, these morphologies remain almost uniform amongst species (e.g. in the case of the *Heterocypris* representatives) and

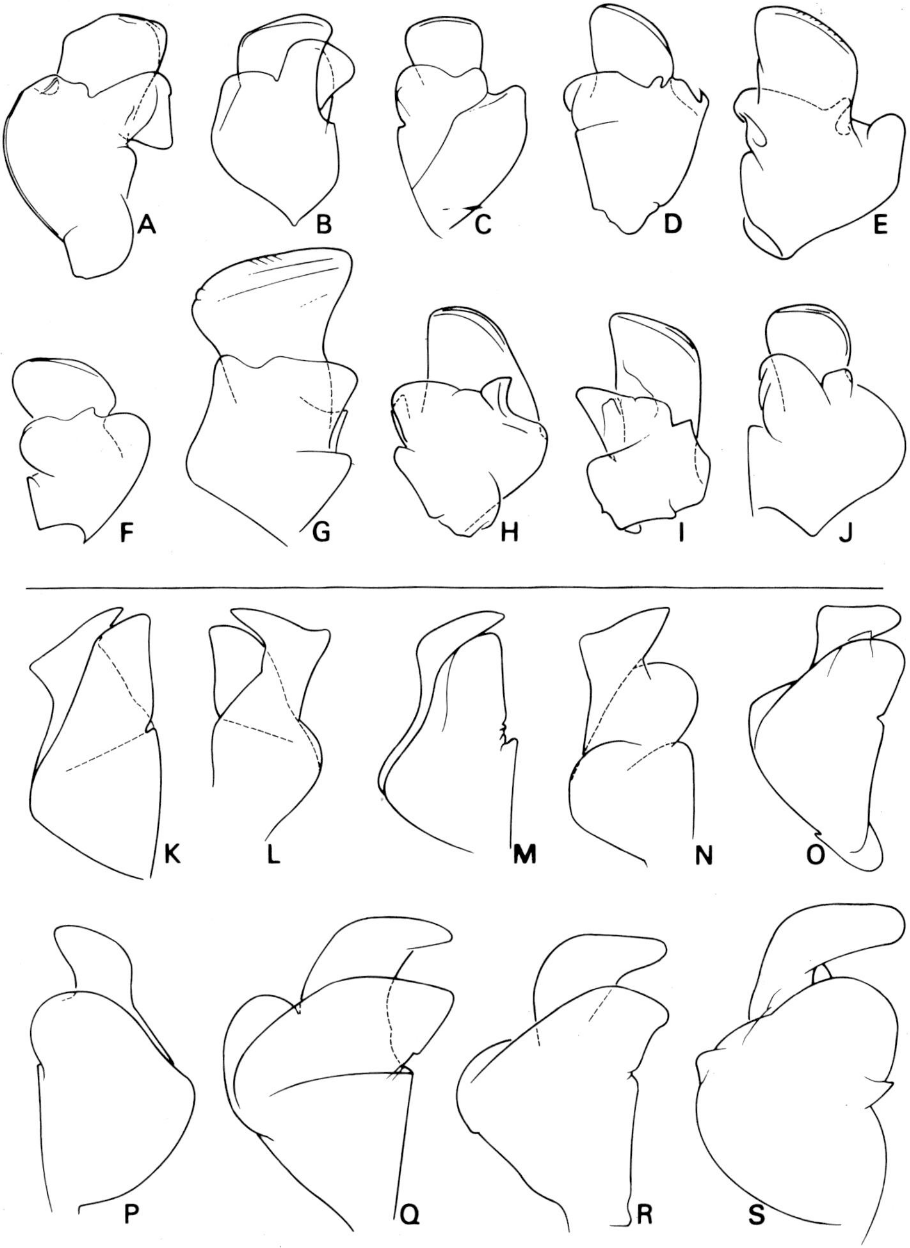

Fig. 10.9 Hemipenes outlines in different groups of Cyprididae, showing large variability in one genus (*Sclerocypris* – A-J), and relative uniformity within and amongst subfamilies (Cyprinotinae (K-P) and Cypridinae (Q-S)).
Large variability: A= *Sclerocypris exerta*; B= *S. coomansi*; C= *S. clavularis*; D= *S. demoori*; E= *S. zelaznyi*; F= *S. major*; G= *S. dayae*; H= *S. sarsi*; I= *S. dedeckkeri*; J= *S. dumonti*.
Relative uniformity in Cyprinotinae: K= *Heterocypris margaritae*; L= *H. antillensis*; M= *H. incongruens*; N= *H. reptans*; O= *H. giesbrechti*; P= *Hemicypris intermedia*; in Cyprididae: Q= *Cypris puberoides*; R= *Pseudocypris bouvieri*; S= *Ramotha hirta*. Not to scale.
(A-J redrawn after Martens 1986, 1988; K, L after Broodbakker 1982; N after Martens 1996b; O, P after Martens 1984; Q-S after Martens 1990b, 1992)

sometimes even amongst genera (*Cypris, Pseudocypris, Ramotha*) (Fig. 10.9). Again, such differences can arise as a consequence of organismal constraints which act during the various developmental stages.

Morphological structures developed for other purposes can become, at a later stage, adaptations to the current use of the living organism (Gould & Vrba 1982). The reduction of the right pairs of the male thoracic legs as compared to the left side and/or the symmetrical legs of the females in various cytherid genera like *Cytherissa, Cyprideis* (Danielopol & Tétart 1990) and/or *Bicornucythere* (Abe & Vannier 1991), can be viewed in a first stage as distortion of the normal developmental program: the right part suffered an arrested evolution of the limbs (which is a non-adaptational feature, these structures became deprived of their original function as walking legs) while the left part developed normally. At a later evolutionary stage, the distal extremities of these abnormal male thoracopods, which come in contact with the female during the mating process, became used for intersexual communication and potentially developed new morphological structures, most probably through direct sexual selection. Such morphological traits, which were originally exaptations, became later on adaptations participating actively to the mating process as in the case described by Abe & Vannier (1991) for *Bicornucythere bisanensis*.

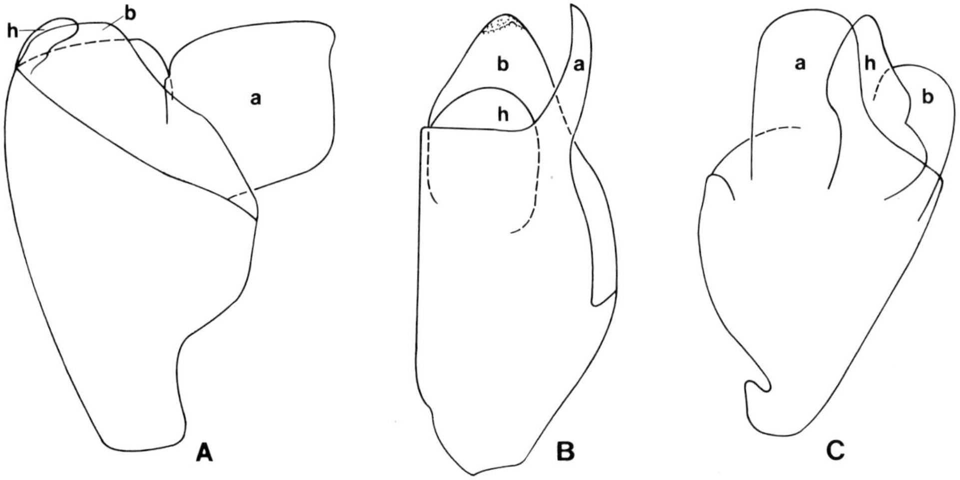

Fig. 10.10 General aspect of Candoninae hemipenes; A= *Candona neglecta,* medial view*;* B= *Mixtacandona pietrosanii,* lateral view*;* C= *Pseudocandona* aff. *compressa,* lateral view. Abbreviations used: a,b, h= distal lobes of the hemipenis (A,C redrawn after Danielopol 1969; B after Danielopol & Cvetkov 1979).

Brood care

Introduction

We are interested in the relationship between brood care and reproductive mode, primarily because it has been suggested in general terms that brooding lineages may be pre-adapted to the maintenance of parthenogenetic strains, and more specifically that brood selection could allow brooding parthenogens to avoid Müller's ratchet and the accumulation of parasites, two of the detrimental effects of long-term parthenogenesis (chapter 1, Lively & Johnson 1994). Different combinations of sexual reproduction, parthenogenesis, brood care and egg-laying exist in living Ostracoda and are often recognisable (by virtue of sexual dimorphism of the shell) in fossils (but see discussion in chapter 6). Ostracods therefore offer a unique opportunity to study the relationship between brood care and reproductive mode.

Brood care in ostracods takes different forms and has evolved independently in different lineages. We define brood care as *any form of parental care of eggs or juveniles once the eggs have been expelled from the uterus, whether this brood care takes place inside or outside the mother's carapace, and whether it involves eggs, developing embryos or juvenile instars*. In the following sections, we describe and discuss the different brood care strategies exhibited by living ostracods, the evidence of brood care in fossil (and in some cases extinct) groups, and the reproductive advantages of species with brood care, compared to those whose reproductive success is influenced solely by genetically pre-determined environmental tolerances.

The occurrence of brood care in ostracods

Many ostracods lay their eggs and pay no further attention to them – the future success of those eggs (and, if they hatch, the juvenile instars) is at the mercy of the environmental factors of which they are more or less tolerant. With regard to extant ostracods, all the Cypridoidea and Bairdioidea, the majority of Cytheroidea and maybe one genus of the Darwinuloidea (*Microdarwinula*), as well as most Halocypridea, fall into this category. Indeed, the reproductive and dispersive success of many cypridoideans is enhanced by their ability to lay desiccation-resistant eggs which can be transferred between habitats by wind, birds or other animals. With regard to *Microdarwinula*, it should be pointed out that in fact *Microdarwinula* does not have an externally visible brood pouch like that of *Darwinula*, although it is probable that brood care nevertheless takes place.

In ostracods which exhibit brood care, the adult females retain their eggs, and in some cases the first two or three instars, within a brood space that in modern species is usually formed by an expansion of the posterior part of the carapace. In this category are the other four extant darwinulid genera, several cytheroidean genera, at least one halocypridid genus and all of the Myodocopida and Platycopida (Cohen & Morin 1990). Such brood care is also recognized in fully extinct groups, such as kloedenellids, lomatopisthiids, monotiopleurids and leperditellids (Lundin *et al.* 1995, see below). In all of these cases, the brood space or brood chamber is formed by the enlargement of the posterior part of the domicilium (the internal space enclosed by the carapace, within which the body and appendages of the animal are housed). This gives rise to distinctive sexual dimorphism in some groups, with the inflated posterior of the female carapace contrasting with the more slender appearance of the male. An extreme example is the living limnocytherid *Metacypris cordata*, in which the female is virtually heart-shaped in dorsal view while the male is almost indistinguishable from the A-1 juvenile stage. By analogy, it is relatively easy to recognize the existence of brood care in sexually dimorphic fossil taxa (but see chapter 6), including extinct genera such as the Jurassic-Cretaceous *Theriosynoecum*. One group of ostracods in the Palaeozoic, the Beyrichioidea, apparently formed brood chambers from bulbous lateral extensions (crumina) on the outside of each valve. That the distinctive sexual dimorphism of this extinct group was related to brood care has been supported by the discovery of juvenile shells inside the crumina of some specimens (Spjeldnaes 1951, Martinsson 1962), although this does not rule out the possibility that these features also had another function such as buoyancy (Kesling 1957, Adamczak 1990 and below).

In extant podocopid lineages, brooding is absent in Bairdioidea and Cypridioidea and most common in Darwinuloidea and Cytheroidea. In the latter, brood care is known to occur in Xestoleberidae (e.g., *Xestoleberis*), Leptocytheridae (*Tanella*), many Cytherideidae (including *Cyprideis*, *Mesocyprideis*, *Romecytheridea*, *Tanganyikacythere* and *Kavalacythereis*; it is unknown in *Cytherissa*, however) and many Limnocytheridae (*Metacypris*, *Gomphocythere*, *Cytheridella*, *Elpidium*, *Kovalevskiella*). In the Cytherideinae, the extant brooder *Cyprideis torosa* is a member of a brooding lineage which extends back to the Jurassic (Colin & Carbonel 1990), but Van Harten (1990) suggested, largely on the basis of observations of reduced sexual dimorphism, that many of its Neogene congeners may have secondarily lost this ability.

The nature of brood care in ostracods

Jaanusson (1985), discussing the functional morphology of platycopids, recognized the following types of brood care:

1. Brood care of juvenile instars in anteroventral or ventral pouches (crumina) – only known in the extinct Beyrichioidea.
2. '*Cyprideis* type': eggs and the first two or three juvenile instars are brooded in the posterodorsal part of the domicilium, with eggs being loose/free in the cavity.
3. '*Cytherella* type': eggs retained in the posterior part of the domicilium, but glued to the inner mantle epithelium with some secretion (cytherellid and cavellinid platycopes). In all *Cytherella abyssorum* studied by Jaanusson (1985), four eggs were found in one valve and three in the other, achieving the tightest possible packing of the relatively large eggs. He pointed out, however, that this brood care concerns only eggs – there is no evidence for brooding of juvenile instars in cytherellids.

We note that in the '*Cyprideis* type' the expansion of the domicilium to form the brood pouch tends to be situated posterodorsally, and the eggs (before hatching, at least) are held in place on a kind of platform on the back of the soft body, above the thoracic legs. In this characteristic, *Cyprideis* and its close relatives differ from brooders such as *Xestoleberis* and *Metacypris*, in which the brood space inflation of the carapace is mainly posterolateral and posteroventral (see 5 below).

To these three we would add four further types, the last two of which are somewhat speculative:

4. 'Myodocopid type': the eggs are loose within the brood pouch, but in most cases there is probably no brooding of juvenile instars. Elofson (1941) observed that juveniles of *Philomedes globosus*, as soon as they hatched, tried to leave the brood space, which was empty within a few hours. He regarded Skogsberg's earlier (1920) observation of first stage juveniles remaining within the brood chamber as mistaken.
5. '*Metacypris* type': eggs and the first two or three juvenile instars are retained loose/free within the posterior part of the domicilium; the brood pouch is mainly posterior and posteroventral in Cytheroidea such as the Xestoleberdidae and Timiriaseviinae, more posterior and lateral in Darwinuloidea.
6. '*Loculicytheretta* type': the function of the loculi in genera such as *Loculicytheretta* and *Heptaloculites* is still unknown (Athersuch & Bonaduce 1976). That they are chambers for brood care has been ruled out on the grounds that there is no internal connection to the domicilium. We do not consider it impossible, however, that juveniles could crawl into them after hatching. Or perhaps the male could transfer the eggs to the female's loculi during copulation. A brooding function was considered (and rejected) for the loculi of the Devonian hollinid *Ctenoloculina cicatricosa* by Kesling (1952), but is still not ruled out for such Palaeozoic forms with externally-opening cavities such as loculi and dolons (Sarv 1966) (Fig. 10.11).
7. 'Nesting': as yet there is little evidence that any ostracods care for their eggs once they have been laid, although many cypridoideans (in which brooding is unknown) apparently take care to deposit (and glue in place) their eggs on suitable substrates (sand grains, moulted ostracod valves, bits of vegetation) rather than simply release them at random. Adult females of *Eucypris virens* in laboratory cultures return at frequent intervals to previously-deposited egg clusters, apparently to clean them or prevent them from being covered by sediment (A. Baltanas, pers. comm.). *Cypria ophtalmica* has been observed to glue clusters of eggs to the umbilicus of the gastropod *Gyraulus crista*, a behaviour which may be regarded as a form of parental care (Janz 1992).

Myodocopid and platycopid eggs are all brooded at the same developmental stage; some halocyprids and all brooding cytheroids and darwinuloids simultaneously carry eggs in different developmental stages (Cohen & Morin 1990). It is beneficial in an animal brooding early juvenile instars ('*Cyprideis* type' and '*Metacypris* type'), that at any one time the eggs or juveniles should be at different stages of development. The first instars to hatch and then moult will clearly occupy more of the brood space, and it is necessary for them to vacate the brood space to make room for those that hatch and moult later. In ostracods that brood only the eggs ('*Cytherella* type', 'Myodocopid type') these can be, and apparently are, all at approximately the same stage of development – they hatch within a few hours of each other and the juveniles immediately leave the brood space.

But many branchiopods manage without doing this.

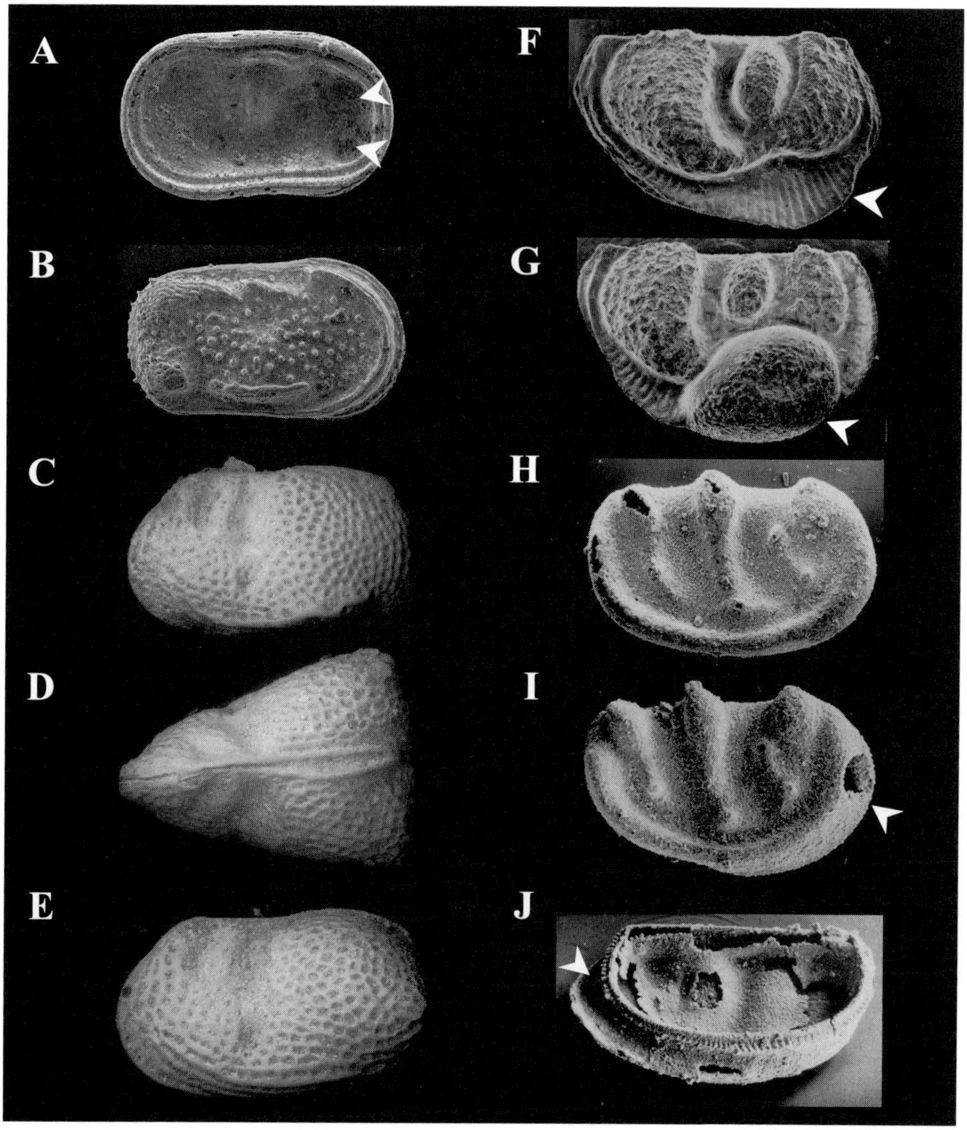

Fig. 10.11 Brood care in fossil ostracods
A,B. *Cytherelloidea granulosa* (Platycopida), adult female, from the Upper Cretaceous Chalk of SE England; A= internal view of right valve, showing the internal loculi which house the eggs (arrowed); B= external view of right valve. Length approx. 0.7 mm. C-E. *Theriosynoecum striatum* (Podocopida, Cytheroidea, Limnocytheridae), from the Lower Cretaceous Purbeck deposits of S. England; C= external view of male left valve; D= dorsal view of female carapace showing inflated posterior brood space; E= external view of female left valve. Length approx. 0.6 mm.
F, G. *Equicastanea lappacea* (Palaeocopida, Beyrichioidea), from the Silurian of England; F= external view of male right valve showing the wide frill or velum (arrowed); G= external view of female right valve showing the crumina (brood pouch). Length approx. 1.5 mm. H-J. *Homeokiesowia epicopa* (Palaeocopida, Beyrichioidea), from the Ordovician of Wales; H= external view of male right valve; I= external view of female right valve showing expanded dolon (arrowed); J= internal view of female right valve showing expanded dolon (arrowed). Length approx. 2.0 mm. (A-E kindly provided by Ian Slipper; G-J kindly provided by David Siveter).

Brood care in extinct groups

In some palaeocopid ostracods, distinctive sexual dimorphism is seen in the adult stage, with females of some genera possessing a large, bulbous swelling in the antero-ventral part of each valve. In Beyrichioidea, these swellings open internally into the domicilium and are termed cruminae or brood pouches, but in certain other groups dolons or 'false brood pouches' are found in which the openings are outside the contact margin of the valves. The dolon may be seen as an evolutionary precursor of the crumina; in some taxa the dolons form an almost totally enclosed space when the valves are closed, while in others (subfamily Wehrliinae) they are fringed with a fence of spines (e.g., in *Bilobatia*) which might serve to retain eggs or juveniles (although there is no evidence for this) while still allowing them contact with the water circulating outside the mother's carapace (Schallreuter 1982). According to Kesling (1957), the dolon evolved in the Middle Ordovician, in the piretellids, from the broad convex frills of their ancestors, the Lower Ordovician eurychilinids. The late Ordovician – Silurian beyrichioids are then seen as descendants of the piretellids in which the incurving outer margin of the dolon eventually fused with the valve margin and a new aperture opened directly into the domicilium, giving rise to the crumina (Fig. 10.11).

Kesling (1957) reviewed the many interpretations and speculations concerning the function of the crumina. Several authors had referred to these as 'brood pouches' but without any concrete evidence that they were actually used for brooding eggs or juveniles. A report of eggs in the 'false brood pouch' of *Piretella reticulata* by Kummerow (1933) was later considered to be erroneous, the features described actually being nodes on the outer surface of the pouch (Triebel 1943 *fide* Kesling 1957). Schmidt (1941) considered that the pouches might be gas-filled buoyancy aids, or that the pouched dimorph was the male and the pouches were for storing sperm, as did Triebel (1941, see also Kesling 1957). It was only when valves and carapaces of juvenile instars were found within the crumina of *Beyrichia kloedini* and *Beyrichia jonesii* (Hessland 1949, Spjeldnaes 1951) that it became clear that they were probably brood pouches (Kesling 1957). Even now, however, an alternative (or perhaps double) function as buoyancy aids is not ruled out by some authors (Adamczak 1990). There is as yet no firm evidence that the dolon was actually used for brood care.

A problem with the brood-pouch interpretation of the crumina relates to how the function evolved, as opposed to how the structure evolved (from convex frills via dolons as outlined above; but see later discussion below). Kesling (1957) thought that the brooding function might have begun quite fortuitously when juveniles took advantage of the protection offered by these swellings (which presumably had a different primary function). Spjeldnaes (1951) supported the brooding function by finding juveniles inside the crumina, but was concerned about the implications of this discovery for the orientation of the adult carapace, suggesting that since all modern brooding ostracods do so in the posterior half of the carapace, the crumina must have been a posterior rather than an anterior swelling. This idea finds little support today, however, and this leaves us with the question of how eggs and/or juveniles entered the crumina.

The following speculative account shows one possible evolutionary pathway to beyrichioid brood care. Suppose that in the Lower Ordovician, eurychilinids laid their eggs freely on the sea-floor, but either stayed near them or returned to them periodically to keep them from being buried or perhaps to guard them from predators (some modern cypridoideans may do the same – see above). Then, in certain environments, it became advantageous for the hatched juveniles to enjoy the same protection, and for the adults to develop ventrally-directed frills which would shield the young beneath, rather like a bird sitting on a clutch of eggs. Note that these frills are found in both sexes (brood care does not have to be a female prerogative, nor is it in birds), but evolved into sexually dimorphic features. The latter can occur by sexual selection (see the first part of this chapter). The development of a dolon in the piretellids would then improve the protection of the brood and allow the adult to move from one place to another.

This speculative interpretation shows how the brood pouch could have developed in the anterior position, since the juveniles could move into it of their own accord once they had hatched. It is also possible, however, that the eggs were first actively transferred to the dolon (outside the carapace) or crumina (inside the carapace – an altogether safer route) by the thoracic appendages of the mother. We know that such manipulation is possible from our observations of living adult female *Darwinula stevensoni*, which can use its legs to select and eject moulted valves from the brood space (Horne *et al.* in press, Fig. 10.12). We might regard a posterior brood chamber as a more efficient means of brood care, as adopted by several ostracod lineages, but the beyrichioids apparently did not have this option. Evolution by natural selection does not create 'fully-formed' major new features, but must act to modify and improve the efficiency of existing structures. If we accept the assumption (almost certainly unprovable) that certain early palaeocopids cared for the eggs they had laid (external brood care), then the evolutionary pathway frills-dolons-cruminae is seen as a progressive adaptive change, working with 'available materials' towards the internalisation of brood care of eggs and juveniles, probably in response to increasing environmental pressures such as lower oxygen levels or more effective predators.

This hypothesis also satisfies the observation that while posterior-domiciliar brood care (the more 'logical' position) appears to have evolved separately in more than one cytheroidean lineage (e.g., Xestoleberididae, Timiriaseviinae, Cytherideidae), anterior-domiciliar (i.e., cruminal) brood care has only evolved once. An objection to the hypothesis is that the features we call cruminae may actually have evolved in different ways in different beyrichiacean lineages (Schallreuter 1993). Some authors have suggested that the closure of the dolon and secondary appearance of an internal aperture is only one way of producing a crumina: Schallreuter (1989) cited Martinsson's (1962) explanation of cruminal origin by the expansion of tubules in the velum (i.e., a separate evolutionary event, not a continuation of dolon development) and suggested a third possibility, the inward migration of the antral opening (i.e., the dolon) across the valve margin (see also discussion by Adamczak 1990). Schallreuter's model was based on observations of the morphology of the earliest known cruminate ostracod, *Fallaticella schaeferi* (Upper Ordovician). A problem with Martinsson's model is the lack of an evident function for an expansion of the tubules in the first place, since the crumina would have to be relatively large to become functional as a brood chamber. In other words, of what use is a half-formed crumina? The frill-dolon-crumina hypothesis seems to be the more robust one, since it allows selection to produce incremental adaptation, while the tubule-expansion hypothesis requires either an improbably large evolutionary 'jump' or an as yet undetermined preadaptation for purposes other than brood care.

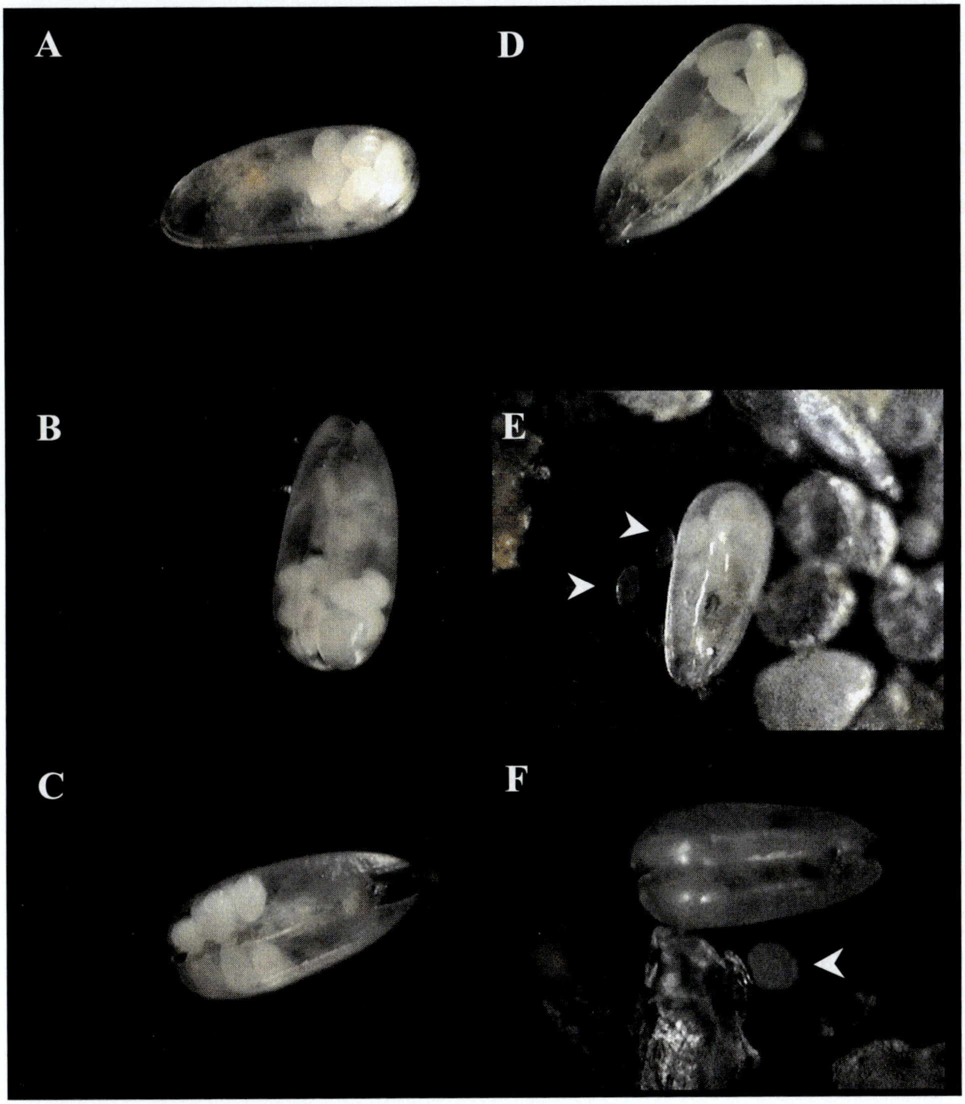

Fig. 10.12 Brood care in *Darwinula stevensoni*
A= lateral view (left side) of an adult female with ten eggs (five or six can be seen) in the brood pouch. B= oblique ventral view of an adult female with ten eggs in the brood pouch. C= ventral view of an adult female with eggs and juvenile instars in the brood pouch. D= dorsal view of an adult female with six juveniles in the brood pouch. E= an adult female just after ejecting two moulted juvenile valves (arrowed) from the brood pouch by lying on her back and kicking them out with her thoracic legs; the adult has just turned over and is seen in lateral view. F= an adult female which has just turned over (seen now in dorsal view) after ejecting a dead, partially-developed embryo (arrowed) from the brood pouch. (Adult females approx. 0.7 mm long. All from original high-resolution video recordings by David Horne and Friederike Mösslacher.)

Several groups of Palaeozoic ostracods exhibit a sexual dimorphism of the posterior part of the carapace, referred to as posterior domiciliar dimorphism (Guber & Jaanusson 1964) or domatial dimorphism (Lundin *et al.* 1995). According to the latter authors, domatial dimorphism takes the form of a special posterior domiciliar space for egg/brood care in females, and is found in extant ostracods such as cytherellids and the podocopid *Cyprideis* lineage as well as in extinct Palaeozoic groups such as kloedenellids, monotiopleurids and leperditellids. As Lundin *et al.* (*loc. cit.*) pointed out, the term 'domatial dimorphism' was coined by Jaanusson (1985) who distinguished between a domatial space for egg/brood care (delimited by a thickening, folding or internal loculation of the shell), as in, for example, the platycopid genus *Cytherelloidea*, and a non-domatial space for egg/brood care, as in, for example, the podocopid genus *Cyprideis*. Lundin and his colleagues expressed reservations about this view but accepted it for the purposes of their paper. We do not regard this distinction as a useful one. Certainly, in many platycopids, a thickening of the shell forms an interior ridge, or limen, which apparently separates the domatium from the rest of the domicilium. In the genus *Cytherelloidea* the eggs are housed in internal loculi, dish-shaped depressions in the wall of the domatium. But in many species of *Cytherella* there appears to be no trace of a limen, yet all are believed to brood their eggs. Moreover, a domatium delineated by a limen is known in at least one member of the podocopid *Cyprideis* lineage, *Mesocyprideis irsacae*, although most of its close relatives (including well-known brooders such as *Cyprideis torosa*) lack such a feature. Therefore, Jaanusson's division of domiciliar dimorphism into domatial and non-domatial types seems to be not very significant. Of course, this in no way invalidates Lundin *et al.*'s recognition of brood care in leperditellids and monotiopleurids on the basis of the presence of a limen or loculi. The danger remains that in groups which have no living representatives for direct comparison, we may mistakenly equate posterior domiciliar dimorphism with brood care, when in fact the posteriorly inflated dimorph could just as easily have been the male of the species (a problem that Lundin *et al.* were clearly aware of). Many podocopid genera exhibit a striking sexual dimorphism in which the male carapace is larger and more inflated posteriorly than that of the female, in order to accommodate the relatively large copulatory appendages (for example, in the marine/ brackish genus *Cytherura*). As with the beyrichioids, we can never be entirely certain of brood care unless we actually find examples of juveniles in the supposed brood space, as Adamczak (1991) did in the kloedenellid *Poloniella tertia* (another extinct group). Morphology alone is a pointer to, but not proof of, brood care. In any case, the true biological distinction between Jaanusson's 'domatial' platycopid brood care and 'non-domatial' *Cyprideis*-type brood care is that the former brood only eggs, which are glued to the inside of the adult female's domatium, while the latter carry loose eggs and also the first two or three juvenile moult stages. Perhaps the term *domatium* should be applied to a space in the posterior domicilium within which eggs are attached (as in platycopids), while the morphologically-similar space in brooding podocopids which houses loose eggs and juveniles could be referred to as a *marsupium*, a term coined by Athersuch (1976) for the brood space of the genus *Xestoleberis*. In many fossil cases, however, it would be impossible to decide which of the terms was appropriate.

The evolutionary significance of brood care in ostracods

Brood care may be regarded as a mechanism to improve the survival chances of eggs and/or juveniles in an unstable environment, that is to say one in which conditions become temporarily adverse (Van Harten 1990), or to protect them from predators. Eggs and juveniles in the brood space exist in a microenvironment within which significant parameters are maintained within survivable limits by the actions of the adults, while ambient conditions outside may exceed those limits (e.g., salinity, dissolved oxygen concentration). When subjected to low oxygen levels, adult females of *Metacypris cordata* are known to increase the rate at which they beat their branchial plates, thus ensuring that eggs in the brood space are adequately ventilated (Danielopol *et al.* 1996). Horne & Rosenfeld (in Jarvis *et al.* 1988), considering why platycopids apparently survived the so-called Cenomanian-Turonian Oceanic Anoxic Event while many contemporary podocopids did not, suggested that the former group's habit of brood care gave them the advantage in dysaerobic conditions.

Van Harten (1990) suggested that brood care 'evolved as an adaptive response to specific ecological problems posed by the environment, and then turned out to be an extremely advantageous preadaptation to transport by air'. For species such as *Cyprideis torosa* and *Metacypris cordata* which inhabit the shallow margins of water bodies, brood care facilitates effective dispersal by intermediaries such as wading birds and waterfowl (McKenzie & Hussainy 1968, Sandberg & Plusquellec 1974, Van Harten 1990, Danielopol *et al.* 1996). However, if widespread dispersal of a brooding, sexually reproducing species is to succeed as an evolutionary strategy, genetic mixing must be maintained. Simply transporting a single adult female containing eggs or juveniles of both sexes may be enough to start a population, but the chances of continued population growth would be enhanced if greater genetic variability could be maintained by mating between individuals that were not all descended from a single parent. This means that anything that improves the chances of many brooding females being transported and deposited together, particularly if by doing so different populations are mixed, is likely to be advantageous. High population densities, such as are characteristic of both migrating flocks of birds and ostracods such as *Cyprideis*, would achieve this. However, most non-marine brooders occur in relatively stable (permanent) lacustrine situations, where the selective advantage could be related to the limitation of fecundity in a highly competitive environment (see chapter 17). As discussed earlier, energy economy is a significant selection currency (Diamond 1993), particularly in habitats with increased competition. Limiting fecundity saves energy, but requires the survival chances of the limited offspring to be optimised, something that can be achieved with brood care.

Brooding and reproductive mode, brood selection

We stated at the beginning of this section that the excellent fossil record of the Ostracoda offers a unique opportunity to appraise the suggestion that brooding lineages may be pre-adapted to maintenance of parthenogenetic strains. In fact, this idea seems to obtain little support from the Ostracoda, since the majority of parthenogenetic forms (mostly non-marine Cypridoidea) are not brooders and many brooding genera are bisexual (e.g., *Cyprideis*, *Xestoleberis*, *Tanella*, *Metacypris*,

Cytherella). Moreover, the majority of parthenogenetic lineages in the Cypridoidea are considered to be relatively young and belong to taxa which also exhibit sexual reproduction, albeit in geographically restricted areas (chapter 5). In persistent brooding lineages, there is no evidence of repeated origin of parthenogenetic strains, although in the Timiriaseviinae parthenogenesis has arisen once in one genus (the Tertiary-Recent *Kovalevskiella*) (Carbonel *et al.* 1986) and possibly in one species of *Frambocythere* at the Cretaceous-Tertiary boundary (Tambareau *et al.* 1997). Persistent and obligate parthenogenesis may be confined to the Darwinuloidea, a group which apparently abandoned sex some time in the Mesozoic (but see discussion in Chapters 6 and 17). *Darwinula stevensoni* is an ostracod with brood care which is normally found in the shallower parts of permanent freshwater lakes living on sandy or muddy substrates rich in organic detritus. It often attains extremely high population densities (e.g., 160,000 individuals / m^2 recorded in an oligotrophic lake in Finland by Ranta 1979), and has survived for at least 20 Myr (Straub 1952). One possible way for such an ancient asexual to have survived for so long in spite of all the theoretical objections against longevity of asexual lineages (chapter 1) may be by 'brood selection', suggested as a theoretical possibility by Lively & Johnson (1994 – *if brooding species can replace or selectively abort defective offspring, they can increase the mean fitness of the brood).* Such brood selection is known to be at least physically and behaviourally possible in *Darwinula stevensoni* (Horne *et al.* in press, Fig. 10.12).

General conclusion

Sexual behaviour of ostracods is diverse. Both sexes play active roles during the mating process. The male's activity consists of getting access to the female genitalia in order to transmit the sperm. The female has the capacity to recognize conspecific males and further on to select them and/or their inseminated spermatozoa. The complex sexual behaviour of various ostracod groups matches the complexity and diversity of the morphological structures used during the successive stages of the mating process. Especially the male's copulatory module (with special emphasis on the hemipenes) of the benthic podocopids is very developed in those ostracod lineages where signalling behaviour through direct contact to female could evolve. For the marine planctonic myodocopids it appears that the sensorial organs used for signalling and recognition at distance developed strongly. The morphology of the female genitalia, at least in the case of the benthic podocopids, displays a rather complicated structure and gamete selection (for fit spermatozoa) can occur within the female during the post-mating phase.

Various processes can be envisaged for the evolution of the diverse male copulatory organ of the ostracods. Evolutionary scenarios based on sexual selection can amply explain mechanistically most of the data pertaining to the diversity of the male's dimorphic morphological traits. In this respect, it offers a clear answer to the question asked by Butlin & Menozzi (in press) and mentioned at the beginning of this chapter, *viz.* which process caused the unusual diversity and complexity of the copulatory organs of the ostracods. Going deeper into the details of the general process of sexual selection, we noted that especially the female choice hypothesis, in its modern variant, Eberhard's *cryptic choice model* (Eberhard 1996), provides an elegant explanation not only for the complex structure of male genital traits, but also for some of the secondary sexual characters of the ostracod species. There are further strong arguments for the active role played by the combined action of natural and sexual selection during the development of male copulatory module as a whole. Following the ideas of the organismic evolvability developed by Gerhart & Kirschner (1997) one could, possibly, envisage a clade selection process for the development of whole copulatory modules. Finally, there is evidence that some of the male dimorphic traits are due to stochastic events and/or to organismal developmental constraints. The maintenance of such traits in the organism during long periods of time could theoretically be due to a stabilising selection process on these traits, as discussed by Ridley (1996) and/or Waxman & Peck (1998).

Brood care (seven different types were identified) has arisen independently in several marine and non-marine ostracod lineages and diverse carapace shapes acting as brood pouches are documented. The ability of the female ostracods to brood eggs and/or juveniles within the carapace may serve to protect them from adverse environmental conditions and predators. Brood care may be advantageous for the dispersal of ostracods such as Timiriaseviinae, which might not have desiccation-resistant eggs. The variety of brood care solutions and carapace brood pouches which evolved within unrelated ostracod groups is one of the most remarkable peculiarities of the reproductive behaviour of the Ostracoda. This contrasts with the restrained number of solutions developed within the other main Crustacea lineages (see Schram 1986). Brood selection, previously suggested as a theoretical aid to long-term survival in a parthenogenetic lineage, has now been shown to be possible in an ancient asexual *Darwinula stevensoni.*

References

Abe, K. & Vannier, J. 1991. Mating behaviour in the podocopid ostracode *Bicornucythere bisanensis* (Okubo, 1975): rotation of a female by a male with asymetric fifth limbs. *J. Crustacean Biol.*, **11:** 250-260.

Adamczak, F.J. 1990. The crumina in *Craspedobolbina* Kummerow 1924 (Palaeocopa, Ostracoda). *Cour. Forsch. Inst. Senckenberg*, **123:** 303-314.

Adamczak, F.J. 1991. Kloedenellids: morphology and relation to non-myodocopid ostracods. *J. Paleontol.*, **65:** 255-267.

Andersson, M. 1994. *Sexual selection*, Princeton Univ. Press: 599 pp.

Athersuch, J. 1976. The genus *Xestoleberis* (Crustacea: Ostracoda) with particular reference to Recent Mediterranean species. *Pubbl. Staz. Zool. Napoli*, **40:** 282-243.

Athersuch, J. & Bonaduce, G. 1976. *Loculicytheretta pavonia* (Brady). *Pubbl. Staz. Zool. Napoli*, **40:** 349-364.

Bell, G. 1997. *The basics of selection,* Chapman & Hall: 378 pp.

Bonaduce, G. & Danielopol, D.L. 1988. To see and not to be seen: The evolutionary problems of the Ostracoda Xestoleberididae. *in*: Hanai, T. (ed), *Evolutionary biology of Ostracoda, its fundamentals and applications,* Elsevier Kodansha: 375-396.

Boxshall, G.A. (ed) 1998. Mating biology of copepod crustaceans. *Phil. Trans. R. soc. Lond. B.* **353**: 667-815.

Broodbakker, N.W. 1982. The genus *Heterocypris* (Crustacea, Ostracoda) in the West Indies. Part. I. Taxonomic characters. *Bijdr. Dierk.*, **52:** 207-227.

Broodbakker, N.W. 1983. The genus *Strandesia* and other Cypricercini (Crustacea, Ostracoda) in the West Indies. Part. I. Taxonomy. *Bijdr. Dierk.*, **53:** 327-368.

Broodbakker, N. & Danielopol, D.L. 1982. The chaetotaxy of Cypridacea (Crustacea, Ostracoda) limbs: proposals for a descriptive model. *Bijdr. Dierk.*, **52:** 103-120.
Butlin, R.K., Schön, I. & Martens, K. (in press). Asexual reproduction in non-marine ostracods. *Heredity.*
Butlin, R. & Menozzi, P. (in press). Open questions in evolutionary ecology: Do ostracods have the answers? *in*: Horne, D.J. & Martens, K. (eds), Proc. 13 Int. Symp. Ostracodologists. *Hydrobiologia.*
Carbonel, P., Colin, J.-P., Danielopol, D.L. & Londeix, L. 1986. *Kovalevskiella* (Ostracoda, Timiriaseviinae), genre à mode de vie benthique depuis l'Oligocène, son adaptation à la vie interstitielle. *Géobios*, **19:** 677-687.
Cohen, A.C. & Morin, J.G. 1990. Patterns of reproduction in ostracodes: a review. *J. Crustacean Biol.*, **10:** 184-211.
Colin, J.-P. & Carbonel, P. 1990. Phylogenetic affinities of *Cytherissa* with other Cytherideinae (*Vernoniella, Fabanella, Neocyprideis, Cyprideis*); a paleontological approach. *Bull. Inst. Geol. Bassin d'Aquitaine, Bordeaux*, **47**:83-95.
Danielopol, D.L. 1969. Recherches sur la morphologie de l'organe copulateur mâle chez quelques Ostracodes du genre *Candona* (Fam. Cyprididae Baird). *in*: J.W. Neale (ed), *The taxonomy, morphology and ecology of Recent Ostracoda,* Oliver & Boyd: 135-153.
Danielopol, D.L. 1972. Supplementary data on the morphology of *Neonesidea* and remarks on the systematic position of the family Bairdiidae (Ostracoda, Podocopida). *Proc. Biol. Soc. Washington*, **85:** 39-48.
Danielopol, D.L. 1975. Quelques remarques sur la diversification morphologique de trois nouvelles espèces d'Elpidium (Ostracoda) à Cuba. *in*: Swain, F. (ed), *Biology and paleontology of Ostracoda, Bull. Amer. Paleont.*, **65:** 47- 60.
Danielopol, D.L. 1976. Supplementary data on *Pusella botosaneanui* Dan. (Ostracoda, Bairdiidae). *Vie Milieu*, **26:** 261-273.
Danielopol, D.L. 1977. Recherches sur les Ostracodes Entocytheridae. Données sur *Sphaeromicola cebennica juberthiei* nov. ssp. et *Sphaeromicola cirolanae* Rioja. *Int. J. Speleol.*, **9:** 21-41.
Danielopol, D.L. 1978. Über Herkunft und Morphologie der süßwasserhypogäischen Candoninae (Crustacea, Ostracoda). *Sitz. Ber. Österr. Akad. Wiss., Math.-Nat.* Kl. Abt. I, **187:** 1-162.
Danielopol, D.L. 1980a. Sur la biologie de quelques Ostracodes Candoninae épigés et hypogés d'Europe. *Bull. Mus. natn. Hist. nat., Paris*, 4e ser, **2A:** 471-506.
Danielopol, D.L. 1980b. On the carapace shape of some european freshwater interstitial Candoninae (Ostracoda). *Proc. Biol. Soc. Wash.*, **93:** 743-756.
Danielopol, D.L. 1985. Stygobiont ostracods from inland subterranean waters. *in*: Botosaneanu, L. (ed), *Stygofauna Mundi*, Brill: 265-274.
Danielopol, D.L. & Cvetkov, L. 1979. Trois nouvelles espèces du genre *Mixtacandona* (Ostracoda, Cyprididae, Candoninae). *Hydrobiologia*, **67:** 249-266.
Danielopol, D.L. & Horne, D.J. 1996. Some observations on sexual behaviour and the morphology of the male copulatory appendages in *Metacypris cordata* (Ostracoda, Timiriaseviinae). *in:* Keen, M.C. (ed), *Proc. 2nd Europ. Ostracodologists Meeting Brit. micropal. Soc.*: 171-174.
Danielopol, D.L. Horne, D.J. & Wood, R.N. 1996. Notes on the ecology of *Metacypris cordata* (Ostracoda, Timiriaseviinae); why does it not colonise groundwater habitats? *Ibid*: 175-180.
Danielopol, D.L., Martens, K. & Casale, L. 1990. Revision of the genus *Leucocythere* Kaufmann, 1892 (Crustacea, Ostracoda, Limnocytheridae), with the description of a new species and two new tribes. *Bull. k. belg. Inst. Natuurw., Biol.*, **59** (1989)**:** 63-94.
Danielopol, D.L. & Tétart, J. 1990. Morphology of *Cytherissa* and *Cyprideis*: (Ostracoda): supplementary data on the appendages and the caryotype. *in*: Danielopol, D.L., Carbonel, P. & Colin, J.P., (eds), *Cytherissa* (Ostracoda), the Drosophila of Paleolimnology. *Bull. Inst. Geol. Bassin d'Aquitaine*, **47-48:** 55-68.
Darwin, C.R. 1859. *On the origin of species*. J. Murray.
Darwin, C.R. 1871. *The descent of man and selection in relation to sex*. J. Murray.
Diamond, J. 1993. Evolutionary physiology. *in*: Boyd, C.A.R. & Noble, D. (eds), *The logic of life. The challenge of integrative physiology,* Oxford Univ. Press: 89-111.
Eberhard, W.G. 1985. *Sexual selection and animal genitalia,* Harvard Univ. Press: 244 pp.
Eberhard, W.G. 1996. *Female control: sexual selection by cryptic female choice,* Princeton Univ. Press: 501 pp.

Elofson, O. 1941. Zur Kenntnis der marinen Ostracoden Schwedens mit besonderer Berücksichtigung des Skagerraks. *Zool. Bidr. Upps.*, **19:** 217-534, figs 1-52. (Translated from German into English, Israel Program for Scientific Translation, Jerusalem, 1969).

Fox, H.M. 1966. Ostracods from the environs of Pallanza. *Mem. Ist. Ital. Idrobiol.*, **20:** 25-39.

Gerhart, J. & Kirschner, M. 1997, *Cells, embryos and evolution*, Blackwell: 642 pp.

Gonzalez Mozo, M.E., Martens, K. & Baltanás, A. 1996. A taxonomic revision of european *Herpetocypris* Brady and Norman, 1889 (Crustacea, Ostracoda). *Bull. k. belg. Inst. Natuurw., Biol.*, **66:** 93-132.

Gould, S.J. & Lewontin, R.C. 1979. The spandrels of San marco and the panglossian paradigm: a critique of the adaptationist program. *Proc. R. Soc. B*, **205:** 581-598.

Gould, S.J. & Vrba, E.S. 1982. Exaptation – a missing term in the science of form. *Paleobiology*, **8:** 4-15.

Guber, A.L. & Jaanusson, V. 1964. Ordovician ostracodes with posterior domiciliar dimorphism. *Bull. Geol. Inst. Univ. Uppsala*, **43:** 1-43, pls I-VI.

Hart, C.W. & Hart, D.G. 1969. The functional morphology of entocyterid ostracod copulatory appendages, with a discussion of possible homologues in other ostracods. *in*: Neale, J.W. (ed), *The taxonomy, morphology and ecology of Recent Ostracoda*, Oliver & Boyd: 154-167.

Hart, D.G. & Hart, C.W. 1974. *The ostracod family Entocytheridae*. Monogr. **18**, Acad. Nat. Sci. of Philadelphia: 239 pp.

Harten, D. Van, 1990. The Neogene evolutionary radiation in *Cyprideis* Jones (Ostracoda: Cytheracea) in the Mediterranean area and the Paratethys. *Cour. Forsch.-Inst. Senckenberg*, **123:**191-198.

Hartmann, G. 1968. *Ostracoda,* Dr. H.G. Bronns Klassen und Ordnungen des Tierreichs, **5** Arthropoda, 1. Abt. Crustacea: 409-568.

Hartmann, G. 1975. *Ostracoda,* Dr. H.G. Bronns Klassen und Ordnungen des Tierreichs, **5** Arthropoda, 1. Abt. Crustacea: 569-786.

Hartmann, G. & Guillaume, M.-C. 1996. Classe des Ostracodes (Ostracoda Latreille, 1802). *in*: P.-P. Grassé (ed), Traité de Zoologie, **7** – Crustacés, F. 2, Masson, Paris: 755 – 839.

Hessland, I. 1949. Investigations of the Lower Ordovician of the Siljan District, Sweden, I, Lower Ordovician ostracodes of the Siljan District, Sweden. *Bull. Geol. Inst. Upps.,* **33**: 97-408.

Horne, D.J., Martens, K. & Mösslacher, F. (in press). A short note: Is there brooding selection in *Darwinula stevensoni*? *in*: Crasquin, S., Braccini, E. & Lethiers, F. (eds), Proceedings of the 3rd European Ostracodologists Meeting. *Bull. Centr. Recherch. Explor.-Prod. Elf-Aquitaine.*

Jaanusson, V. 1985. Functional morphology of the shell in platycope ostracodes – a study of arrested evolution. *Lethaia*, **18:** 3-84.

Janz, H. 1992. Eine fakultative Beziehung zwischen *Cypria ophthalmica* (Jurine) (Ostracoda) und *Gyraulus crista* (L.)(Gastropoda) und ihre mögliche biologische Bedeutung. *Stuttgarter Beitr. Naturk.*, Ser. A, **476:**1-11.

Jarvis, I., Carson, G., Cooper, M.K.E., Hart, M.B., Leary, P.N., Tocher, B.A., Horne, D.J. & Rosenfeld, A. 1988. Microfossil assemblages and the Cenomanian-Turonian (late Cretaceous) oceanic anoxic event. *Cret. Res.*, **9:** 1-103.

Kamiya, T. 1988. Morphological and ethological adaptations of Ostracoda to microhabitats in *Zostera* beds. *in*: Hanai, T. *et al.* (eds), *Evolutionary biology of Ostracoda, its fundamentals and applications,* Elsevier Kodansha: 303-318.

Kesling, R.V. 1952. A study of *Ctenoloculina cicatricosa* (Warthin). *Contr. Mus. Paleontol. Univ. Michigan*, **9:** 290.

Kesling, R.V. 1957. Origin of Beyrichiid ostracods. *Contr. Mus. Paleontol. Univ. Michigan*, **14:** 7-80.

Kesling, R.V. 1965. Anatomy and dimorphism of adult *Candona suburbana* Hoff. *in*: Kesling, R.V., Darby, D.G., Smith, R.N. & Hall, D.D. (eds), *Four reports of ostracod investigations. Mus. Paleont. Univ. Michigan,* Ann Arbor: 1-56.

Klie, W. 1938. Ostracoda, Muschelkrebse. *in*: Dahl, F. (ed), *Die Tierwelt Deutschlands und der angrenzenden Meeresteile*, Gustav Fischer: 230 pp.

Kummerow, E. 1933. Zur Palaeobiologie der Ostracoden und Trilobiten. *Cbl. Mineral. Geol., Palaeontol.*, (1933) B., **1**: 42-53

Lively, C.M. & Johnson, S.G. 1994. Brooding and the evolution of parthenogenesis: strategy models and evidence from aquatic invertebrates. *Proc. R. Soc. Lond.* B, **256:** 89-95.

Löffler, H. & Danielopol, D.L. 1978. Ostracoda. *in*: Illies, J. (ed), *Limnofauna Europaea,* Gustav Fischer:196-208.

Lundin, R.F., Williams, M. & Siveter, D.J. 1995. Domatial dimorphism occurs in leperditellid and monotiopleurid ostracodes. *J. Paleont.*, **69:** 886-896.
McGregor, D.L. & Kesling, R.V. 1969. Copulatory adaptations in ostracods. Part 2. adaptations in living ostracods. *Contr. Mus. Paleont. Michigan*, **22:** 221-231.
McKenzie, K.G. & Hussainy, S.U. 1968. Relevance of a freshwater cytherid (Crustacea, ostracoda) to the continental drift hypothesis. *Nature*, **220:** 806-808.
Martens, K. 1984. On the freshwater ostracods (Crustacea, Ostracoda) of the Sudan, with special reference to the Red Sea Hills, including a description of a new species. *Hydrobiologia* **110**: 137-161.
Martens, K. 1986. Taxonomic revision of the subfamily Megalocypridinae Rome, 1965 (Crustacea, Ostracoda). *Verh. konink. Acad. Wetensch. Lett. Schone K., Kl. Wetensch.*, **48 (174):** 81 pp., 64 figs.
Martens, K. 1987. Homology and functional morphology of the sexual dimorphism in the antenna of *Sclerocypris* Sars, 1924. (Crustacea, Ostracoda, Megalocypridinae). *Bijdr. Dierk.*, **57:** 183-190.
Martens, K. 1988 Seven new species and two new subspecies of *Sclerocypris* Sars, 1924 from Africa, with new records of some other Megalocypridinids (Crustacea, Ostracoda). *Hydrobiologia* **162**: 243-273.
Martens, K. 1990a. Revision of African *Limnocythere* s.s. Brady, 1867 (Crustacea, Ostracoda) with special reference to the Eastern Rift Valley Lakes: morphology, taxonomy, evolution and (palaeo-) ecology. *Arch. Hydrobiol.*, Suppl. **83:** 453-524.
Martens, K. 1990b. Revision of African Cypridini. I. The genera *Cypris* O.F. Müller, *Pseudocypris* Daday and *Globocypris* Klie. *Bull. k. belg. Inst. Natuurw., Biol.*, **60**: 127-172.
Martens, K. 1991. A new and enigmatic representative of *Leucocythere* Kaufmann, 1892 from the Eastern Cape province (South Africa) (Crustacea, Ostracoda). *J. nat. Hist.*, **25:** 583-596.
Martens, K. 1992. Taxonomic revision of African Cypridini. Part II. Description of *Ramotha* gen.nov. *Ann. s. afr. Mus.* **102**: 91-130.
Martens, K. 1994. Summary of the morphology, taxonomy and the distribution of *Limnocythere inopinata* (Baird, 1843) (Ostracoda, Limnocytherinae). *in*: Horne, D.J. & Martens, K. (eds), *The evolutionary ecology of reproductive modes in non-marine Ostracoda*, Greenwich Univ. Press: 17-22.
Martens, K. 1996a. On *Korannacythere* gen. nov. (Crustacea, Ostracoda), a new genus of temporary pool limnocytherids from southern Africa, with the description of three new species and a generic reassessment of the Limnocytherinae. *Bull. k. belg. Inst. Natuurw., Biol.*, **66:** 51-72.
Martens K. 1996b. On *Heterocypris reptans* (Kaufmann, 1900) (Crustacea, Ostracoda), a new species for the fauna of Israel. *Isr. J. Zool.* **42**: 287-291.
Martens, K. (in press). On the origin and development of different recognition systems in the Limnocytherinae (Crustacea, Ostracoda, Cytherocopina): chance or selection? *in*: Horne, D.J. & Martens, K. (eds), Proc. 13 Int. Symp. Ostracodologists. *Hydrobiologia*.
Martinsson, A. 1962. Ostracodes of the family Beyrichiidae from the Silurian of Gotland. *Publ. pal. Inst. Univ. Upps.*, **41**: 369 pp.
Matzke-Karasz, R. 1997. Descriptive nomenclature and external morphology of the Zenker's organs of Cypridoidea (Crustacea, Ostracoda). *Sonderveröffentl. Geol. Inst. Univ. Köln*, **114** (Festschrift Eugen K. Kempf): 295-315.
Morin, J.G. & Cohen, A.C. 1991. Bioluminescent displays, courtship and reproduction in ostracodes. *in*: Bauer, R.T. & Martin, J.W. (eds), *Crustacean sexual biology*, Columbia Univ. Press: 1-16.
Parker, A. 1995. Discovery of functional iridescence and its coevolution with eyes in the phylogeny of Ostracoda (Crustacea). *Proc. R. Soc. B*, **262:** 349-355.
Parker, A. 1997. Mating in Myodocopina (Crustacea, Ostracoda): results from video recordings of a highly iridescent cypridinid. *J. mar. biol. Ass. U.K.*, **77:** 1223-1226.
Paterson, H.E.H. 1985. The recognition concept of species. *in*: Vrba, E.S. (ed), *Species and speciation.* Transvaal Museum Monograph, Pretoria, **4:** 21-29.
Poulsen, E.M. 1973. Ostracoda-Myodocopa. Part. 3B Halocypriformes-Halocyprididae Conchoecinae. *Dana-Report*, **84:** 1-224.
Raff, R.A. 1996. *The shape of life. Genes, development, and the evolution of animal form*, Chicago Univ. Press: 520 pp.
Ranta, E. 1979. Population biology of *Darwinula stevensoni* (Crustacea, Ostracoda) in an oligotrophic lake. *Ann. Zool. Fenn.*, **16:** 28-35.
Ridley, M. 1996. *Evolution* (2nd Edition), Blackwell: 719 pp.

Rogulj, B., Danielopol, D.L., Marmonier, P. & Pospisil, P. 1993. Adaptive morphology, biogeographical distribution and ecology of the species group *Mixtacandona hvarensis* (Ostracoda, Candoninae). *Mem. Biospeleol.*, **20:** 195-207.

Sandberg, P.A. & Plusquellec, P.L. 1974. Notes on the anatomy and passive dispersal of *Cyprideis* (Cytheracea, Ostracoda). *in*: Van den Bold, W.A. (ed), *Geoscience and Man, Volume VI, Ostracoda. The Henry V. Howe memorial volume*, 1-26, School of Geoscience, Louisiana State University, Baton Rouge.

Sarv, L.I. 1966. Sexual dimorphism in Early Palaeozoic Ostracods. *in*: Vyalov, O.S. (ed), *Fossil Ostracoda* (papers from the First All-Union Symposium on Fossil Ostracoda, Lvov, 1963), Izdatel'stvo 'Naukova Dumka, Kiev, 1966. (English translation, Israel Program for Scientific Translations, Jerusalem, 1971): 11-23.

Schallreuter, R.E.L. 1982. On *Bilobatia serralobata* Schallreuter, *Stereo-Atlas Ostracod Shells*, **9:** 9-16.

Schallreuter, R.E.L. 1989. On *Fallaticella schaeferi* Schallreuter. *Stereo-Atlas Ostracod Shells*, **16:** 25-28.

Schallreuter, R.E.L. 1993. The polyphyletic origin of cruminate ostracods. *in*: McKenzie, K.G. & Jones, P.J. (eds), *Ostracoda in the Earth and Life Sciences*, Balkema: 13-21.

Schram, F.R. 1986. *Crustacea*, Oxford Univ. Press: 606 pp.

Schmidt, E.A. 1941. Studien im böhmischen Caradoc (Zahoran-Stufe). 1. Ostrakoden aus den Bohdalec-Schichten und über die Taxonomie der Beyrichiacea. *Abh. Senckenb. Naturf. Gesell.*, **454**: 1-96.

Spjeldnaes, N. 1951. Ontogeny of *Beyrichia jonesi* Boll. *J. Paleontol.*, **25:** 745-755.

Straub, E.B. 1952. Mikropaläontologische Untersuchungen im Tertiär zwischen Ehingen und Ulm a.d. Donau. *Geol. Jb.*, **66:** 433-523.

Tambareau, Y., Hottinger, L., Rodriguez-Lazaro, J., Villatte, J., Babinot, J.-F., Colin, J.-P., Garcia-Zarraga, E., Rocchia, R. & Guerrero, N. 1997. Communautés fossiles benthiques aux alentours de la limite Crétacé/Tertiaire dans les Pyrénées. *Bull. géol. Soc. France*, **168:** 795-804.

Triebel, E. 1941. Zur Morphologie und Ökologie der fossilen Ostracoden. *Senckenbergiana*, **23**: 294-400.

Wagner, G. 1998. Complexity matters. *Science*, **279:**1158-1159.

Waxman, D. & Peck, J.R. 1998. Pleiotropy and the preservation of perfection. *Science*, **279:** 1210-1213.

Wingstrand, K.G. 1988. Comparative spermatology of the Crustacean Entomostraca. 2. Subclass Ostracoda. *Biol. Skr. Kong. Dansk. Vidensk. Selsk.*, **32:** 1-149.

Yin, Y. 1997. Contribution to the morphology and ecology of Ostracoda Limnocytheridae and Candonidae. Comparative studies between Austrian and Chinese Crustacea. PhD thesis, Univ. Vienna, 163 pp.

CHAPTER 11
REPRODUCTIVE MODES AND TAXONOMY

K. Martens, G. Rossetti and A. Baltanás

Abstract

Nomenclature, taxonomy and species concepts of clonal (asexual) lineages have for a long time been hotly debated subjects. We argue that separating the 'asexuals' into different types, facilitates the discussion. In *ancient asexuals*, separate lineages have low intraspecific genetic and morphological variability and clear interspecific genetic and morphological gaps. Their taxonomy is thus (relatively) straightforward, species are conform to morphological, genetic and evolutionary species concepts. No special species concept for these taxa is necessary. In *species with mixed reproduction*, bisexual and asexual populations cluster amongst each other and form an inclusive species, with large intraspecific genetic and morphological variability. Accepting an inclusive approach, such species with mixed reproduction are again conform to morphological, genetic and evolutionary species concepts. Many problems remain, however, with the asexual spin-offs from species with sexual roots, be they advanced mutational lineages or interspecific hybrids. Congeneric lineages are often difficult to identify unambiguously and, for several reasons, most of these clusters must be regarded as evolutionary species complexes for which a separate (theoretical) species concept (e.g. the agamospecies) must be invoked.

Subspecies in clonal taxonomy should be avoided in all categories of asexuals, and this for different reasons. Possible exceptions are identifiable sib clones with special ecological preferences.

Introduction

Taxonomy attempts to classify natural variation and variability, both morphological and genetic (Box 11.1). To this end, taxonomy applies a set of nomenclatorial rules and a number of concepts, relating to the two most important taxonomic levels, *species* and *genera*. These two levels form the basis of all taxonomies and the International Code of Zoological Nomenclature (ICZN) contains rigid rules relating to their application in Zoology. Other, higher levels of classification involve families, orders and phyla and their subdivisions. As an example, the higher classification of the Ostracoda is presented in chapter 3 (Table 3.2).

The present chapter assesses whether similar species concepts are applicable in groups with different reproductive modes. The Ostracoda are a rather unique animal group in that an extensive range of reproductive strategies, from fully sexual, over mixed to fully asexual reproduction occurs. They are thus excellently suited to test whether different reproductive strategies have an effect on morphological and genetic variability and, as a consequence, on the taxonomic interpretation of this variability. These considerations are not only theoretical, they are directly related to practical problems. For example, should sexual and asexual populations of what

Sex and parthenogenesis: evolutionary ecology of reproductive modes in non-marine ostracods, edited by K. Martens, pp. 197–214

could morphologically be considered one and the same species, be allocated different taxonomic ranks, i.e. species or subspecies? Delorme (1971: 55-56) wrote: "Although the similarities between the females of *Limnocythere inopinata* and *L. sappaensis* are overwhelming, it would appear that there is sufficient cause to allow the species *L. sappaensis* to stand and maintain *L. inopinata* in a parthenogenetic state." Krstić (1987) referred a fossil sexual population to a new subspecies, *Limnocythere inopinata pleistocenica*.

Other authors (Dobbin 1941, Petkovski 1959) unequivocally referred their syngamic populations to *L. inopinata*. Also, what species concepts are available to translate extant variability in fully asexual groups into a classification? The present discussion will thus deal with what is arguably one of the most difficult aspects in modern systematics: the concept of clonal taxonomy.

Box 11.1 Taxonomy, Systematics and Classification

Definitions of these fundamental concepts differ according to different authors. Most commonly accepted are those of Mayr & Ashlock (1991), slightly modified here.

TAXONOMY is the theory and practice of classifying organisms. A special part of taxonomy is NOMENCLATURE, which sets down the rules for the vehicular language used in naming organisms.

SYSTEMATICS is the scientific study of the kinds and diversity of organisms and of any relationships among them, applying the taxonomic rules.

CLASSIFICATION is the resulting ordering of entities in groups, whose members share certain (non-homologous) traits (mostly morphological) due to common descent.

Species concepts

Mayden (1997) recognised 25 different species concepts from the vast literature that has been devoted to this fundamental problem (Box 11.2). The ideal species concept should be universally valid, easily applicable and theoretically significant (Hull 1997). However, these three aims tend to conflict with one another. For example, theoretically significant concepts tend not to score very high on applicability; generally valid concepts tend to have less theoretical background. The morphological species concept is probably the most universally applicable one, but lacks theoretical background unless other conceptual aspects (mate recognition, phylogeny,...) are merged with it.

Leaving aside the universality, classifications should reflect natural phylogenies as closely as possible, but must at the same time be workable tools, also for non-specialists. These are the two (often conflicting) tasks of the taxonomists. In the first, systematics finds its own definition as an independent biological science; in the second, it provides vital tools for all other biological disciplines. This intrinsic dichotomy has often been identified (Mayr 1963, 1969 and others) as a main source of confusion. In an attempt to link pattern to process (Chandler & Gromko 1989), both the mode of speciation (*Evolutionary species*) and of reproduction (*Biological species*) have repeatedly been used to define 'species concepts'. In reality, however, 99% of nominal species are defined on morphology only (*Linnean or Morphological species*), and, as material available to the taxonomist generally consists of preserved specimens only, nothing is known about the processes that lead to their origin. Many nominal taxa (species, genera) in, for example, the Ostracoda, are furthermore described from fossils, where no biological tests are possible. It thus appears that only a minority of described taxa has any bearing on process-related (evolutionary) species concepts, and it would thus seem logical that taxonomy should base its decisions primarily on patterns (morphology) and worry less about the underlying processes. We demonstrate, using non-marine ostracods as an example, that this is a misconception and that the mode of reproduction has a significant bearing upon evolutionary processes, most importantly on speciation, so that different modes of reproduction justify the application of different species concepts.

Box 11.2 Species concepts

King (1995) stated that "it is no exaggeration to say that the history of science is littered with discarded species concepts" (1995: 7). Mayden (1997) distinguished no less than 25 different species concepts. The main concepts, relevant to the present discussion, are here briefly described.

Linnean (Morphological) Species Concept: this is an operational concept, where morphological distinctiveness is assumed to be heritable and serves as a surrogate for lineage independence. In the Linnean concept, morphology is assumed to be constant. Later versions accept a degree of intraspecific morphological variability and accept that taxa are characterised by gaps in the variability range.

Biological Species Concepts (termed sexual species concepts by Bell 1997) are applicable only to sexually reproducing, recent taxa. Included are:

- *Isolation Concept:* ...a group of interbreeding natural populations that is reproductively isolated from other such groups (Mayr & Ashlock 1991).
- *Recognition Concept:* ...that most inclusive population of individual, biparental organisms which share a common fertilization system (Paterson 1993).
- *Cohesion Concept:* ...the most inclusive group of organisms having the potential for phenotypic cohesion through intrinsic cohesion mechanisms (Templeton 1989).
- *Agamospecies Concept:* (including microspecies, paraspecies, pseudospecies, semispecies): refers specifically to taxa which do not conduct sexual reproduction; mostly apomictic parthenogens, although they can be part of a species complex in which sexual reproduction exists.

Evolutionary Species Concepts include:

- *Genetic Species Concept:* ...a genetic unit consisting of a large, intercommunicating gene pool (Mayr 1969).
- *Genotypic cluster definition:* ...clusters of monotypic or polytypic biological entities, identified using morphology or genetics, forming groups of individuals that have few or no intermediates when in contact (Mallet 1995).
- *Evolutionary Species Concepts:...* a single lineage of ancestor-descendant populations, which maintains its identity from other such lineages and which has its own evolutionary tendencies and historical fate (Wiley 1978, modified after Simpson 1961).
- *Ecological Species Concept:* ...a lineage (or a closely related set of lineages) which occupies an adaptive zone, minimally different from that of any other lineage in its range and which evolves separately from all lineages outside its range (Van Valen 1976).
- *Phylogenetic Species Concept:* ...the smallest biological entities that are diagnosable and/or monophyletic (Cracraft 1983).
- *Cladistic (Hennigian) Species Concept:* ...reproductively isolated natural populations or groups of natural populations (that) originate via the dissolution of the stem species in a speciation event and cease to exist either through extinction or speciation (Meier & Willmann 1997).
- *Successional (palaeo-, chrono-) Species Concept:* relies on divergence through time as determined from studying fossil assemblages; distinction between successional species is an arbitrary delineation in time or strata based on divergent morphologies or gaps in morphologies or time (Mayden 1997).

Sexual reproduction and taxonomy

Before dealing with the problems of clonal taxonomy, it is important to assess how morphology of sexual taxa is reflected in their taxonomy. The preceding chapter has dealt with the importance of behaviour and mate recognition in the maintenance and evolution of sexually reproducing taxa. We will here assess the importance of variability in the copulatory complex (see chapters 4 and 10) for the taxonomy of these lineages, i.e. does higher male morphological variability bias recognised taxonomic diversity?

It is obvious that (specific) diversity is much higher in Cytheroidea and Cypridoidea, both groups with either sexual or mixed reproduction, when compared to the ancient asexual Darwinuloidea (chapter 3, Fig. 3.4). Do these significantly higher diversities reflect a reality, or are they a bias resulting from the fact that the greater differences in male morphology allow distinction of species, which in asexual taxa would remain cryptic, because all females look alike? One way to test this is to check whether females of species described primarily on male morphology, are indeed indistinguishable.

The taxonomic diversification of the Limnocytherinae is described in chapter 3. It is clear from the descriptions that most specific features (in this case: features used to diagnose species) are in the male morphology, especially in the copulatory complex of the hemipenis. These different morphologies are largely the result of sexual selection, as is illustrated in chapter 10. A detailed analysis of inter- and intraspecific variability in hemipenis morphology in part of this lineage (the African species of *Limnocythere s.s.)* was conducted by Martens (1990a, b) in several taxonomically different populations of the Ethiopian Shala-Awassa basins. As these lakes are relict basins of one large palaeolake, it was assumed that these taxa constitute a monophyletic (or at least paraphyletic) flock. The taxonomy of the flock was primarily determined relying on micro-morphology of male copulatory appendages (Fig. 4.10), where (infrasubspecific) variability within populations was nearly non-existent and inter (sub-)specific differences were constant. Surprisingly, when female valve morphology was re-assessed, clear differences in shape are also found between the different (sub-) species. This shows that species defined on sexual (male) characters are not necessarily a bias of the presence of these males. There are of course situations were females of different species (described on male morphologies) are morphologically indistinguishable, but these are exceptions rather than the rule. Therefore, we accept that higher diversities in Cypridoidea and Cytheroidea, as compared to Darwinuloidea, do reflect real situations.

The validity of the present test hinges upon the distinction between the Darwinuloidea as an ancient asexual group with no sexual reproduction for about 100 Myr (see chapter 3) as compared to asexual taxa clustered around species with a sexual root in Cypridoidea and Cytheroidea. We argue that at some stage sexual reproduction has played a role in the evolution of the latter clusters.

Clonal taxonomy: a brief overview

Introduction

The nomenclature and taxonomy of organisms reproducing in a non-Mendelian way (clonal, vegetative, asexual, etc.) are long-standing problems, which have generated often heated debates and for which various (and sometimes conflicting) solutions have been proposed (review in Foottit 1997). The inability to clarify the taxonomic position of clusters of asexually reproducing populations hampers the present discussions on the evolution of sexual and asexual reproduction and their implications on speciation and biodiversity. Classical papers (e.g. Lynch 1984) repeatedly refer to 'races', clearly avoiding the discussion whether or not bisexual and sexual lineages are conspecific or not. As a consequence, the important term 'hybridisation' becomes also confused (Box 11.3).

A large part of this confusion is due to the fact that 'clonal species' is a polyphyletic concept, which refers to very different types of uniparental organisms. Most important is the degree to which they remain (phylogenetically) associated with bisexuality.

Box 11.3 Sexual and asexual species and populations

Sexual species are species from which only fully sexual populations are known. *Asexual species* are those from which only all-female populations are known (e.g. Darwinulidae). In the species with *mixed reproduction, parthenogenetic, sexual* and *mixed* populations (with both parthenogenetic and sexually reproducing females) occur and it is at all times important to mention which type of population is discussed in such taxa.

Sometimes, sexual populations are discovered of species that were until then deemed fully asexual (e.g. Martens & Meisch (1985) on *Potamocypris villosa*). Sexual populations in such taxa have a geographically restricted distribution (termed *geographical parthenogenesis* by Vandel 1928, see chapter 5), or occur in specific habitats only (for example, the bisexual populations of *Cytherissa lacustris* appear to be restricted to certain long-lived lakes).

In theory, *sexual populations* have a tertiary sex ratio close to 50:50, while *populations with mixed reproduction* have higher female ratios. However, deviations of tertiary sex-ratios in a population can be caused by several factors (e.g. environmental factors causing differential mortality amongst sexes, seasonal differences in male and female occurrences, etc. (see discussion in chapter 2, Abe 1990). Determining the reproductive strategy present in a given population is therefore not always straightforward. Sometimes *rare males* occur. These are isolated male specimens amongst hundreds or thousands of females. It must be stressed that the occurrence of such specimens is not sufficient to consider such populations to have sexual, or even mixed, reproduction.

Clones are reproductive series of female individuals that are genetically identical. Even modern molecular techniques mostly analyse only part of the genome, so that the actual number of clones in a given population is difficult to identify. As we further assume that small genetic differences are not necessarily directly translated into different morphologies, we use the term *sibling clones* or *sib clones* to indicate groups or clusters of clones which all have similar morphologies.

The main difficulties with clonal nomenclature

The following are the main problems with clonal nomenclature cited in the literature. The first three result from rigidly applying one or more of the evolutionary species concepts; the fourth is a problem of recognisability and relates to the morphological species concept.

Clonal lineages can be polyphyletic, as they can originate from multiple hybridisation or mutational events. Every clone resulting from a single hybridisational or mutational event must thus receive specific rank, even if they are genetically and morphologically indistinguishable (Cole 1985).

Most species concepts require that a species has a monophyletic origin. But asexual lineages can have polyphyletic origins. If P_1 is an asexual species which originated from the hybridisation of bisexual species B_1 x B_2, then such (interspecific) hybridisations can occur at different times and in different populations, yet the resulting hybrids will be morphologically indistinguishable, regardless of where and when the hybridisation event took place. Documented evidence for this exists in several fish and lizard groups (Walker 1986, Echelle 1990, Crisp & Weston 1993, Moritz & Heideman 1993). The same is true for mutational clones. If P_1 is the parental species of the mutational species P_2 (in ostracods, for example, causing simple valve reversal – see below), then such mutations can occur at different times in different populations, but again the resulting mutational clones will be morphologically indistinguishable. The probability of such occurrence can be low, but it does exist. Several authors, rigidly following the evolutionary species concepts, conclude that asexual organisms can have no species (e.g. Ghiselin 1987). Others (Echelle 1990) recognise the intrinsic value of asexual lineages as different reproductive entities, but insist on communicating that non-Mendelian complexes can have multiple hybrid (or mutational) origins and thus do not represent single historical entities. Echelle (loc.cit.) proposed to use the term 'species-complex' for such clusters of indistinguishable taxa with different histories.

All clones are by definition reproductively isolated. Every single clone should thus be allocated specific rank, even if they are not or hardly morphologically recognisable.

Another pitfall for clonal taxonomy is the 'reproductive isolation' required by certain evolutionary species concepts. There are two (contradictory) problems with this. Firstly, clones are by definition reproductively isolated. Following this logic, every single clone should be given specific rank. As certain asexual lineages consist of hundreds, if not thousands, of genetically identifiable clones, nearly all authors agree that to name all of these is a practical impossibility (Enghoff 1976, Ghiselin 1987, Gosling 1994), although Walker (1986) still suggests to name all clones with informal codes (this already happens for certain organisms which are intensively used for genetic research, such as *Drosophila*, *Caenorhabditis*, etc.). To counter this problem, different types of taxonomic levels have been suggested: paraspecies (Mayr 1970), agamospecies (Ghiselin 1984), etc. However, whereas these concepts may be theoretically sound, they are hardly or not applicable. The use of subspecific rank for clonal lineages is instinctively, but correctly, rejected by Enghoff (1976) (see below).

Secondly, there is increasing evidence that in asexuals with bisexual root, occasional back-crossing between males and asexual females does occur (chapter 15). Such hybridisations mostly create new (polyploid) clones. Although the genetical integrity of the parental clone remains intact, this no longer qualifies as absolute reproductive isolation. Likewise, it has been suggested that sexual parasites should not be allocated specific rank as they are not strictly reproductively isolated (Walker 1986), but this was extensively challenged by Echelle (1990).

As differences in ploidy reflect different historical and evolutionary backgrounds and different evolutionary potential, they should be treated as different species.

This is a special case of the previous problem. Polyploid lineages can have different origins, but nearly always originate instantaneously and as a result of hybridisation, either intra- or interspecific. Cole (1985) argues, based on the evolutionary species concepts, that such hybridisation events mark the origin of a historical, evolutionary entity, which should thus be taxonomically recognised.

The ruling theory on speciation dictates that groups of uniparentally reproducing animals consistently offer more taxonomical difficulties than related groups of bisexuals.

A final problem deals with recognisability of clonal lineages. Holman (1987) tested this hypothesis by comparing the number of published synonymous names for monogonont rotifers, with mixed sexual and asexual reproduction, and bdelloid rotifers, with obligate parthenogenesis. Holman's results indicated that, in fact, the opposite is true, and that bdelloid species were more consistently recognised than monogononts. This problem is directly related to the confusion of different types of asexuals: Holman was not comparing sexuals with asexuals, but groups with mixed reproduction (amongst which asexuals with bisexual roots – monogononts) with the ancient asexuals in bdelloids (Birky & Gilbert 1971). Holman's test still validly shows that ancient asexuals can produce stable and morphologically recognisable lineages, a result which is corroborated by our revision of the extant Darwinuloidea (Rossetti & Martens 1998).

The above problems will be discussed for the different types of asexual lineages occurring in non-marine ostracods: ancient asexuals, taxa with mixed reproduction, asexual spin-offs and interspecific hybrids. In the first two categories, species are largely conform to morphological, genetic and evolutionary concepts. Only in the latter two categories are special concepts, such as the agamospecies, necessary.

Ancient asexuals

The only two examples of genuine ancient asexuals groups, reported in the literature, are darwinulid ostracods and bdelloid rotifers. Both were termed 'ancient asexuals scandals' by Judson & Normark (1996), because, according to the ruling evolutionary theory, there are several genetic mechanisms which should make persistence of fully asexual animals over geological time frames impossible (see chapter 1). Other putative ancient asexuals should actually be referred to other categories, as they are phylogenetically closely associated with sexuality (e.g. *Artemia parthenogenetica*).

The Darwinulidae have very limited extant diversity (c 28 extant species and 5 genera – Rossetti & Martens 1998), but they do qualify as ancient asexuals, as the fossil record has shown the absence of sexual reproduction in this group at least for the past 100 Myr (chapters 3, 6). Historical records of two putative males in the literature (Brady & Robertson 1870, Turner 1895) are either false or refer at most to an atavistic specimen (Rossetti & Martens 1996, Schön *et al.* 1998), while there is no evidence for automixis.

As reproduction-based concepts cannot be applied in this group, the darwinulid species are tested to the evolutionary, morphological and genetic species concepts, in order to evaluate their theoretical content and their applicability.

Material and methods

Intraspecific morphological variability was assessed in several specimens of four populations of the ubiquitous and nearly cosmopolitan species *Darwinula stevensoni* (from Finland, Belgium, Italy, South Africa); valve and soft part morphology of five specimens of each populations were examined in detail (Rossetti & Martens 1996, chapter 9). Intraspecific genetic variability was assessed with starch-gel electrophoresis of allozymes (chapter 15) and both mitochondrial and nuclear DNA (chapter 16) in numerous populations of the same species; the latter two techniques also provided interspecific distances between *D. stevensoni* and 3 other darwinulid species. Interspecific morphological differences were assessed by partly revising the genus world-wide, based on both literature, new collections and museum-specimens (Rossetti & Martens 1996, 1998; Martens *et al.* 1997, Martens & Rossetti 1997).

Results

Intraspecific morphological variability of both valves and soft parts in several geographically and climatically distant populations of *D. stevensoni* is found to be minimal or non-existing. Only size varies significantly between populations and this environmentally-cued effect can be attributed to the differences in ambient temperatures during juvenile development. The number and shape of muscle scars also vary, but this both within and between populations. The valve shape and chaetotaxy of limbs are remarkably constant. One female from the Italian population has aberrant T_1, but this specimen is considered a teratological specimen (Rossetti & Martens 1996). Nuclear genetic variability amongst these populations is either non-existent (DNA – Schön *et al.* 1998, chapter 16) or minimal (allozymes – chapter 15); only mitochondrial DNA shows a reasonable, albeit still restricted, variability (Schön *et al. loc.cit.*). Other darwinulid species also seem to have very low intraspecific variability (although this is less intensively investigated). Morphological differences between recognised species are small, but constant (Rossetti & Martens 1998); this is corroborated by the limited DNA data on interspecific differences (Schön *et al. loc.cit.*).

Discussion

From a *morphological point of view*, all conditions are fulfilled to consider the various morphologically different strains as species: they are easily recognisable, as intraspecific variability is limited and interspecific morphological gaps are clear and constant and, important in the case of Ostracoda, mostly present in both valve and soft part features. As was already pointed out by Holman (1987) for the bdelloid rotifers, these genuine ancient asexuals offer less taxonomic difficulties than groups with mixed reproduction. *Genetically*, the situation is similar, as within-species variability is lower than between-species variability, although only a limited sample set has thus far been tested.

From an *evolutionary and phylogenetic point of view*, the situation is less obvious, especially with regard to the first problem, the potential polyphyletic origin of the different lineages. However, there are several arguments favouring the recognition of darwinulid species as evolutionary entities with monophyletic origin. Firstly, as was cited above, most species can be recognised by a mixture of features on both valves and (several) limbs. The probability that a large set of combined mutations is successfully repeated is low. Secondly, the interspecific genetic differences indicate a considerable divergence between the various lineages and this is corroborated by fossil evidence: extant species can be quite old (20 My or more – Straub 1952). Apart from the fact that this, too, favours the survival of only few clones from a sib clone cluster, it could also explain why extant lineages are morphologically and genetically discrete: potential intermediate lineages will have been subjected to directional selection and will have become extinct. Thirdly, morphologically identical lineages occur on different continents; which again shows the antiquity of these various lineages (given that darwinulids are bad dispersers, as they do not have the drought-resistant eggs present in both other non-marine lineages, the Cypridoidea and the Cytheroidea). All this points towards the fact that, although isolated mutations can occur repeatedly, most likely the extant lineages each have a monophyletic origin. Potential exceptions do exist, e.g. the South African *Alicenula inversa* (Martens & Rossetti 1997), which differs from *A. serricaudata* only in valve reversal, a morphological difference suspected to be related to a fairly simple genetic alteration (see below). Finally, all lineages are perfectly reproductively isolated, as extensive allozyme screening of hundreds of specimens of *Darwinula stevensoni* has not detected polyploidy (chapter 15). Darwinulids thus seem largely conform to the evolutionary species concepts.

Morphological stability and speciation

The low morphological and genetic intraspecific variability in *D. stevensoni* is remarkable, the more so because this species can be considered ubiquitous. This species at least is conform to the hypothesis that asexuals should have a general purpose genotype (Lynch 1984). In this, it differs drastically from the strategies in asexuals in clusters with mixed reproduction, where every species comprises many different clonal lineages, with different morphologies and ecological requirements (chapters 14, 17 and below), although the existence of sib clones with large ecological tolerances in these clusters cannot be completely excluded.

Equally surprising is the interspecific stability in darwinulids, and apparently also in the other ancient asexual lineage, the bdelloid rotifers (Holman 1987). The low extant diversity (as compared to the other non-marine ostracod lineages), finally, is striking and indicates that speciation in this group is a slow process. The absence of intraspecific hybridisation of asexual females with males, explains both the apparent absence of polyploidy and the paucity of clonal lineages. All evolution in these lineages has to occur through mutation and natural selection only.

Conclusion

Darwinulids have good taxonomic species. Apart from the fact that a polyphyletic origin of some taxa cannot completely be excluded, most evidence seems to indicate that these morphological species are monophyletic and can also be considered valid evolutionary species. As Coyne (1994) pointed out, the fact that also ancient asexuals are organised in discrete units contributes towards our understanding why organic forms in general are discontinuous.

Species with mixed reproduction

These are species comprising parthenogenetically and syngamically reproducing populations, while also populations with mixed reproduction can exist. Bisexual populations in fact often have both types of females (bisexual and parthenogenetic) living sympatrically; such populations mostly have a geographically limited distribution. Note that there is at present no information on the factors which cause females to be either sexual or asexual. Circumstantial evidence seems to indicate that no or few reversals of the reproductive mode within the life span of one individual female can occur, but no experimental data exist on this (chapter 2).

Material and methods

Species with mixed reproductive modes are found in most of the Cyprididae and, to a lesser extent, in the Cytheroidea. Intraspecific (genetic and morphological) variability is assessed in *Eucypris virens*, which occurs during winter in larger, grassy temporary habitats and can be found all over Europe and North America in parthenogenetic populations and in bisexual/mixed populations in the southern part of the area (around the Mediterranean in Europe). We have investigated bisexual populations from Spain (Extremadura) and Italy (Sicily) and asexual populations from the same localities, as well as several other populations in western and northern Europe. The techniques used include morphometric valve shape analyses (chapter 8), starch-gel electrophoresis of allozymes (chapter 15) and sequencing of both mitochondrial and nuclear DNA (chapter 16).

Results

Bisexual (and mixed) populations of *E. virens* cluster within and amongst the parthenogenetic populations, and this with regard to morphometric analyses of valve outlines (chapter 8), allozyme starch gel electrophoresis (chapter 15) and both mitochondrial and nuclear DNA-sequences (Schön *et al.* in prep., chapter 16). Sexual and clonal strains thus seem to show a genetic coherence, which is further exemplified by the fact that hybridisation between males and parthenogenetic females does occur and results in new clonal lineages (chapter 15).

Discussion

The first practical question to be answered here is: do bisexual strains and associated parthenogenetic lineages belong to the same species? As was cited in the introduction, morphologically similar parthenogenetic and bisexual populations have in the past indeed been referred to as different nominal species or subspecies, for no reason other than their different reproductive mode. Secondly, it must be discussed whether some or all clonal lineages must be allocated taxonomic rank?

From a *taxonomic (morphological) point of view*, the above-cited morphological coherence offers sufficient support to include bisexual and parthenogenetic populations within one and the same taxonomic species. The continuous mixture of both reproductive modes makes that this species has a wide morphological variability, which originates from the syngamically reproducing pool and from the sum of the different morphologies of the many associated clonal lineages. The number of the latter increases, among other mechanisms, through recurrent hybridisations between parthenogenetic females and males. Even if isolated parthenogenetic populations show (recognisable) disjunct morphologies, it follows from investigation of larger numbers of samples that such morphologies mostly fit into a wider and continuous intraspecific variability range. There is thus a problem with regard to morphological recognisability of different lineages in this case, the solution being to consider taxa with mixed reproductive strategies to be widely inclusive. The above-cited *genetic coherence* again offers strong support to the view that parthenogenetic and bisexual populations should all be considered part of one inclusive taxon, the species.

From an *evolutionary and phylogenetic point of view*, the problem of the polyphyletic origin is solved if these clonal lineages are considered part of the same species, as multiple generation of genetically identical lineages is an intraspecific process, which can potentially also occur, albeit with lower probability, in sexual taxa. The same is true for the third problem (polyploidy), which then simply becomes an intraspecific mechanism increasing the variability. The recurrent intraspecific hybridisation largely neutralises the problem of reproductive isolation, even when the parental clone retains its genetic integrity. In short, also from an evolutionary point of view, there seem to be sufficient arguments to consider such taxa with mixed reproductive modes as inclusive species.

Intraspecific hybridisation

Why intraspecific hybridisation does not occur more frequently within such species, is an intriguing question in itself. Either the production of such hybrids is advantageous or at least neutral, and then one would expect such events to occur more often. Or these hybridisations have negative effects, and then reproductive isolation is expected to be more complete. Following the present observations, these hybridisations most likely do have positive effects, but only if they are allowed at low frequencies, as they do increase the genetic variability within the species, but should not threaten the integrity of the sexually reproducing strains, for example by diverting too much reproductive effort from sexual to parthenogenetic females. Males in ostracods indeed attempt to copulate with all females, regardless of their reproductive mode (chapter10), but if most parthenogenetic females will resolutely refuse copulation at an early stage, less effort will be lost for the bisexual reproduction. These observations raise questions of the efficiency and functionality of sexual selection within species with mixed reproduction, an exciting avenue for future research.

Conclusion

In parthenogenetic lineages with mixed reproduction, the clonal lineages form a genetic and morphologically coherent cluster with their bisexual roots. According to morphological, genetic and evolutionary species concepts, they should therefore be considered to belong all in one, inclusive species. Note that the observed genetic variability within *Eucypris virens* is very high, to the extent that it reaches levels of variability normally observed between species with sexual reproduction (Schön *et al.* in prep., chapter 16).

Asexual spin-offs and interspecific hybrids

Amongst the groups which comprise species with mixed reproduction, there are species, seemingly fully asexual, which have been described as separate taxa. Amongst the genera which comprise species with mixed reproduction, subfamilies also hold genera with (apparently) only fully asexual species. It is of course possible that bisexual populations of such species either exist and remain to be discovered, or have recently become extinct. Nevertheless, it seems unlikely that this would be the case for all (seemingly) asexual spin-offs.

Material and methods

The taxonomic situation in the Eucypridini is at present insufficiently stable to allow assessment of interspecific (but intrageneric) variability. Therefore, the related cypridinid genus *Herpetocypris* was revised and morphological species definitions tested in this genus (Gonzalez Mozo *et al.* 1996). *Herpetocypris* is an ideal test-case for this level of assessment, as there is no taxonomic confusion related to the genus and at least one European species within the genus comprises both bisexuals and asexually reproducing lineages.

Results

Prior to the revision by Gonzalez Mozo *et al.* (1996), the genus *Herpetocypris* in Europe consisted of about 20 nominal (sub-)species. Several bisexual populations around the Mediterranean were known, all other European populations were entirely asexual. The revision was performed exclusively on morphology, but was based on type-materials of nearly all extant nominal taxa. In a first step, it was determined that all bisexual populations belong to only one species, *Herpetocypris brevicaudata*, which furthermore has many, morphologicaly variable, parthenogenetic populations in the whole of Europe. This variability thus constitutes a continuous pool, and no disjunct units could be identified. The situation in this species is thus very similar to that in *Eucypris virens*.

Apart from *H. brevicaudata*, which has both reproductive modes, three clusters of a total of 6 (sub-)species, all exclusively asexual, were retained in European *Herpetocypris*: *H. reptans* (with subspecies *H. r. curvata*), the *H. mateusorum-intermedia* group and the *H. chevreuxi-helenae* cluster. Other nominal species were sunk into synonymy of these senior taxa. All species have a wide distribution (mostly also outside Europe), only *H. mateusorum* appears to be endemic to the western part of the Iberian peninsula.

Discussion

This category of asexuals presents problems for all types of species concepts. Such lineages are similar to those found in vertebrates (lizards, fish) and on which most discussion in the past has focused. Their morphological distinctness can be very variable, which further illustrates the intrinsic necessity to separate these populations from the species with sexual roots, as the latter already comprise an unusually large genetic and morphological variability if the inclusive concept discussed above is applied. The main problems, however, are how morphological species boundaries must be defined and how such species must be considered from an evolutionary point of view.

Whether or not different nominal taxa belong to the same *morphological species* was, in the revision of *Herpetocypris*, decided upon comparing type materials, i.e. the specimens used for the original description. Most sib clones could be referred to the genetically coherent parent cluster (see above). Some, however, were more divergent from this cluster and show morphological (and most likely also genetic, but this remains to be tested in this genus) gaps with the whole of the 'central' cluster, and previous ostracod taxonomists have translated this into the description of new (sub-)species. Differences between most of these species are small, but rather constant (e.g. in *H. reptans)*. However, such decisions are mostly arbitrary and subjective. For example, the morphological differences between *H. chevreuxi* and *H. helenae* are small and subtle and intra- and interspecific variabilities in this case are comparable. Because of this, taxonomists have in the past repeatedly taken opposite decisions. Also the large number of synonyms of *H. helenae* indicates that this taxonomic situation is unclear (method of Holman 1987).

With regard to the *evolutionary and phylogenetic* concepts, the problem of polyphyletic origin can at least partly be weakened by arguments similar to those used for ancient asexuals.

Specific spin-offs which are morphologically recognised as different from a species with mixed reproduction mostly differ in several character states. These differences are either the result of interspecific hybridisation or of accumulation of mutations or of a combination of both. In the former case, the probability of repeated occurrence of such events is relatively high, and the logic of Echelle (1990) for parthenogens should be followed: such taxonomic species are evolutionary species complexes. In the latter two cases, repeatability has a low probability and such mutational (or combined) spin-offs can actually be considered to have a monophyletic origin and be considered good evolutionary species. However, it will often be impossible to decide on the origin of such asexuals without screening the entire genus or subfamily with genetic methods.

The second and third problems (reproductive isolation and levels of polyploidy) are in favour of considering congeneric spin-offs as separate from the sexual root, but leave uncertainty as to how they, themselves clusters of groups of sib clones as exemplified by observed variability in, for example, the *H. chevreuxi-helenae* complex, can be defined as separate and inclusive species. This question requires a combined set of data on genetic and morphological variability amongst such spin-offs, which is at present unavailable.

Conclusion

Asexual lineages, which are separated from the sexual cluster by a morphological (and most likely also genetic) gap, can be allocated specific (in some cases even generic) rank. Such taxa comply with most species concepts to some extent, although many problems remain. Morphological recognition of congeneric lineages is not always unambiguous and these taxonomic species, themselves often being clusters of clonal lineages, mostly have to be considered evolutionary species complexes, maybe conform to the agamospecies concept of which Ghiselin said that they are like "... heaps of leaves that have fallen of the tree that gave rise to them." (1984: 213). We suggest to apply morphological species concepts based on subjective assessment of morphology and to consider such taxa evolutionary species complexes, following Echelle (1990).

General discussion

We have shown that at least some types of asexual lineages do not create taxonomic and nomenclatorial problems. *Ancient asexuals* can conveniently be separated in discrete morphological units, which are largely congruent with genetic units. Naming such units species is furthermore in line with most evolutionary species concepts, as they are almost invariably monophyletic. *Asexual lineages closely clustered amongst sexual lineages* must be incorporated with those lineage in one species. Again, this reflects an underlying morphological and genetic reality, while such species also constitute monophyletic, evolutionary units.

Asexual spin-offs, separated from the species with mixed reproduction at least by a morphological gap, and *interspecific hybrids* do create the taxonomic problems extensively discussed in the literature on clonal vertebrates (Echelle 1990 a.o.) and here, either morphospecies or agamospecies concepts must be applied as has been suggested for such cases.

To distinguish between intraspecific sib clones and asexual spin-offs will require a subjective decision when only limited material is available. However, the latter populations can more easily and more objectively be separated when the intraspecific variability range of the species with mixed reproduction becomes more extensively documented.

Speciation

Fig. 11.1 presents a hypothetical scheme of evolution and speciation in the different types of asexuals occurring in non-marine ostracods. This is of course a simplified scenario and patterns can become more confused by several other factors: it is not at all certain that the bisexual taxon is indeed the root of such genera; generally, more than one species with mixed reproduction occurs; cladogenesis occurs world-wide, so the taxonomy of genera should be assessed on that, and not on a continental scale; sexuality could be regained in certain asexual spin-offs; etc. Nevertheless, the above scenario could summarise the common modes of speciation in at least the Cyprididae, which is by far the most successful non-marine ostracod lineage in the world.

Clonal subspecies

One issue has consistently been avoided in the above discussion, i.e. the use of infra-specific taxa, more in particular the subspecies, in clonal taxonomy. It would indeed seem a practical solution, to institutionalise a category of 'clonal subspecies', parallel to geographical and ecological subspecies. There are several arguments against this.

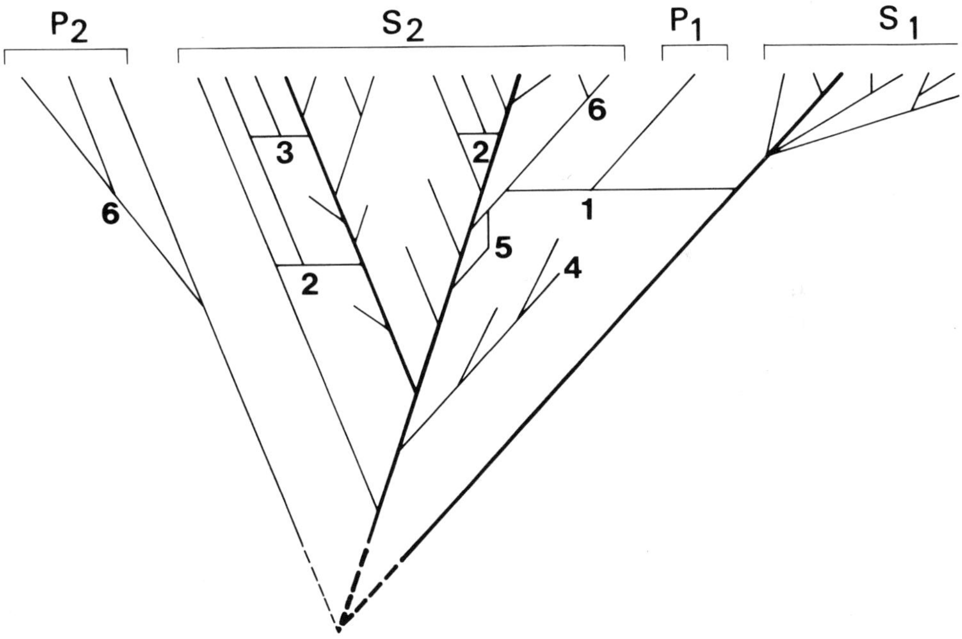

Fig. 11.1 Hypothetical scheme of evolution and speciation in a genus with taxa with mixed reproduction. S1 and S2: species with mixed reproduction. P1: interspecific hybrid. P2: asexual spin-off. Figures refer to different processes: 1. interspecific hybridisation. 2. intraspecific hybridisation (resulting clone = 3N). 3. intraspecific hybridisation with polyploid clone (resulting clone= 4N). 4. extinct clonal lineages. 5. polyphyletic origin of a clone (intraspecific). 6. recent branching of clones through mutation.

Enghoff (1976) argues that if the specific level in clonal taxonomy is still hotly debated, the discussion should not be further confused by prematurely introducing another level. He provides the following arguments: (1) such lineages are fully reproductively isolated; (2) they do not comply with the requirements for geographical or ecological subspecies; (3) there are no (geographical or ecological) hybrid zones.

We will now briefly discuss the potential use of subspecies in the three categories of asexuals cited above.

Ancient asexuals

Martens & Rossetti (1997) described *Alivenula inversa* from South Africa, a species known from several populations in the former Eastern Transvaal and distinguishable from *A. serricaudata* only by the reversed valve overlap. These populations comply with all arguments to be regarded as species. Firstly, only individuals with RV>LV have been found, and the reversed valve-overlap is thus a stable character, almost certainly genetically fixed and thus useful and sufficient to delimit a new taxon. Secondly, the idea that many of these clonal lineages in Darwinulidae are geographically restricted is falsified: for example, there are no morphological differences between African and South American *A. serricaudata* and *Penthesilenula brasiliensis* is found in South America, Europe and Africa (Martens *et al.* 1997, Rossetti & Martens 1998). Finally, these lineages are fully reproductively isolated and no hybrids are possible. This new lineage was thus allocated specific rank. Most other darwinulid species are distinguishable by a set of different character states and present even less taxonomic problems. The use of subspecies in this group should be avoided.

Species with mixed reproduction and spin-offs

Given the continuity of the range of variability in both types of asexuals, also here the use of subspecies should be avoided, unless it is decided to allocate this rank to each individual clone. We have argued extensively against this solution above.

Exceptions

In the revision of *Herpetocypris*, the subspecies *Herpetocypris reptans curvata* was retained. This was done for practical reasons, mostly because the position of the population from which it was described remains uncertain. The valve morphology of *H. r. curvata* is clearly different from the real *H. reptans*, but the soft part morphology is identical. This taxon has never been found again since its original description and it is not certain whether this is a genetically stable morphology, or an ecophenotypic exception. It could be a mutational novelty, or a hybrid, for example between *H. brevicaudata* and *H. reptans*. The subspecific status of *H. reptans curvata* is retained, because there are at present no new data which provoke alteration of the taxonomy, in which case the original status should be conserved. We do not suggest to introduce the subspecific level to illustrate taxonomic uncertainty, we only say that it is to the advantage of stability, when taxonomic levels are changed only when new data require this.

A second potential exception deals with practical recognition of certain clones with typical morphology and ecological requirements. Indicator species (of Ostracoda) are used for palaeoenvironmental reconstructions. If this recognition could be refined to recognisable clonal lineages, for which narrower ranges of environmental variables are known, then this would greatly improve the technique. As this situation coincides with the notion of ecological subspecies, the use of a subspecific trinomen for such practical applications can be considered.

References

Abe, K. 1990. What the sex ratio tells us: a case from marine ostracods. *in*: Whatley, R.C. & Maybury, C. (eds), *Ostracoda and Global Events*, Chapman & Hall: 175-186.

Bell, G. 1997. *Selection. The mechanism of evolution*, Chapman & Hall: 699 pp.

Birky, C.W. Jr. & Gilbert, J.J. 1971. Parthenogenesis in rotifers: the control of sexual and asexual reproduction. *Am. Zool.,* **1:** 245-66.

Brady, G.S. & Robertson, D. 1870. The Ostracoda and Foraminifera of Tidal Rivers. *Ann. Magazine Nat. Hist., ser.* **4, 6:** 1-33, 307-309.

Chandler, C.R. & Gromko, M.H. 1989. On the relationship between species concepts and speciation processes. *Syst. Zool.*, **38:** 116-125.

Cole, C.J. 1985. Taxonomy of parthenogenetic species of hybrid origin. *Syst. Zool.*, **34:** 359-363.

Coyne, J.A. 1994. Ernst Mayr and the origin of species. *Evolution*, **48:** 19-30.

Cracraft, J. 1983. Species concepts and speciation analysis. *in*: *Current Ornithology*, **1:** 159-187, Plenum Press.

Crisp, M.D. & Weston, P.H. 1993. Geographic and ontogenetic variation in morphology of Australian waratahs (Telopea, Proteaceae). *Syst. Biol.*, **42:** 49-76.

Delorme, L.D. 1971. Freshwater ostracodes of Canada. Part V. Families Limnocytheridae, Loxoconchidae. *Can. J. Zool.*, **49:** 43-64, pls. 1-19.

Dobbin, C.N. 1941. Fresh-water Ostracoda from Washington and other western localities. *Univ. Washington Publ. Biol.*, **4 (3):** 175-245, 14 plts.

Echelle, A.A. 1990. Nomenclature and non-Mendelian (clonal) vertebrates. *Syst. Zool.*, **39:** 70-78.

Enghoff, H. 1976. Taxonomic problems in parthenogenetic animals. *Zool. Scr.*, **5:** 103-104.

Foottit, R.G. 1997. Recognition of parthenogenetic insect species. *in*: Claridge, M.F., Dawah, H.A. & Wilson, M.R. (eds), *Species, the units of diversity,* Chapman & Hall: 291-308.

Ghiselin, M.T. 1974. A radical solution to the species problem. *Syst. Zool.*, **23:** 536-544.

Ghiselin, M.T. 1984. Narrow approaches to phylogeny: a review of nine books on cladism. *in*: Dawkins, R. & Ridley, M. (eds), *Oxford surveys in Evolutionary Biology*: 209-222.

Ghiselin, M.T. 1987. Species concepts, individuality and objectivity. *Biol. Philos.*, **2:** 127-143.

Gosling, E.M. 1994. Speciation and wide-scale genetic differentiation. *in*: Beaumont, A.R. (ed), *Genetics and evolution of aquatic organisms*, Chapman & Hall: 1-14.

Gonzalez Mozo, M., Martens, K. & Baltanás, A. 1996. A taxonomic revision of European *Herpetocypris* Brady & Norman, 1889 (Crustacea, Ostracoda). *Bull. k. belg. Inst. Natuurw., Biol.*, **66:** 93-132.

Holman, E.W. 1987. Recognisability of sexual and asexual species of rotifers. *Syst. Zool.*, **36:** 381-386.

Hull, D.L. 1997. The ideal species concept – and why we can't get it. *in*: Claridge, M.F., Dawah, H.A. & Wilson, M.R. (eds), *Species, the units of diversity,* Chapman & Hall: 357-380.

Judson, P.O. & Normark, B.B. 1996. Ancient asexual scandals. *Trends Ecol. Evol.,* **11:** 41-46.

King, M. 1995. *Species evolution. The role of chromosome change,* Cambr. Univ. Press: 336 pp.

Krstic, N. 1987. Three new ostracode genera from lacustrine Miocene. *Bull. Acad. serb. Sci. Arts, Cl. Sci. nat. math., Sci. nat.*, **27:** 129-139 + 12 plts.

Lynch, M. 1984. Destabilizing hybridization, general-purpose genotypes and geographic parthenogenesis. *Quart. Rev. Biol.*, **59:** 257-290.

Mallet, J. 1995. A species definition for the modern synthesis. *Trends Ecol. Evol.*, **10:** 294-299.

Martens, K. 1990a. Revision of African *Limnocythere* s.s. Brady, 1867 (Crustacea, Ostracoda) with special reference to the Eastern Rift Valley Lakes: morphology, taxonomy, evolution and (palaeo-) ecology. *Arch. Hydrobiol. Suppl.*, **83 (4):** 453-524.

Martens, K. 1990b. Speciation and evolution of the genus *Limnocythere* Brady, 1867 s.s. in the East African Galla and Awassa basins. *Cour. Forschungsinst. Senckenberg*, **123:** 87-95.
Martens, K. & Meisch, C. 1985. Description of the male of *Potamocypris villosa* (Jurine, 1820) (Crustacea, Ostracoda). *Hydrobiologia*, **127:** 9-15.
Martens, K. & Rossetti, G. 1997. On two new species of *Darwinula* BRADY & ROBERTSON, 1885 (Crustacea, Ostracoda) from South African dolomitic springs. *Bull. k. belg. Inst. Natuurw., Biol.*, **67:** 57-66.
Martens, K., Rossetti, G. & Fuhrmann, R. 1997. Pleistocene and Recent species of the family Darwinulidae Brady & Norman, 1889 (Crustacea, Ostracoda) in Europe. *Hydrobiologia*, **357:** 99-116.
Mayden, R.L. 1997. A hierarchy of species concepts: the denouement in the saga of the species problem. *in*: Claridge, M.F., Dawah, H.A. & Wilson, M.R. (eds), *Species, the units of diversity*, Chapman & Hall: 381-424.
Mayr, E. 1963. *Animal species and evolution,* Belknap Press Harvard Univ. Press.
Mayr, E. 1969. *Principles of systematic zoology,* McGraw-Hill.
Mayr, E. 1970. *Populations, species and evolution,* Belknap Press Harvard Univ. Press: 453 pp.
Mayr, E. & Ashlock, P.D. 1991. *Principles of systematic zoology,* McGraw-Hill.
Meier, R. & R. Willmann, 1997. The Hennigian species concept. *in*: Wheeler, Q.D. & Meier, R. (eds), *Species concepts and phylogenetic theory: a debate,* Columb. Univ. Press.
Moritz, C. & Heideman, A. 1993. The origin and evolution of parthenogenesis in *Heteronotia binoei* (Gekkonidae): reciprocal origins and diverse mitochondrial DNA in western populations. *Syst. Biol.*, **42:** 293-306.
Paterson, H.E.H. 1993. Animal species and sexual selection. *in*: Lees, D.R. & Edwards, D. (eds), *Evolutionary patterns and processes, Linn. Soc. Symp. Ser.*, **14:** 209-228.
Petkovski, T.K. 1959. Beiträge zur Kenntnis der Ostracoden-Fauna Jugoslawiens (V). Hidrobiologi, *Publications of the Hydrobiological Research Institute, Faculty of Science, University of Istanbul, Seri B.*, **4 (4):** 158-165.
Rossetti, G. & Martens, K. 1996. Redescription and morphological variability of *Darwinula stevensoni* (Brady & Robertson, 1870) (Crustacea, Ostracoda). *Bull. k. belg. Inst. Natuurw. Biol.*, **66:** 73-92.
Rossetti, G. & Martens, K. 1998. Taxonomic revision of extant Darwinulidae, with the description of two new genera. *Bull. k. belg. Inst. Natuurw. Biol.*, **68**.
Schön, I., Butlin, R.K., Griffiths, H.I. & Martens, K. 1998. Slow evolution in an ancient asexual ostracod. *Proc. R. Soc. London B*, **265:** 235-242.
Simpson, G.G. 1961. *Principles of animal taxonomy,* Columbia Univ. Press.
Straub, E.B. 1952. Mikropaläontologische Untersuchungen im Tertiär zwischen Ehingen und Ulm a.d. Donau. *Geol. Jb.*, **66:** 433-523.
Templeton, A.T. 1989. The meaning of species and speciation: a generic perspective. *in*: Ott, D. & Endler, J.A. (eds), *Speciation and its consequences*, Sinauer Assoc.: 3-27.
Turner, C.H., 1895. Freshwater Ostracoda of the United States. *Minnesota Geol. Nat. Hist. Surv., Zool. series*, **2:** 277-337.
Vandel, A. 1928. La parthénogénèse géographique. *Bulletin Biol. Franc. Belg.*, **62:** 164-281.
Van Valen, L. 1976. Ecological species, multispecies and oaks. *Taxon*, **25:** 233-239.
Walker, J.M. 1986. The taxonomy of parthenogenetic species of hybrid origin: cloned hybrid populations of *Cnemidophorus* (Sauria, Teiidae). *Syst. Zool.*, **35:** 427-440.
Wiley, E.O. 1978. The evolutionary species concept reconsidered. *Syst. Zool.*, **27:** 17-26.

CHAPTER 12

POPULATION DYNAMICS, LIFE HISTORIES AND REPRODUCTIVE MODES

W. Geiger

Abstract

On the base of existing data, the importance of reproductive mode and habitat type on population dynamic parameters like maximum and mean abundances, numbers of abundance peaks, development times and fertility is explored. The results from inter- and intra-habitat comparisons only support hypotheses which predict the success of bisexuality in predictable habitats like the littoral and profundal of permanent lakes or the groundwater. They could not demonstrate the predicted dominance of parthenogenesis in harsh, pioneer environments. In this type of environment, the acquisition of some kind of resting stage is by far more important than the mode of reproduction. In general, environmental conditions are acting on parthenogenetic and bisexual populations in a similar way, thus determining to a large extent population dynamic variables.

Introduction

When comparing the evolutionary success of parthenogenetic and bisexual reproduction, often reference to differences in life history strategies and population dynamics is made (Hughes 1989, Carvalho 1994). In general, asexual reproduction coincides with conditions favouring population growth and with habitats where the influence of abiotic factors is higher than that of biological interactions (Bell 1982, Lynch 1984, Jackson 1985). Rapid population growth is a consequence of the fact that no males have to be produced, resulting roughly in twice the number of reproducing offspring than for a bisexual species in similar conditions (Maynard Smith 1978). On the other hand, sex should be superior in producing new genotypes and genetic diversity (see chapter 1). Therefore, clonal organisms are supposed to be good colonisers and exploiters of harsh, pioneer habitats, but to be at a disadvantage when habitats are fine-grained and complex (Williams 1975, Bell 1982). Long term replacement of parthenogens by bisexuals in stable habitats, where biotic interactions dominate, is thought to be inevitable as the genetic diversity generated by clonal organisms is not sufficient to cope with the permanent process of niche development and the need for rapid adaptation (Bell 1982, Stearns 1987).

In a comparison of life histories, one has to distinguish between variables which are measured on individuals (fertility, development time) and those measured on populations (population growth rates). Whilst we might not detect any difference at

Sex and parthenogenesis: evolutionary ecology of reproductive modes in non-marine ostracods, edited by K. Martens, pp. 215–228

the first level, the effects of parthenogenesis on total population growth can be tremendous (Hughes 1989).

In freshwater ostracods many species have made the transition from bisexual reproduction to apomictic parthenogenesis (Bell 1982 and chapter 1). Both types of reproduction are represented in a wide variety of water bodies and a considerable amount of literature exists on their taxonomy and geographical distribution. The present chapter attempts to analyse the ecological information on ostracods in the light of the above-cited hypotheses and therefore concentrates on the empirical evidence available. It does not aim to invalidate existing theories nor to erect new ones, but rather intends to show agreement and disagreement between theory and reality. As an anecdotal reminiscence of the European project, on which the present volume is based, the terms b-pops and p-pops will be used throughout the paper for bisexual and parthenogenetic populations. These terms proved not only to be useful abbreviations but also should underline the dynamic atmosphere in which the present study was carried out.

The state of the art – a re-analysis of the existing data

When reviewing the existing literature on population dynamics and life histories in freshwater ostracods, one is confronted with the problem that most of the studies were designed for different purposes and that they are most commonly cited only as total ostracods without any reference to individual species. Only two studies (Herman *et al.* 1983, Geiger 1990) include the necessary information (stage specific abundance, birth and death rates and fecundity) to calculate key variables in population dynamic studies and life history analyses like r (intrinsic rate of population increase), R_0 (net productive rate) or generation time.

Nevertheless, data exist for 32 species with 99 populations from 25 studies providing at least partial information on parameters like maximum and mean abundances, lifespan, development times, fecundity or survival. They represent a subset of the European Database NODE (see chapter 5) including only studies which have been conducted over at least one year or were performed in the laboratory. The most complete data are found in studies where both field and laboratory investigations have been combined. Unfortunately, their number is very low (7%). 16% are carried out entirely in the laboratory, causing various problems when extrapolating these data to field populations. The remaining 77% are field studies, most of them do not distinguish between developmental stages and are therefore not providing any information about development times nor fecundity. Keeping this in mind, we tried to detect the general patterns in population dynamics and life history and the possible influence of the reproductive mode. We opted for a minimal approach, selecting the most frequently stated parameters, knowing well that they are not exactly the kind of data needed for this type of analysis but being constrained to use what is available. An overview of the parameters included in this analysis and the number of populations studied is presented in table 1.

Qualitative variables were analysed with χ^2-tests. Statistical significance of differences between quantitative variables has been assessed with ANOVA at a significance level of 0.05.

Table 12.1 Population parameters, N (number of populations) and ranges of given values.

Information available	N	Range	References
Habitat type	99		all references
littoral	72		1, 2, 3, 5, 7-14, 16-19, 23
profundal	6		6, 7, 15, 20
rivers	2		4
subterranean	6		4
brackish	11		5, 21, 22
Mean density (Ind x m^{-2})	75	65 - 259,000	1, 2, 3, 5, 6, 8, 9, 10, 12, 13, 14, 15, 17, 19, 20, 22
Max. density (Ind x m^{-2})	37	200 - 1.800,000	1, 2, 6, 8, 9, 10, 12, 13, 14, 15, 17, 19, 20, 22
Number of density maxima	30	1 - 2	1, 2, 6, 8, 9, 10, 12, 13, 14, 18, 19, 20, 22
Life span (days)	15	28.8 - 1280	8, 11, 13, 14, 16, 17, 18, 20
Juvenile development time	28	12.8 - 800	4, 6, 6, 7, 8, 9, 10, 11, 12, 14, 17, 18, 20, 21
Number of eggs/female	18	1 - 40	4, 6, 7, 12, 13, 17, 18, 20
Total number of eggs laid	17	8 - 290	4, 6, 7, 11, 14, 16, 18
Type of egg laying	34		4, 6, 7, 8, 9, 11, 12, 13, 14, 16, 17, 18, 19, 20,
continuous	5		
pulsed	29		
Survival (%)	9	37 - 100	7, 9, 13, 14, 20
Number of generations (per year)	38	0.25 - 3	6, 8, 9, 10, 12, 13, 14, 17, 18, 19, 20, 21, 22
Resting egg production	36		all references
yes	21		
no	15		
Mode of reproduction	99		all references
bisexual	56		
parthenogenetic	43		
Type of study	99		all references
field	77		1, 2, 3, 5, 6, 8, 9, 10, 12, 13, 15, 17, 19, 20,
laboratory	15		21, 22
field and laboratory	7		4, 6, 7, 11, 16, 7, 14, 18, 23

References: 1) Alcorlo unpubl. data, 2) Baltanás 1994, 3) Benzie 1989, 4) Danielopol 1980, 5) Delling 1981, 6) Geiger 1990, 7) Geiger *et al.* in prep., 8) Griffiths 1995, 9) Herman *et al.* 1983, 10) Mallwitz 1984, 11) Marín 1984, 12) Martens & Tudorancea 1991, 13) McGregor 1969, 14) McLay 1978a, b, 15) Newrkla 1985, 16) Otero *et al.* in press, 17) Ranta 1979, 18) Semenova 1973, 19) Strayer 1988, 20) Thaler 1977, 21) Theisen 1966, 22) Wolfram 1993, 23) Yin 1997.

Environmental stability versus instability

The concept of environmental stability and its assessment not being a straightforward task and seldom undertaken even for simple variables as temperature (Stearns 1981), we were looking for an operational definition which allows testing of the above-cited hypotheses taking the data available in the literature. Therefore, we prefer to call stability "predictability of environmental changes on a time scale appropriate to the organism and within their adaptation limits", rather than the absence of change. Unstable would therefore be equal to stochastic changes and/or changes which exceed the range normally encountered by the organism. No abiotic and even less so biotic factors remain constant over long time periods, but it makes a big difference if these changes can be foreseen or if they occur in a stochastic, catastrophic way (e.g. the 'capricious environments' quoted in Bell 1982).

Often, predictable environments go together with habitat complexity in space and intensity of abiotic and biotic interactions, other important components of existing hypotheses concerning the advantages of sex (see also chapter 1 for their complete listing).

We are aware that this is a very subjective definition of stability, but in referring to what the organisms experience, has the advantage to be applicable to all kinds of environments. Stability can be very different, as we will see, when dealing with living and reproducing organisms or with their respective reproductive propagules.

In a similar way we look at stability in time from the point of view of the organism rather than from that of the observer. For organisms without resting stages the longevity of the single water body is of higher importance than for organisms with resting stages. For the latter, the duration (temporal stability) of the type of water body (e.g. temporary pond) is the key parameter as metapopulation processes become more important than population dynamics as such.

In such a concept, ancient lakes like Lake Baikal or Tanganyika with their high proportion of endemic, bisexual species, pose the problem if we are confronted with examples for the long-term success of bisexuality over parthenogenetic reproduction or, alternately, for the importance of bisexuality and longevity of the water body for speciation. On the base of what is more extensively discussed in other chapters, we consider these water bodies as good examples for the pre-requisite of bisexuality and of the necessary longevity of the habitat for species without resting stages for speciation. When talking about temporal stability (= longevity) we therefore have to distinguish between the age of a single water body and the age of water body type depending on the potential to produce resting stages or not. However, this discrimination is somewhat weakened by the fact that also non-resting egg-producing organisms might represent metapopulations rather than isolated populations subjected to the fate of the water body they are living in.

Comparison of bisexual and parthenogenetic populations in habitats of different degree of permanency

Bisexual ostracod populations should have an advantage when exploiting stable, saturated habitats like the profundal zone of lakes or subterranean waters and to a lesser degree in habitats with unpredictable fluctuations, high incidences of disturbance and abiotic factors as dominating selective agents like ephemeral ponds or new, man-made habitats. They should occupy an intermediary position in environments which are fluctuating with regard to abiotic parameters in a predictable way which is the case in the littoral of lakes or in permanent ponds. The reverse should be true for parthenogenetic species (Bell 1982, Chaplin *et al.* 1994).

Habitat type and reproductive mode

From our analysis – a more general picture using the entire European database NODE is presented in chapter 5 – no statistically significant relationship between habitat type and reproductive mode can be found. There are more bisexual populations in subterranean waters and a slight dominance in the lake littoral and shallow permanent water bodies, but the number of populations investigated is too low to allow firm conclusions. Similarly, if we pool habitats into only two categories – stable and unpredictably fluctuating water bodies – results indicate a higher percentage of parthenogens in the latter but, again, without statistical significance.

Resting stages and reproductive mode

Many ostracod species developed some kind of resting stage which allow them to overcome unfavourable periods like draughts and freezing and which also facilitate dispersion. This can be, like in Cypridoidea, a resting egg or some larval stage (Cyprididae, Candoninae resp.). Drought-resistant resting stages in Cytheroidea are rare (Martens 1989, 1996) but only Darwinuloides seem to lack them completely.

The lack of correlation between habitat type and mode of reproduction at the population level could be better explained (i.e. without confronting the existing hypotheses) if the advantage of parthenogenesis in recolonising disturbed areas is outweighed by a higher frequency of bisexual species producing resting stages. Our data show that the opposite is the case. There are significantly ($p<0.01$) more parthenogenetic species with resting stages (57%) than bisexual (25%).

A closer look to the habitats classified as unstable reveals that there are two kinds of instability: rapid and unpredictable changes in environmental factors (mainly salinity) and the predictable disappearance of the habitats due to seasonal drying. Whilst the first are inhabited by populations of a few widely distributed asexual and bisexual species without resting stages, the latter harbour populations of many species with both modes of reproduction but exclusively with resting stages. Strong, unpredictable fluctuations which affect living organisms allow only for a few tolerant species independent of their mode of reproduction, whilst predictable seasonal disappearance of the habitat under otherwise less constraining conditions permits a much higher number of species, provided they can overcome the dry period by some kind of resting stage.

Combining habitat type and the potential or producing resting stages, the results clearly indicate the importance of resting stages to successfully colonise unstable habitats (Fig.12.1). Whilst the predicted dominance of parthenogenetic populations in harsh, unstable environment could not be corroborated, there are some indications that this could be true for populations of species which do not produce resting stages (Fig. 12.1). This is based on the fact that the number of bisexual populations without resting stages is drastically reduced in unstable habitats and much higher than those with parthenogenetic reproduction in habitats classified as stable, the number of parthenogenetic populations not differing between the two habitat types. Therefore, we might find here some support for the success of bisexuality in stable environments but only if we consider populations of species without resting eggs.

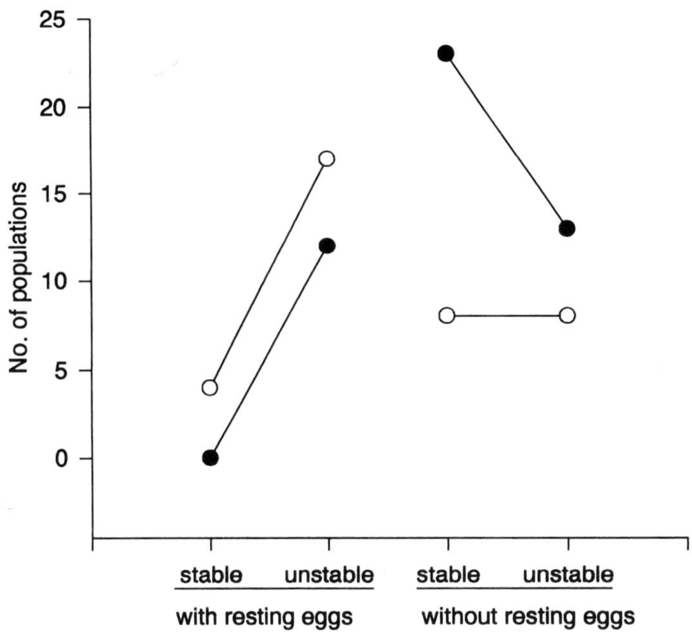

Fig. 12.1 Frequency of parthenogenetic (white dots) and bisexual (black dots) populations in stable and unstable habitats and their capability to produce resting stages.

All that was said above holds also true when switching from populations to species. There are more parthenogenetic species in fluctuating and more bisexuals in stable environments, but again without statistical significance ($p=0.072$). The advantages of parthenogens in dispersal abilities and the fact that they can start a population from a single individual are not sufficient to result in a statistically significant dominance in ephemeral, harsh or frequently disturbed habitats as predicted by many ecological hypotheses concerning the evolution of parthenogenesis.

Abundance, life history parameters and reproductive mode

When analysing quantitative data on population dynamics or life histories, one is confronted with the extreme scarcity of available data. Most of the data reported from field surveys include only mean and maximum densities, both of them strongly habitat-dependent, and therefore of limited use in a comparative approach. Other data, like survival and fecundity, are much less frequently stated and are mainly gathered from laboratory experiments. Therefore, conclusions drawn from such analyses have to be taken with care.

Mean annual densities of b-pops and p-pops are somewhat higher in the former, but again this is not statistically significant. Most frequent densities are for both reproductive modes between 10,000 and 100,000 ind/m^2, with a frequency maximum of around 20,000 ind./m^2. The distribution of mean abundances between 4 different abundance classes (<100, 100-1000, 1000-10,000, >10,000) is identical for both reproductive modes. Maximum abundances are also higher in b-pops, but due to the extreme variability, this difference is also not significant. P-pops appear to be more homogeneous with regard to maximum abundances.

Lifespan tends to be longer in p-pops due to a longer juvenile development. Fecundity in terms of egg numbers per egg-carrying female does not differ between b-pops and p-pops. None of these results, however, are statistically significant.

Due to the heterogeneity of the data set, and to the fact that ostracods might adopt different strategies in different habitats, the outcome of the overall approach – the between-habitat comparison – did not provide any support or falsification of existing hypotheses.

Within-habitat comparison

A within-habitat comparison should avoid variability due to different environmental factors and should therefore reveal intrinsic properties better than the between habitat approach. Habitats can be classified by many factors, the most interesting with regard to modes of reproduction being 'stability' as defined in the introductory section. For this approach we are forced to concentrate on single, complete studies rather than be able to provide a comprehensive review.

Predictable habitats

THE LAKE LITTORAL

Mallwitz (1984) in his study on the Schmalsee littoral (Germany) demonstrated the dominance of bisexual species in this type of habitat. Only 4 out of the 23 species reproduce parthenogenetically in the study area and two of them are very low in numbers. Species richness is high and environmental variables (except temperature) vary little and if they do then it is in a predictable way. Incidence of disturbances is low, wave action due to the small size of the lake minimal. High substrate diversity allows for rich niche differentiation, a picture frequently encountered in lake littoral zones. None of the species belongs to those commonly producing resting eggs. In summer, fast-developing candonid species dominate whilst slower developers reach their maximum abundance in winter or spring. Mallwitz's study provides considerable support for the hypothesis that sexual reproduction is favoured under predictably changing environmental conditions and rich development of niches. In contrast, wind-exposed shores of larger lakes or large shallow lakes are often dominated by parthenogenetic species. Unpublished data from the author comparing a highly structured littoral within a sheltered bay and a wind-exposed, frequently disturbed shore habitat showed an impoverished ostracod community dominated by a single parthenogenetic population in the latter, whereas the former is rich in species and bisexual forms are abundant. In a similar way is the open lake ostracod fauna of Neusiedlersee, a large, shallow, wind-exposed lake in the east of Austria, dominated by the parthenogenetic populations of *Limnocythere inopinata* and *Ilyocypris bradyi* and the sheltered shore by bisexual species (Wolfram 1993). Summarising these findings, the above studies show that in low disturbance habitats with regularly changing environmental conditions, bisexual species are at an advantage and that whenever environmental change is rapid and unpredictable, resulting in a disturbance of the entire community, parthenogens do dominate.

THE PROFUNDAL OF LAKES

If the above-mentioned findings hold true for the littoral, they must do so even more in the profundal zone of lakes, defined as the lake bottom area without primary production (i.e. below the euphotic zone). This habitat is generally, together with groundwater systems, considered to be the most stable amongst freshwater environments.

The profundal ostracod fauna of temperate lakes is generally characterised by low species richness. None of the assemblages reviewed in this chapter is composed of more than 8 species and communities are surprisingly similar throughout continental Europe. In all studies included in this analysis, bisexual populations are dominating. Most of the communities include a single parthenogenetic representative (*Cytherissa lacustris* or *Candona candida*) which in terms of individual numbers is often the most abundant. In most lakes this community was established very soon after deglaciation and has remained unchanged throughout time (Danielopol *et al.* 1993). All species have a long life cycle (up to 2 years) and low fecundity. Mean densities are lower than in the littoral zone, with significantly less variation reflecting the lack of seasonality and constant environmental conditions. The two species frequently encountered with parthenogenetic populations in the profundal of lakes (*Cytherissa lacustris* and *Candona candida*) are both cold-adapted species showing, when occurring in shallower areas, a clear winter seasonality (Geiger 1990, Griffiths 1995). While *C. lacustris* is exclusively parthenogenetic, at least some of the populations of *C. candida* have males (Griffiths 1993, Semenova 1973). No differences in habitat characteristics of b-pops and p-pops can be detected.

In order to explain the dominance of bisexual populations in the profundal of lakes, several hypotheses have been erected:

1) The number of species is restricted due to the scarcity of coarse-grained niches. This has been proposed by Sarvala (1986) based on data on cyclopoid and harpacticoid copepods. Once all niches are occupied, no further increase in species is possible. This affects parthenogenetic and bisexual species equally. However, in some sexual species inhabiting the profundal of lakes, micro-evolutionary processes, resulting in species clusters, have been observed – *Candona neglecta* in Loch Ness (Griffiths 1995); a *Fabaeformiscandona* species close to *Fabaeformiscandona caudata* in pre-alpine and Polish lakes (Danielopol *et al.* 1988, Namiotko 1992) and *Candona studeri* in Lake Neuchâtel (Kaufmann 1900) – which deserve closer examination in the future. If these lakes would persist long enough, such processes might circumvent the problem of the lack of coarse-grained niches by producing new species with fine-grained adaptations similar to what can be observed in ancient lakes (Martens 1994, 1997). In this context, only bisexuality can produce rapidly enough the genetic diversity necessary for 'short-term' speciation.
2) The overall scarcity of the profundal habitat as compared to shallow water bodies and problems associated with its colonisation. There is no doubt that many more favourable factors have to coincide for a successful colonisation of deep water habitats, but again, this should affect species independently of their reproductive mode (unless they developed different dispersal strategies which is up to now not demonstrated for profundal species) and therefore does not provide an answer to the dominance of bisexual populations in this habitat.

3) The specific ecological conditions of the lake profundal (low oxygen, poor quality food) require the acqusition of physiological adaptations which can be generated more rapidly by bisexuals. In other words: we should expect much more bisexuals pre-adapted to colonise this area. However, we know that parthenogenesis can produce ecological specialists and, on the other hand, highly tolerant generalists (chapter 12), but we still do not know at what rate this is occurring. If rates of producing physiological adaptations are very different between reproductive modes, this could serve as a possible explanation of the dominance of bisexuality in this type of habitat.

GROUNDWATER

The ecology of groundwater ostracods is poorly known. Especially scarce are quantitative field data due to sampling logistics. Groundwater ostracods are overwhelmingly bisexual, species with parthenogenetic populations and affinities to the groundwater being restricted to near-surface areas (Danielopol 1991). As these layers are strongly influenced by the physical environment of the adjacent water bodies (rivers, lakes), we find there, besides some true elements of a hypogean fauna, mainly epigean species with similar biological characteristics as those living in the littoral of lakes or in the hyporheal of rivers (Rouch & Danielopol 1997). In contrast, in the phreatic zone, species number and mean densities are lower than in epigean habitats comparable to those found in the profundal zone of oligotrophic lakes (Rouch & Danielopol 1997) even if data are difficult to compare and vary between geographic areas (higher species numbers in Southern Europe).

We can regard the zone of the groundwater environment which lacks short-term environmental changes as the endpoint of the stability scale. Ostracods occurring exclusively in groundwater have long development (up to 400 days), low fecundity (2 to 3 eggs) but larger egg size as compared to epigean species (Danielopol 1980). A similar trend can be observed in groundwater populations of epigean species – both parthenogenetic and bisexual. The environmental conditions (low temperatures, low oxygen and low predatory pressure) are by far more influencing life history parameters than is the reproductive mode as such. Concerning the hypotheses explaining the dominance of bisexuality in the environmentally stable zone of the groundwater, the same factors listed above for the profundal zone of lakes might also apply here.

Temporary habitats

Temporary aquatic environments can be found throughout Europe, from salt pans like Playa lakes in Spain, rice fields, floated meadows, ditches or road tracks in more temperate climates up to the arctic, where ponds are forming only during short periods of time, just long enough for a single reproductive cycle. These are the habitats where parthenogens should be at an advantage, with their capacity of fast colonisation, fast reproduction and high densities, the classical r-strategists, inhabiting areas with a high incidence of disturbances and low biotic interaction and with abiotic factors ruling the scene.

Temporary freshwater habitats

In spite of their high frequency of occurrence, extensive studies on temporary freshwater ponds are relatively rare. Limnologists, especially in Europe, were much more attracted to larger lakes than to the ubiquitous small water bodies.

One of the most complete studies is the one of McLay (1978a, b) on the succession and competition between four ostracod species in a pool formed by the filling of car tracks with water in the United States. Temperature is the dominant abiotic factor displaying both a depth and a longitudinal gradient. Of the four ostracod species in the pond, two reproduced bisexually and two parthenogenetically. All of them were fast developing, had moderate to high fecundities (70 eggs/female) and showed distinct temporal patterns. Sexual *Candona decora* overcame the drying period in a juvenile stage and therefore had an advantage when the pond was filling. Development times of the other three species differed so that egg-laying periods did not coincide. They formed resting eggs and, depending on the duration of the inundation period, they produced one or two generations. None of the measured quantitative life history parameters differed significantly between species with bisexual and parthenogenetic reproduction. Once again, environmental parameters were most important, more so than the reproductive mode.

In both reproductive modes, species have evolved life history features which allow them to survive under highly unpredictable conditions: resting stages, short development times, high fecundity and the potential for multiple generations. In theory, such ponds can harbour a considerable number of species as a function of longevity and habitat complexity, but temporal succession of species seems to be more frequent. This is true not only for species but also for clones of parthenogenetic species. Clones of *Heterocypris incongruens* in Italian rice fields succeed each other according to their temperature preferences throughout the inundation period of the fields (Rossi & Menozzi 1990). As this kind of water bodies is colonised by developing resting stages from the sediments and to a much lesser degree by immigration from adjacent permanent water bodies, the advantage of parthenogenetic species in colonising only exists in newly created habitats but not during the periodic filling of the ponds.

Temporary saline habitats

Situated at the extreme end of the instability scale are temporary, saline habitats which harbour few species with sometimes extremely high abundances. *Cyprideis torosa* can reach in such environments abundances of up to 1,800,000 individuals/m^2 (Herman *et al.* 1983) and densities of *Limnocythere inopinata* in Austrian astatic ponds and abundances of *Heterocypris barbara* in Spanish Playa lakes (Geiger unpubl.) are in the same order of magnitude. Poorly diversified food sources and niches, together with the strong impact of abiotic conditions (salinity and temperature), allow only few species to survive. Among such specialists we find both species reproducing bisexually and parthenogenetically. *Cyprideis torosa,* and *Prionocypris aragonica*, on the one side, and *Limnocythere inopinata, Ilyocypris* and *Heterocypris* species, on the other. Development is fast and energetic demands are furthermore increased by osmoregulatory stress. Under ideal conditions, as many as 5 generations per inundation cycle can be produced (Yin 1997).

First arrival does not always pay – colonisation and reproductive mode

There is considerable evidence that parthenogenetic species are the first colonisers of man-made water bodies like reservoirs or gravel pit lakes. Being the first arrived does, however, not necessarily mean that this prevents the colonisation by other species when conditions become favourable. Two examples might illustrate this: in two Austrian gravel pit lakes (age less than 30 years), the parthenogenetic populations of *Limnocythere inopinata* are the only ostracods present. Both populations appeared to be monoclonal and therefore support the idea that a single or a few individuals are at the origin of these populations. During the study period, one population disappeared simply because its habitat changed: the sandy bottom became overgrown by a dense mat of macrophytes, thus creating anoxic conditions above the sediment (Geiger in prep.). As a consequence, phytophilous, bisexual ostracod species appeared, most probably from the nearby macrophyte rich ponds. A simple change in habitat explains what could have been interpreted as a short-term replacement of a parthenogenetic species by bisexuals due to competitive superiority.

The second example is taken from the littoral of Lake Mondsee (Austria) where the colonisation history has been investigated (Danielopol *et al.* 1993): parthenogenetic populations were colonising the shallow part of the lake as late as in the Subatlantic, when temperature increased and macrophytes developed. They could readily utilise the newly-available niches and establish stable populations. The apparent replacement of the pre-existing, predominantly bisexual community is due to their temperature and substrate preferences and not to competition between bisexual and parthenogenetic populations. Indeed, the original post-glacial littoral community moved to deeper areas and nowadays forms the typical sublittoral assemblage.

Conclusions

Within- and between-habitat comparisons stress the importance of the acquisition of resting stages and the predictability and complexity of the habitat as major factors determining the occurrence, numbers and life history strategies of ostracod species. Concordance with existing theory was only detected where it predicts the success of bisexuality in habitats with predictable environmental conditions. The lack of significance in the results of the overall analysis is mainly due to the fact that bisexuals are more successful than predicted in harsh, unpredictable environments. Parthenogenetic and bisexual populations of the same species are co-occurring in such habitats (*Eucypris virens*) or, if they are not, ecological differences between their respective habitats cannot be detected (*Candona candida, Heterocypris barbara*). Parthenogenetic and bisexual populations within a given habitat and belonging to different species show a high degree of similarity with regard to life history parameters and abundances. The expression of their potentialities depends much more on the environmental set-up in terms of abiotic and biotic interactions than on the reproductive mode.

Perspectives for future work

When looking for data on population dynamics and life histories of ostracods, I was surprised how many papers have been published and how little information could be extracted from them. Whilst mean and maximum abundances are frequently mentioned, survival and fecundity data are extremely rare and, if existing, are derived from laboratory experiments. Quantitative data taken at time intervals corresponding to the life cycle of the species with enough replicates to ensure reliability are urgently needed. The analyses should include all developmental stages and be conducted on the whole ostracod community. Fecundity data should not only include the number of eggs observed within the ovaries, but also the total number of eggs laid or the number of egg depositions. With these data, life tables could be constructed and life history parameters such as intrinsic rates of population increase, generation times and mortality schedules calculated.

There is special need for this type of studies in temporary environments which are often neglected by limnologists. The assessment of 'stability' is best done following the method given by Stearns (1981). For this, one needs time series of environmental data which are best approximated by long-term records of meteorological data, as water level, salinity and temperature in small water bodies are related to the regional climate. The stability of food supply is much more difficult to assess. It requires the identification of potential food items and the monitoring of their dynamics. A considerable amount of work and therefore seldom undertaken. Last, but not least, the stability of biotic interactions should be estimated by identifying the number, nature and intensity of links between compartments and by evaluating the number and dynamics of available niches. Such sampling programs must be designed in a way to test the specific hypotheses concerning the short-term ecological benefits of the different reproductive modes.

Beside the need for new data, new questions became apparent from this analysis:

Why are parthenogenetic ostracods not much more successful than bisexuals, given their advantage of rapid colonisation and the fact that they are lacking all kind of costs associated with sexuality (Lewis 1987), especially when population size is large enough to circumvent potentially negative genetic effects (see chapter 1)?

How do parthenogenetic species survive in highly structured, rapidly changing environments, where they are supposed to be outcompeted by bisexual species?

How do p-pops and b-pops of one and the same species coexist in the same water body (see also the excellent review on coexistence of clones by Sebens & Thorne 1985)?

Could it be possible that the advantages of parthenogenesis predicted by theory are simply not fully expressed in complex environments with high intensities of biotic interactions? Some indications from as yet unpublished competition experiments show that a parthenogenetic species displays much faster population growth when kept alone than in the presence of a bisexual species, whereas the population growth rate of the latter remains unchanged. Similarly, both parthenogenetic and bisexual populations have higher population growth rates when alone than when they occur in multi-species assemblages (*Cyprideis torosa*, *Limnocythere inopinata, Heterocypris incongruens*).

Once it is accepted that parthenogenetic ostracods have the necessary genetic potential for evolution and that they are not an evolutionary dead end, why should they not develop under similar selective forces similar adaptations and life history strategies as bisexual species?

References

Baltanás, A. 1994. Taxonomy and ecology of *Eucypris virens* (Ostracoda, Cyprididae). *in:* Horne, D.J. & Martens, K. (eds), *The evolutionary ecology of reproductive modes in non-marine Ostracoda,* Greenwich Univ. Press: 9-16.

Bell, G. 1982. *The masterpiece of nature,* Croom Helm: 625 pp.

Benzie, J.A.H. 1989. The distribution and habitat preference of ostracods (Crustacea: Ostracoda) in a coastal sand-dune Lake, Loch of Strathbeg, north-east Scotland. *Freshw. Biol.*, **22:** 309-321.

Carvalho, G.R. 1994. Evolutionary genetics of aquatic clonal invertebrates: Concepts, problems and prospects. *in:* Beaumont, A.R. (ed), *Genetics and evolution of aquatic organisms*, Chapman & Hall: 291-323.

Chaplin, J.A., Havel, J.E. & Hebert, P.D.N. 1994. Sex and ostracods. *Trends Ecol. Evol.*, **9:** 435-439.

Danielopol, D.L. 1980. Sur la biologie de quelques Ostracodes Candoninae épigés et hypogés d'Europe. *Bull. Mus. natn. Hist. nat., Paris, 4ème ser.*, **2:** 471-506.

Danielopol, D.L. 1991. Spatial distribution and dispersal of interstitial Crustacea in alluvial sediments of a backwater of the Danube at Vienna. *Stygologia*, **6:** 97-110.

Danielopol, D.L., Geiger, W., Tölderer-Farmer, M., Orellana, C.P. & Terrat, M.N. 1988. In search of *Cypris* and *Cythere* – A report of the evolutionary ecological project on limnic Ostracoda from the Mondsee (Austria). *in*: Hanai, T., Ikeya, N. & Ishizaki K. (eds), *Evolutionary biology of Ostracoda,* Kodansha: 485-500.

Danielopol, D.L., Handl, M. & Yin, Y. 1993. Benthic ostracods in the pre-alpine deep lake Mondsee. Notes on their origin and distribution. *In:* McKenzie, K.G. & Jones, P.J. (eds), *Ostracoda in the Earth and Life Sciences*, Balkema: 465-480.

Delling, D. 1981. Eutrophierungseffekte an Ostrakodengemeinschaften oligohaliner Gewässer Norddeutschlands. PhD thesis, Univ. Kiel.

Geiger, W. 1990. Field and laboratory studies on the life cycle of *Cytherissa lacustris* (SARS) (Crustacea, Ostracoda) with special emphasis on the role of temperature. *Bull. Inst. Géol. Bassin d'Aquitaine, Bordeaux*, **47:** 191-208.

Griffiths, H. 1993. The spatial distribution of benthic ostracods in the profundal zone of Loch Ness. *Scot. Naturalist*, **105:** 137-147.

Griffiths, H. 1995. The application of freshwater Ostracoda to the reconstruction of Late Quaternary Palaeoenvironments in North-western Europe. PhD thesis, Univ. Cardiff.

Herman, P.M.J., Heip, C. & Vranken, G. 1983. The production of *Cyprideis torosa* Jones 1850 (Crustacea, Ostracoda). *Oecologia*, **58:** 326-331.

Hughes, R.N. 1989. *A functional biology of clonal animals,* Chapman & Hall: 331pp.

Jackson, J.B.C. 1985. Distribution and ecology of clonal and aclonal benthic invertebrates. *in:* Jackson, J.B.C., Buss, L.W. & Cook R.E. (eds), *Population biology and evolution of clonal organisms,* Yale Univ. Press: 297-355.

Kaufmann, A. 1900. Cypriden und Darwinuliden der Schweiz. *Rev. suisse Zool.,* **8(3):** 209-421.

Lewis Jr., W.M. 1987. The cost of sex. *in:* Stearns, S.C. (ed), *The evolution of sex and its consequences,* Birkhäuser Verlag: 33-57.

Lynch, M. 1984. Destabilizing hybridization, general-purpose genotypes and geographic parthenogenesis. *Quart. Rev. Biol.*, **59:** 257-290.

McGregor, D.L. 1969. The reproductive potential, life history and parasitism of the freshwater ostracod *Darwinula stevensoni. in:* Neale, J.W. (ed), *The taxonomy, morphology and ecology of recent Ostracoda,* Oliver & Boyd: 194-221.

McLay, C.L. 1978a. Comparative observations on the ecology of four species of ostracods living in a temporary freshwater puddle. *Can. J. Zool.*, **56:** 663-675.

McLay, C.L. 1978b. The population biology of *Cyprinotus carolinensis* and *Herpetocypris reptans* (Crustacea, Ostracoda). *Can. J. Zool.*, **56:** 1170-1179.

Mallwitz, J. 1984. Untersuchungen zur Ökologie litoraler Ostracoden im Schmal- und Lüttauersee (Schleswig-Holstein, BRD). *Arch. Hydrobiol.*, **100:** 311-339.
Marín, J.A. 1984. Estudio del desarrollo de los ostrácodos *Eucypris aragonica* y *Heterocypris salina* en cultivo de barro. *Limnética* **1:** 345-354.
Martens, K. 1989. *Ovambocythere milani* gen. n., sp. n. (Crustacea, Ostracoda), an African Limnocytherid raised from dried mud. *Rev. Zool. afric.*, **103:** 379-388.
Martens, K. 1994. Ostracod speciation in ancient lakes: a review. *in*: Martens, K., Goddeeris, B. & Coulter, G. (eds), *Speciation in Ancient Lakes*, *Adv. Limnol.*, **44:** 203-222.
Martens, K. 1996. On *Korannacythere* nov.gen. (Crustacea, Ostracoda) a new genus of temporary pool limnocytherids from southern Africa, with the description of three new species and a generic reassessment of the Limnocytherinae. *Bull. k. Belg. Inst. Natuurw. Biol.*, **66:** 51-72.
Martens, K. 1997. Speciation in Ancient lakes (review). *Trends Ecol. Evol.*, **12:** 177-182.
Martens, K. & Tudorancea, C. 1991. Seasonality and spatial distribution of the ostracods of Lake Zwai, Ethiopia (Crustacea, Ostracoda). *Freshw. Biol.*, **25:** 233-241.
Maynard Smith, J. 1978. *The evolution of sex,* Cambridge Univ. Press.
Namiotko, T. 1992. On a collection of subfossil and recent Ostracoda found in deep lake habitats from North-Eastern Poland – Rare and little known species. *in: Symposium on glacial relict and rare Crustacea.* Marine Biology Center, Pol. Acad. Sci.
Newrkla, P. 1985. Respiration of *Cytherissa lacustris* (Ostracoda) at different temperatures and its tolerance toward temperature and oxygen concentration. *Oecologia,* **67:** 250-254.
Otero, M., Rossi, V., Baltanás, A. & Menozzi, P. (in press). Effect of genotype and photoperiod on diapause strategies in *Eucypris virens* (Jurine, 1820) (Crustacea: Ostracoda). *Arch. Hydrobiol.*
Ranta, E. 1979. Population biology of *Darwinula stevensoni* (Crustacea, Ostracoda) in an oligotrophic lake. *Ann. Zool. Fennici*, **16:** 28-35.
Rossi, V. & Menozzi, P. 1990. The clonal ecology of *Heterocypris incongruens* (Crustacea: Ostracoda). *Oikos*, **57:** 388-398.
Rouch, R. & Danielopol, D.L. 1997. Species richness of Microcrustacea in subterranean freshwater habitats. Comparative analysis and approximate evaluation. *Int. Rev. Ges. Hydrobiol.*, **82:** 121-145.
Sarvala, J. 1986. Patterns of benthic copepod assemblages in oligotrophic lakes. *Ann. Zool. Fenn.*, **23:** 101-130.
Sebens, K.P. & Thorne, B.L. 1985. Coexistence of clones, clonal diversity and the effects of disturbance. *in:* Jackson, J.B.C., Buss, L.W. & Cook, R.E. (eds), *Population biology and evolution of clonal organisms,* Yale Univ. Press: 357-398.
Semenova, L.M. 1973. Key features in the biology of *Candona candida* (Ostracoda, Crustacea). *Hydrobiol. J.*, **15:** 21-24.
Stearns, S.C. 1981. On measuring fluctuating environments: Predictability, constancy, and contingency. *Ecology*, **62:** 185-199.
Stearns, S.C. 1987. Why sex evolved and the differences it makes. *in:* Stearns, S.C. (ed), *The evolution of sex and its consequences,* Birkhäuser Verlag: 15-27.
Strayer, D. 1988. Life history of a lacustrine ostracod. *Hydrobiologia*, **160:** 189-191.
Thaler, B. 1977. Die benthischen Crustaceen des Piburger Sees (Ötztal, Tirol). PhD thesis, Univ. Innsbruck.
Theisen, B.F. 1966. The life history of seven species of Ostracods from a Danish brackish-water locality. *Meddel. Danmarks Fisk. Havundersogelser.*, **4:** 215-270.
Williams, G.C. 1975. *Sex and evolution,* Princeton University Press.
Wolfram, G. 1993. Untersuchung der benthischen Lebensgemeinschaft des Neusiedlersees. Dipl. Thesis, Univ. Vienna: 99pp.
Yin, Y. 1997. Contribution to the morphology and ecology of Ostracoda Limnocytheridae and Candonidae. Comparative studies between Austrian and Chinese Crustacea. PhD thesis, Univ. Vienna: 163pp.

CHAPTER 13

OSTRACOD POPULATIONS AS METAPOPULATIONS

A. Baltanás

Abstract

Population structures and dynamics of many ostracod species can be described as metapopulations, living in patches (separate waterbodies) which are connected through passive dispersal. A simple metapopulation model was adopted for a single-species system with mixed reproduction, where sexual populations are competitively superior and asexuals show higher colonisation rates. If colonisation success of a single patch is supposed to depend on geographic distance, the model provides amongst other results the typical distribution pattern of geographical parthenogenesis, with a widespread distribution of asexuals, combined with a restricted, southern distribution of sexuals.

Introduction

For a given species, suitable habitats are not distributed in a continuous range but appear in patches. Some of these patches are occupied by the species and some are not. Quite frequently, the pattern of occupancy is not fixed but permanently changing with the species going extinct in some patches and (re)colonising others. The dynamics among patches were already valued as important by some population ecologists in the fifties (Andrewartha & Birch 1954). However, the attention paid to the spatial pattern, as it was developed in the turmoil of the controversy 'density-dependent *vs.* density-independent factors', was soon forgotten. It was reborn in the '60s with the theory of island biogeography (MacArthur & Wilson 1967). But it was Richard Levins (1969) who finally introduced the term 'metapopulation' in the ecological jargon and who set the basis for a general spatially structured approach to populations (and communities) ecology.

A metapopulation was originally defined as 'a series of local populations that are linked via dispersal' (Levins 1969). Thus, the concept of metapopulation concerns the spatial structure of populations. Can ostracod species be approached as metapopulations?

Ostracod faunas are ill-known for many places. There are many species that have been found just once in their so-called type locality (see chapter 4). However, it is reasonable to assume that most of those species have broader distribution ranges than currently known. Indeed, we know already of many ostracod species widely distributed at a regional, continental or even global scale. In those cases, suitable

Sex and parthenogenesis: evolutionary ecology of reproductive modes in non-marine ostracods, edited by K. Martens, pp. 229–241

habitats for ostracods (i.e. mostly water bodies) are patchily distributed and surrounded by a dry hostile matrix. That is the rule for most aquatic organisms. Each of those patches can harbour local populations of ostracod species with their own dynamics: a stable population, a population in decline running to extinction or a successful population growing in numbers of individuals from one year to the next. Of course, some patches remain empty in spite of the fact that they fulfil the ecological requirements for species to survive. Occupied and empty patches form a pattern that is not static, as the status of a habitat patch can change. The frequency of these changes depends on colonisation and extinction rates.

Ostracods disperse passively, but, especially cypridids, in a very efficient way. Resistant eggs that can survive desiccation can be transported by wind (McKenzie 1971), fish (Kornicker & Sohn 1971, Victor *et al.* 1979), insects (Fryer 1953), inside the alimentary canal of birds (Proctor 1964, Proctor & Malone 1965, Frith *et al.* 1969), attached to the body of birds (Sandberg & Plusquellec 1974, De Deckker 1977) and by man when trading with plant seeds (Fox 1965, McKenzie & Moroni 1986). The better an ostracod species disperses, the higher its colonisation rate; i.e. the probability that it will reach an empty patch and found a new local population.

There is a plethora of causes that can lead to the extinction of a population. Some are endogenous (low growth rates, inbreeding depression, etc.) and some are exogenous (mainly climatic catastrophes). Many ostracod species live in fluctuating habitats like temporary ponds or the littoral zone of lakes. In those unstable habitats, extinction rates are expected to be higher because of the stronger influence of stochastic events. Natural selection favours strategies that buffer the effect of fluctuations: life cycles synchronised with environmental signals, bet-hedging strategies for reproductive events, high intrinsic growth rates, etc. But, although population extinction rates can be reduced in this way, they will never become zero.

Summarising, many ostracod species can be described as a set of local populations living in habitat patches (water bodies) which are likely to be interconnected through colonisation (passive dispersal). Thus, the metapopulation approach seems adequate to explore the behaviour of ostracod population dynamics.

Metapopulation models

The simplest metapopulation model is that of Levins (1969)

$$\frac{\partial p}{\partial t} = cp\,(1\text{-}p) - ep, \qquad (13.1)$$

where c = colonisation rate; e = extinction rate and p = fraction of patches that are occupied.

The number of patches occupied at equilibrium is

$$\hat{p} = 1 - \frac{e}{c} \qquad (13.2)$$

Levin's classical model requires several assumptions (Hanski & Simberloff 1997):
– patches have equal areas and isolation,
– local populations have uncorrelated dynamics, and
– there is no influence of migrants on the dynamics of local populations.

Most likely, real ostracod metapopulations do not fully satisfy these requirements, especially those concerning the degree of isolation of the habitat patches and the independence of dynamics among local populations. Nevertheless, the model still holds after relaxing some assumptions as long as the notion of local inbreeding populations connected *via* dispersion is not altered (Hanski & Simberloff 1997).

One of the most attractive aspects of Levin's model is that it can be expanded for multi-species systems (Hastings 1980, Nee & May 1992, Tilman 1994). For instance, a system with two interacting species with a competitive hierarchy can be modelled as:

$$\frac{\partial p_1}{\partial t} = c_1 p_1(1-p_1) - e_1 p_1$$

$$\frac{\partial p_2}{\partial t} = c_2 p_2(1-p_1-p_2) - e_2 p_2 - c_1 p_1 p_2 \qquad (13.3)$$

Here, species 1 is a superior competitor which displaces the inferior competitor (species 2) when both co-occur at a site.

The number of patches occupied at equilibrium by both species is:

$$\hat{p}_1 = 1 - \frac{e_1}{c_1}$$

$$\hat{p}_2 = 1 - \hat{p}_1 - \frac{(e_2 + c_1 \hat{p}_1)}{c_2} \qquad (13.4)$$

When both species do not differ in their respective extinction rates ($e_1=e_2$), the model predicts (Fig. 13.1) that for the superior competitor to persist in the landscape, the ratio c_1/e_1 should be greater than 1. The inferior competitor will be outcompeted if its colonisation rate is lower than that of the superior competitor. To have $c_2>c_1$ is a necessary, but not sufficient, condition. The inferior competitor will persist only when e_1/c_1 is larger than e_2/c_2 [or $c_2/e_2> c_1/e_1$]. In other words, a high colonisation rate does not necessarily make the inferior competitor to persist, but the ratio between colonisation rate and extinction rate does. Note that, when both species are present and stable in the system, the fraction of unoccupied patches remains constant and equal to the ratio c_1/c_2.

Let us use this model (13.4) to approach the behaviour of a single-ostracod species system which includes two kinds of reproductive modes, sexual and asexual. For this, it is necessary to add some assumptions to the premises of Levin's classical model (see above):

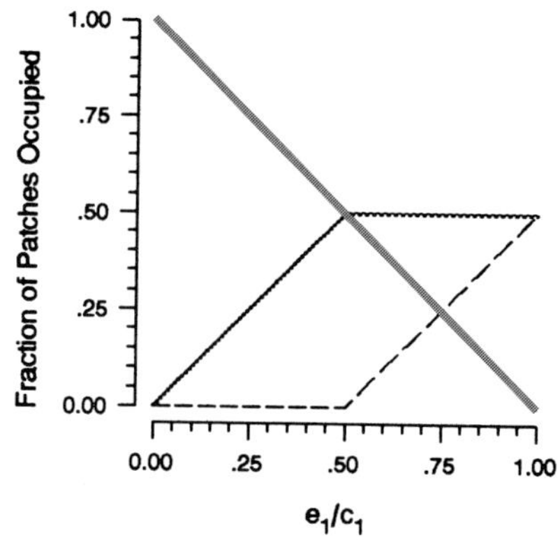

Fig. 13.1 Fraction of patches occupied at equilibrium by the superior (solid line) and inferior (dashed line) competitor. Grey line is the fraction of unoccupied patches. [e_1=e_2 and c_2=2·c_1]

1) Sexual and asexual lineages are fixed and do not change one into the other; i.e. sexual lineages do not regularly produce asexual lineages. This is an unlikely supposition at an evolutionary time scale because there is evidence suggesting that apomictic lineages might have resulted from multiple divergence from sexual species (Chaplin *et al.* 1994, Schön *et al.* in prep.). At an ecological time scale, however, the assumption is quite acceptable.
2) One of the two reproductive modes must be considered competitively superior. This superiority is absolute, meaning that the superior competitor always displaces the inferior competitor from a site and cannot be displaced by it. Thus, habitat patches (water bodies) which are occupied can harbour only one reproductive mode.

There is a widespread belief that parthenogens perform well in situations where sexuals could not succeed (sub-optimal environments), leading to wide distribution ranges that sometimes reach harsh environments like the arctic regions. Success can be achieved by parthenogens either through ecological differentiation between clones (Rossi & Menozzi 1990, 1993, Otero *et al.* in press) or through the production of clones with a 'general purpose genotype' (Lynch 1984). However, evidence of direct competition between reproductive modes is scarce. Chaplin (1993) reported the case of displacement of a sexually reproducing population of *Candonocypris novaezelandiae* by a conspecific parthenogen. Although several ponds in which both sexual and asexual *Eucypris virens* co-occur too (Rossi *et al.* in prep.) have been sampled, the duration of the study period was not long enough to test for changes in the frequency of each type. On the other hand, it has been proposed that sexuals outcompete their asexual conspecifics (or congenerics) in stable environments (Chaplin *et al.* 1994). Most likely, there is not a superior competitor as such but different lineages are better competitors under different circumstances. Despite that, simulations drive our attention to certain features that otherwise would have been neglected.

One model parameter of great importance in the case at hand is colonisation rate. Parthenogens are excellent colonisers. One drought-resistant egg from a parthenogenetically reproducing population would theoretically be enough to found a new population in an empty patch. A sexual population needs, at least, two propagules of opposite sex to reach such a patch and their resulting individuals need to locate each other in a temporally and spatially very diluted environment. Other things being equal, i.e. assuming egg banks of equal size for each population type, the expected colonisation rate of an asexual population is not just twice that of a sexual one but, at least, two times the square root of it. This is so, because for a sexually reproducing population, the propagules that must arrive to the new patch are not just 2 randomly selected out of the total egg-bank, but must be one of each gender, male and female.

Hence,

$$c_{sexual} = (c_{asexual} / 2)^2. \qquad (13.5)$$

Non-linearity in the relationship between both colonisation rates produces a pattern which, contrary to the model explored above, is dependent on the absolute value of c_1 (Fig. 13.2).

For e_1=e_2, and rearranging (9.4) we arrive at

$$\hat{p}_1 = 1 - \frac{e_1}{c_1}$$

$$(13.6)$$

$$\hat{p}_2 = 1 - \hat{p}_1 - \frac{c_1}{c_2}\,(e_1 / c_1 + \hat{p}_1)$$

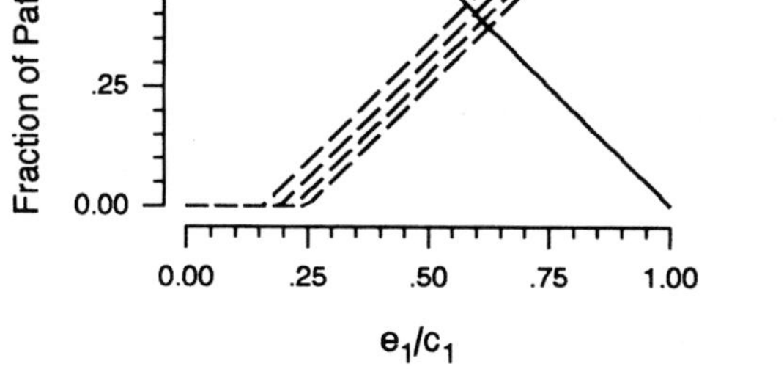

Fig. 13.2 Fraction of patches occupied at equilibrium in a two-reproductive mode system with one type acting as superior competitor (see text). Superior competitor=solid line; inferior competitor= dashed line.

The fraction of patches occupied by the superior competitor depends solely on the ratio e_1/c_1 (the smaller the ratio, the more patches are occupied at the equilibrium). As long as the population size among habitat patches is constant, biological fitness correlates with the number of patches occupied. Thus, it is reasonable to expect that the selective pressures will act towards increasing c_1 and/or lowering e_1. The '-success' of the inferior competitor, estimated as the fraction of patches occupied, depends on both e_1/c_1 and c_1/c_2 (13.6). In a two-species system, an increase of c_1 can be (over)compensated by the inferior competitor increasing its own colonisation ability (c_2). But in a sexual-asexual system like the one presently considered for ostracods, it is difficult to accept that a trait such as colonisation-rate can evolve independently in both kinds. If the relationship between c_1 and c_2 is mechanistically (non-linear) determined as explained above, increases in c_1 will eventually lead to increases in c_2 but with a decrease in the ratio c_1/c_2 and, consequently, a decrease in the number of patches occupied by the inferior competitor. The vicious circle is broken when the two extinction rates are not equal. If the asexuals, which are frequently widely spread over peripheral areas (sub-optimal habitats) of the species distribution range (geographical parthenogenesis), have a lower mean extinction rate, then they can cope much better with their sexual counterparts (Fig. 13.3).

Why do we allow for extinction rates to differ between both kinds of reproductive mode, while at the same time we assume that colonisation rates cannot evolve independently? Asexual lineages originate from sexual populations and, hence, they inherit the traits already present in their ancestors (e.g. lifespan, fecundity, age at maturity, etc.). There is no evidence, likely because of the paucity of population-oriented studies in non-marine ostracods, concerning differences in biological features determining colonisation abilities between sexuals and asexuals. By biological features we mean population size, fecundity, propagules survival rate, etc. We can be certain, however, of some probabilistic effects which set a minimum level for those differences (see 13.6). With regard to extinction rates, geographic patterns of both reproductive modes within species suggest that asexuals have lower extinction rates (widespread distributions, occur in harsh environments, etc.). How can we explain the divergence in life-histories between asexual and sexual lineages? In spite of the absence of recombination, clones in the asexual lineages can diverge significantly as shown by phenotypic selection models (Lynch & Gabriel 1983) and neutral models (Lynch 1994). However, we do not even need to apply such an explanation as long as clones can be independently generated at different times from different individuals (different genotypes) in the sexual lineage. Differentiated clones produced in such way can be subject to natural selection and be pushed towards areas where their ecological tolerance translates into lower extinction rates. Under these premises, the metapopulation model here explored predicts equilibrium states in which asexual lineages, although theoretically inferior in their competitive abilities, can succeed, by occupying a certain fraction of the total habitat patches available.

It is worth stressing again that the speculations developed here do not seek to make precise predictions about ostracod population dynamics. They are used under the philosophy of the so-called 'strategic' models (Holling 1966) which are intended to be theoretical ones which aim to give qualitative general insights (May 1974). So, the model gives us useful guidelines for future research which we can summarise as follows:

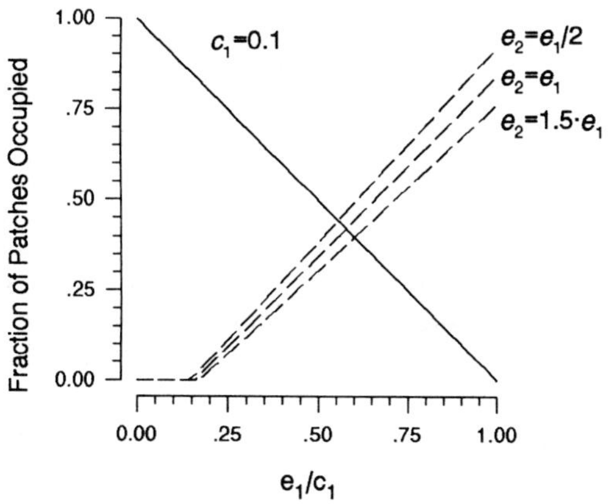

Fig. 13.3 Influence of extinction rates on fraction of patches occupied. Superior competitor=solid line; inferior competitor=dashed line.

- In species with both sexual and asexual reproductive mode, the sexual mode must be the superior competitor in certain habitat/climate types, otherwise, because of its lower colonisation rate, it could not persist.
- The superior competitor is not necessarily always the most successful. Under certain conditions, the inferior competitor can occur in more patches.
- For asexual populations to occupy a larger fraction of the patches available, i.e. with sexual populations less common and with more restricted distribution ranges, there must be a significant difference between extinction rates in both kinds and the ratio of extinction to colonisation rate in the sexual mode must be close to 1.
- Both the superior and the inferior competitors never fill all available patches. There is always a proportion of patches which are empty of either type.
- The influence of extinction rates in metapopulation dynamics of a system like the one described can be greater than that of colonisation rates.

Patches occupied and habitat loss/gain

The whole system can be further complicated by introducing the effects of a changing number of habitat patches. This modification (13.7) was introduced in order to reflect changes due to habitat loss (Nee & May 1992, Tilman 1994). When we consider the nature of the habitats inhabited by many of the non-marine ostracod species with mixed reproduction (ponds, ditches, troughs, littoral zone of lakes, etc.), we can agree that the number of patches which can be occupied by the species is permanently changing. To reflect this in the model, we simply have to add a parameter h for the fraction of the potential total number of patches which are available for the species to colonise (Nee & May 1992, Tilman 1994). The fraction of patches occupied for each species (or reproductive mode lineage) at the equilibrium is:

$$\hat{p}_1 = h - \frac{e_1}{c_1}$$

$$\hat{p}_2 = h - \hat{p}_1 - \frac{(e_2 + c_1\hat{p}_1)}{c_2} \qquad (13.7)$$

The results are similar to those obtained for models with a fixed amount of available patches (Fig. 13.4). It should be noted, nevertheless, that the inferior competitor increases its spatial dominance over the superior competitor when the total number of habitat patches decreases. This fact and the presence of empty patches in the system has some important consequences for conservation that will be further explored in chapter 17.

Spatial explicit models for ostracod metapopulations

The models explored above do not include explicit information with regard to space. All patches are assumed to have an equal probability of being colonised and propagules are thought to be able to reach any patch from any other. Those are unlikely suppositions in the real world, specially if we are dealing with large spatial scales. Although applications of the metapopulation model have been addressed to local-regional scales (Thomas & Hanski 1997, Smith & Gilpin 1997) it is tempting, due to the large distribution ranges of many ostracod species, to extend it at larger scales. In doing so, spatial relationships become increasingly important, as it becomes less probable that an egg produced somewhere in the Mediterranean

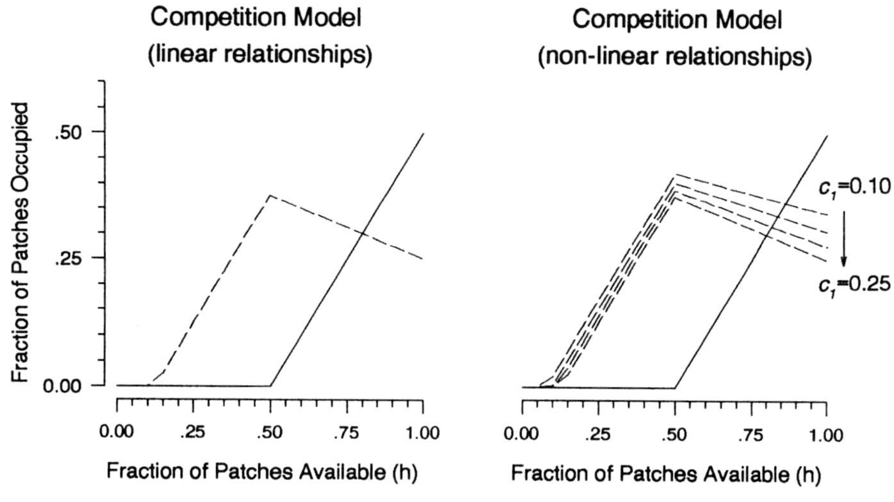

Fig. 13.4 Fraction of patches occupied in two-species systems having a linear (c_2=$4c_1$) and a non-linear relationship ($c_2=2\sqrt{c_1}$) between colonisation rates of both competitors. [e_1= e_2; e_1/c_1=.5]. Superior competitor=solid line; inferior competitor=dashed line.

region can reach with similar likelihood a pond nearby and a ditch in Denmark. To account for such distance dependence and for the structure of a complex territory like Western Europe, we have developed a spatial explicit cellular automaton model (Hassell *et al.* 1991, Hanski & Simberloff 1997). As for the models explored above, this is again a strategic model which pretends to grasp the influence in distribution patterns of some ecological factors given a simple (but realistic) set of assumptions.

Some premises already used in the former approach are maintained here, for instance, the non-linear relationship between colonisation rates in both sexuals and asexuals, or the competitive superiority of the sexuals. For the analysis, Western Europe (*sensu lato*) is divided in 2049 quadrats. Adjacent quadrats have their mid-point 50 km apart. Not all the quadrats are expected to include suitable habitats, only 1000 of them. Those quadrats ready to be used have been chosen at random. In the real world, however, we expect them to be distributed in a contagious way (e.g. the Lake District, the Finnish lakes, Salzkammergut lake area, etc.).

The starting point for each simulation run is a set of 30 quadrats occupied by populations with each kind of reproductive mode, 25 sexuals and 5 asexuals, respectively (Fig. 13.6 A-C). These populations are distributed in the circum-Mediterranean area (south of parallel 48°N) to reflect the likely distribution of these organisms during the glacial age. Although, as far as we know, there are no data concerning extinction rates in non-marine ostracods (see Jeffries 1994 for estimation of turnover in various invertebrate groups inhabiting wetland ponds), we assume it must be low. In a similar way, the colonisation rates used in the model have been selected arbitrarily, given the lack of information. The underlying idea is not to replay what actually happened to ostracod populations in Europe, but to explore the kind of patterns that can emerge from a process which only considers colonisation and extinction as driving forces.

For each simulation, 500 rounds are iterated. Each of these rounds includes a cell-by-cell evaluation of the actual extinction. The remaining populations subsequently act as potential founders. For each empty cell, the likelihood of being colonised from any of the occupied cells is evaluated. This likelihood is function of the number of propagules (resting eggs) produced by each population (which here is assumed to be fixed and equal for all of them), the kind of reproductive mode (which determines the number of propagules needed for the colonisation to succeed) and the distance between both the source and the target cells. Distance is assumed to affect the actual colonisation rate in a non-linear way. Long range distances dramatically decrease the efficiency of the colonisation process (Fig. 13.5).

Some of the possible outputs are displayed in Fig. 13.6 (A-C). Given that extinction rates and propagule availability have been kept constant, differences between them reflect the influence of the distance factor. Slight changes in the dependence of colonisation success on geographic distance result in patterns ranging from a way down to extinction (Fig. 13.6A) to a widespread distribution of the asexuals combined with a restricted southern distribution for the sexuals (Fig. 13.6C) (notice the resemblance of this last pattern with the so-called geographical parthenogenesis). The importance of contingency is also stressed. In Fig. 13.6B, the eastern-biased location of the original populations determines an easterly distribution in the final pattern. It is convenient to remark here that although the original locations (n=5) were randomly placed, such biases are expected when their number is low.

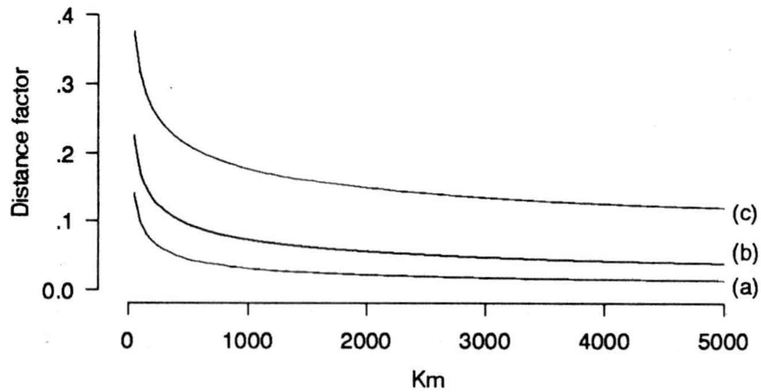

Fig. 13.5 Distance factors (DF) used in the spatial explicit model. (a) DF = $1/dist^{0.5}$; (b) DF = $1/dist^{0.38}$; (c) = $1/dist^{0.25}$.

This simulation shows that, despite being competitively superior, sexual populations can be doomed to extinction due to their lower colonisation rates. Indeed, real sexual populations are likely to overcome this fate because actual colonisation rates must be higher than here assumed. One possible reason for this is that although 2 propagules, each of different gender, are indeed needed for the successful colonisation, both eggs do not necessarily have to be produced in the same habitat cell, as different cells can act as donors to found a new population.

Concluding remarks

Models like the one here presented give us an important message: complex patterns, similar to those actually observed in nature, can result from different kinds of processes. The one we have explored, although simple and even naïve, is able to mimic realistic patterns. It does not include any reference to changes in ecological requirements along the latitudinal gradient, or to demographic variability between populations (both of which are likely to exist). However, by providing an insight into the relative importance of factors such as 'distance effect' or historical contingency, such models indicate relevant variables worth studying if we want to produce testable hypotheses on the biogeography of non-marine ostracods.

Finally, this model also calls for a tighter co-operation between neontology and palaeontology. Many parameters affecting geographic distribution and metapopulation dynamics can only be assessed on living populations. Some others, however, must be traced back in time using palaeoecological techniques. There are not many groups of organisms suitable for such a multiple approach, but non-marine ostracods, with a rich fossil record (at least for some lineages), can provide a major contribution.

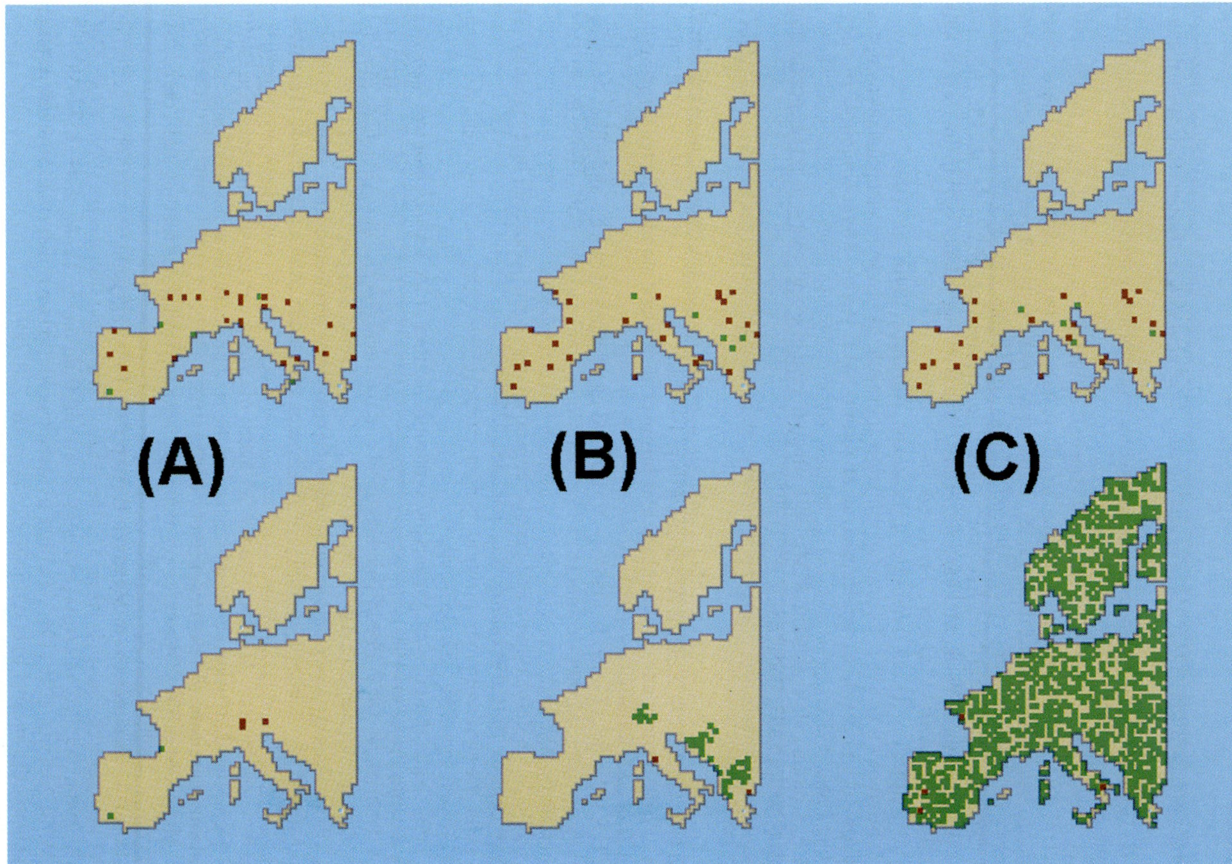

Fig. 13.6 Distribution patterns obtained in a simulation spatially explicit cellular model after 500 rounds. Colonisation and extinction rates are kept constant. Only the distance effect changes: (A) = $1/dist^{0.5}$; (B) = $1/dist^{0.38}$; (C) = $1/dist^{0.25}$.

References

Andrewartha, H.G. & Birch, L.C. 1954. *The distribution and abundance of animals,* Univ. Chicago Press.

Chaplin, J.A. 1993. The local displacement of a sexually reproducing ostracod by a conspecific parthenogen. *Heredity*, **71:** 259-268.

Chaplin, J.A., Havel, J.E. & Hebert, P.D.N. 1994. Sex and ostracods. *Trends Ecol. Evol.*, **9:** 435-439.

De Deckker, P. 1977. The distribution of the "giant" ostracods (family: Cyprididae Baird 1845) endemic to Australia. *in*: Löffler, H. & Danielopol, D.L. (eds), *Aspects of Ecology and Zoogeography of Recent and Fossil Ostracoda,* Junk, The Hague: 285-294.

Fox, H. M. 1965. Ostracod crustacea from ricefields in Italy. *Mem. Ist. Ital. Idrobiol.*, **18:** 205-214.

Frith, H.J., Braithwaite, L.W. & McLean, J.L. 1969. Waterfowl in inland swamp in New South Wales. II. Food. *CSIRO Wild Res.*, **14:** 17-64.

Fryer, G. 1953. Notes on certain freshwater crustaceans. *Naturalist*, **846:** 101-109.

Hanski, I. & Simberloff, D. 1997. The metapopulation approach, its history, conceptual domain, and application to conservation. *in*: Hanski, I. & Gilpin, M.E. (eds), *Metapopulation Biology*, Academic Press: 5-26.

Hassell, M.P., Comins, H.N. & May, R.M. 1991. Spatial structure and chaos in insect population dynamics. *Nature*, **353:** 255-258.

Hastings, A. 1980. Disturbance, coexistence, history, and competition for space. *Theor. Pop. Biol.*, **18:** 363-373.

Holling, C.S. 1966. The strategy of building models of complex ecological systems. *in*: Watt, K.E.F. (ed), *System analysis in ecology,* Academic Press: 195-214.

Jeffries, M. 1994. Invertebrate communities and turnover in wetland ponds affected by drought. *Freshw. Biol.*, **32:** 603-612.

Kornicker, L.S. & Sohn, I.G. 1971. Viability of ostracod eggs ingested by fish and effect of digestive fluids on ostracod shell- ecologic and paleontologic implications. *Bull. Centr. Rech. Pau-SNPA Suppl.,* **5:** 207-237.

Levins, R. 1969. Some demographic and genetic consequences of environmental heterogeneity for biological control. *Bull. Entomolog. Soc. Am.*, **15:** 237-240.

Lynch, M. 1984. Destabilizing hybridization, general-purpose genotypes and geographic parthenogenesis. *Quart. Rev. Biol.*, **59:** 257-290.

Lynch, M. 1994. Neutral models of phenotypic evolution. *in*: Real, L.A. (ed), *Ecological Genetics*, Princeton Univ. Press: 86-108.

Lynch, M & Gabriel, W. 1983. Phenotypic evolution and parthenogenesis. *Am. Nat.*, **122:** 745-764.

MacArthur, R.H. & Wilson, E.O. 1967. *The Theory of Island Biogeography,* Princeton Univ. Press.

May, R.M. 1974. *Stability and Complexity in Model Ecosystems,* Princeton Univ. Press.

McKenzie, K.G. 1971. Paleozoogeography of freshwater Ostracoda. *Bull.Centr. Rech. Pau-SNPA, Suppl.,* **5:** 207-237.

McKenzie, K. G. & Moroni, A. 1986. Man as an agent of crustacean passive dispersal via useful plants- exemplified by ostracoda 'ospiti esteri' of the Italian ricefields ecosystem- and implications arising therefrom. *J. Crustacean Biol.*, **6:** 181-198.

Nee, S. & May, R.M. 1992. Dynamics of metapopulations: habitat destruction and competitive coexistence. *J. Anim. Ecol.*, **61:** 37-40.

Otero M., Rossi, V., Baltanás, A. & Menozzi, P. (in press). Effect of genotype and photoperiod on diapause strategies in *Eucypris virens* (Jurine, 1820) (Crustacea: Ostracoda). *Arch. Hydrobiol.*

Proctor, V.W. 1964. Viability of crustacean eggs recovered from ducks. *Ecology*, **45:** 656-658.

Proctor, V.W. & Malone, C.R. 1965. Further evidence of the passive dispersal of small aquatic organisms via the internal tract of birds. *Ecology*, **46:** 728-729.

Rossi, V. & Menozzi, P. 1990. The clonal ecology of *Heterocypris incongruens* (Ostracoda). *Oikos*, **57:** 388-398.

Rossi, V. & Menozzi, P. 1993. The clonal ecology of *Heterocypris incongruens* (Ostracoda): Life history traits and photoperiod. *Funct. Ecol.*, **7:** 177-182.

Sandberg, P.A. & Plusquellec, P.L. 1974. Notes on the anatomy and passive dispersal of *Cyprideis* (Cytheracea, Ostracoda). *Geosci. Man*, **6:** 1-26.

Smith, A.T. & Gilpin, M.E. 1997. Spatially correlated dynamics in a pika metapopulation. *in*: Hanski, I. & Gilpin, M.E. (eds), *Metapopulation Biology*, Academic Press: 407-428.

Thomas, C.D. & Hanski, I. 1997. Butterfly metapopulations. *in*: Hanski, I. & Gilpin, M.E. (eds), *Metapopulation Biology*, Academic Press: 359-386.

Tilman, D. 1994. Competition and biodiversity in spatially structured habitats. *Ecology*, **75:** 2-16.
Victor, R., Chan, G.L. & Fernando, C.H. 1979. Viability of ostracods passing through the gut of *Catastomus comersoni* (Lacepede) (Pisces, Catostomidae). *Can. J. Zool.*, **57:** 1745-1747.

CHAPTER 14
CLONAL ECOLOGICAL DIVERSITY

W. Geiger, M. Otero and V. Rossi

Abstract

Existing hypotheses concerning maintenance and nature of genetic and ecological diversity in parthenogenetic ostracod populations are reviewed and examined on the base of ecological data of single clones. Ecological diversity goes together with genetic diversity. In species with resting eggs, selection acts on hatching (*Heterocypris incongruens*) or on the proportion of resting eggs produced as a function of photoperiod (*Eucypris virens*), southern clones producing more resting eggs at long day conditions than northern ones. In both cases, a large fraction of the reproductive output is deposited in the sediments and thus withdrawn from selection. This explains, together with a high degree of ecological specialisation, why genetic diversity in these species can be maintained at least over ecological time periods.

In *Limnocythere inopinata,* most likely a species without drought-resistant eggs, environmental instability and high ecological similarity are the main factors for the maintenance of genetic diversity. A complex pattern of generalist and specialised clones along a gradient of habitat permanency is a more realistic picture of what could be interpreted at the species level as a broadly adapted, generalist organism.

Darwinula stevensoni, the oldest parthenogenetic and less genetically diverse species in the study, is the only candidate for what was called a 'general purpose genotype', a feature supposed to be common in parthenogenetic organisms. Taking ostracods as an example, the general problem of testing the 'general purpose genotype' hypothesis is discussed.

Introduction

The existence of high genetic diversity in some lineages of freshwater ostracods providing the necessary background for selection processes as described in chapters 1 and 13 is in striking contradiction to some long lasting hypotheses about parthenogenetic species as evolutionary dead ends. However, when trying to explain genetic diversity we are confronted with two processes: 1) the generation of genetic diversity and 2) its maintenance. Whilst the first is a consequence of genetic mechanisms, the maintenance of genetic diversity in parthenogenetic species has been for a long time the subject of ecological studies (Loaring & Hebert 1981, Jacobs 1990) and a number of hypotheses have been developed:

1. Some clones are confined to peripheral areas away from the species optimum, where abiotic factors predominate. Such clones should have what Lynch (1984) called a 'general purpose genotype' (see Box 14.1. for his definition).

Sex and parthenogenesis: evolutionary ecology of reproductive modes in non-marine ostracods, edited by K. Martens, pp. 243–256

2. The continuous generation of new clones from hybridisation with bisexual conspecifics together with a high degree of niche diversification allows for short-time coexistence of several clones. At a longer run, competition will reduce the number of clones to a few (Bell 1982, Lynch 1983, Carvalho 1994).
3. Clones may succeed each other throughout the growing season (Rossi & Menozzi 1990).
4. Extreme ecological similarity between clones may explain coexistence by neutral stability over a limited time period (Loaring & Hebert 1981).
5. Different clones use micro-niches within a heterogeneous habitat (habitat segregation) (Wilson & Hebert 1992).
6. Similar to the positive effect of intermediate disturbances on species diversity, the frequency of disturbances might also influence the short-time coexistence of clones (Sebens & Thorne 1985).

Like in many fields of ecology, tests of hypotheses with real data are rare. The present chapter tries to fill some of these gaps, to provide data on the ecology of clones within three distinct freshwater ostracod lineages including parthenogenetic populations of four species (*Darwinula stevensoni*, *Limnocythere inopinata*, *Eucypris virens* and *Heterocypris incongruens*) and to analyse them in the light of the above-mentioned hypotheses. The results from other studies (provided they have been performed on a clonal level) will be integrated. In order to avoid lengthy repetitions, the terms 'b-pops' and 'p-pops', referring to bisexual and parthenogenetic populations, will be used throughout the paper.

Clonal ecology of the species

As an understanding of the clonal ecology of the four species investigated is basic to the discussion of the above-mentioned problems, a short summary of the results of several detailed studies carried out during the project, published elsewhere or in preparation, will be given.

Heterocypris incongruens and *Eucypris virens*

Due to the similarity of their ecology, both species will be treated together. However, whilst the data on *Eucypris virens* have been gathered entirely within the project, knowledge on the clonal ecology of *Heterocypris incongruens* has been mainly elaborated by Rossi & Menozzi (1990) and Rossi *et al.* (1996). Both species are cosmopolitan and occur in temporary habitats, but show different temperature preferences. Whilst *H. incongruens* occurs over a wide range of temperatures, *E. virens* is living in cold water ponds with main occurrence in winter. In both species the number of generations is a function of the duration of the habitat and temperature. In man-made water bodies such as the rice fields investigated by Rossi & Menozzi (1990, 1993) where high temperatures and the presence of water coincide, the number of generations achieved by both species is highest (Rossi, unpubl.). Clonal diversity in both species is high (chapter 11). The maximum number of clones has been observed in Italian rice fields. Dispersal occurs through desiccation-resistant resting eggs.

Genotypic differences in resting egg hatching were found for *H. incongruens* clones triggered by temperature and photoperiod providing the base for temporal successions of clones. In *E. virens* the amount of energy invested in resting egg production as a function of photoperiod differs among clones from different latitudes (Fig. 14.1). Whilst all clones increased the overall production of eggs at long day conditions, they differed with regard to the percentage of resting egg production. A clone isolated from an English pond (Bramhope) showed no reaction towards long day conditions, whereas southern clones from Italy and Spain reinforce the production of resting eggs under long day conditions. In these latitudes, long day conditions coincide with the disappearance of the habitats due to drying. Table 14.1 shows differences and common aspects in the clonal ecology of the two species.

In *H. incongruens* the apparently broad tolerance towards temperature is the result of the superimposed abundance of succeeding clones, each of them with narrow temperature ranges. Rossi & Menozzi (1990) could differentiate between winter and summer clones with corresponding temperature optima for juvenile development time, survival and fertility schedules. In addition, photoperiod seems to influence the hatching of resting eggs, thus providing the base for the temporal succession observed. In *E. virens*, as a cold-adapted species, temperature fluctuations are less strong in the field. No succession of clones along a temperature gradient has been observed and clones only differ with regard to photoperiod.

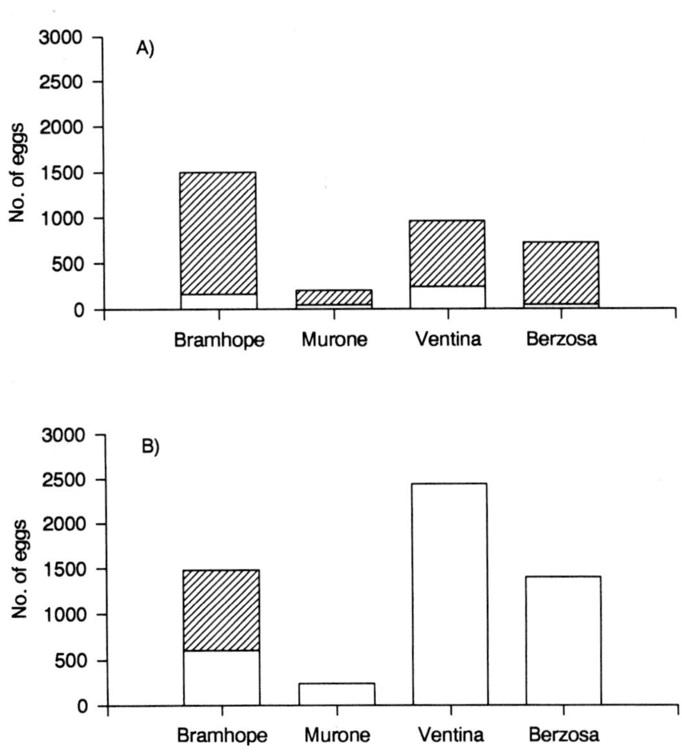

Fig. 14.1 Number of resting and subitaneous eggs produced by clones of *Eucypris virens* under A) short day (6:18 LD) and B) long-day (16:8 LD) photoperiod at 16°C. Unshaded = resting eggs; shaded = subitaneous eggs.

Table 14.1. Ecological information on similarity and differences between populations of *H. incongruens* and *E. virens* investigated with regard to their clonal ecology.

	H. incongruens	***E. virens***
Bisexual populations known	yes	yes
Desiccation-resistant eggs	yes	yes
Habitat	Rice fields	Temporary ponds
Dry phase	None	May-October
Mean densities	High	Low
Temperature range for the species (°C)	0-32	0-16
Clonal differences in:		
Temperature optima	yes	no
Juvenile development	yes	not known
Lifespan	yes (temperature)	no (photoperiod)
Number of eggs	yes (temperature, photoperiod)	yes (photoperiod)
Proportion of resting eggs	yes (temperature, photoperiod)	yes (photoperiod)
Hatching	yes (temperature, photoperiod)	yes (photoperiod)
Survival	yes (temperature)	no (temperature)

The high genetic diversity together with the possibility to produce desiccation-resistant eggs which serve both as a storage of genetic diversity withdrawn from the selection process and as units of dispersion, allow these two species to colonise wide geographic areas. Which clones are successfully established is decided by the environmental conditions prevailing in the specific habitats.

Limnocythere inopinata

The species has a holarctic distribution with bisexual populations in North America (Dobbin 1941) and Southeast Europe (Petkovski 1959). Tétart (1978) described the karyotype of a bisexual population in France without further comments on the locality. A subfossil bisexual population has been described from Mindelsee in Central Europe (Petkovski 1977). Both parthenogenetic and bisexual populations have been found in Lake Qinghai in Tibet (China) (Yin 1997). Both b-pops and p-pops are described from the last Interglacial (Eem) in Northern Germany (Fuhrmann & Pietrzeniuk 1990). Findings of single males in populations with an overwhelming majority of females have been reported (Löffler 1959, 1990). A distribution map of bisexual and parthenogenetic populations of this species in Europe can be found in chapter 5.

The species is described as summer form (Hiller 1972, Mallwitz 1984) and occurs in both permanent and temporary water bodies covering a wide range of salinities up to chloride contents of 2160 mg l^{-1} (Löffler 1959).

Data from the literature (Hiller 1972, Mertiny 1980) and field observations in Mondsee (a medium-sized freshwater lake in the Austrian Alps), Neusiedlersee (a shallow, slightly saline step lake in the east of Austria) and the adjacent astatic ponds of Seewinkel, revealed some substantial differences in the ecology of freshwater and saline populations: besides morphological differences in size and carapace nodation, populations differ in their abundance pattern and their overwintering strategies. Whereas freshwater populations appear in April or May and disappear or reduce to very small numbers in October-November, saline populations are overwintering as adults or late juvenile instars. Below a certain temperature threshold, there is no more development and adults 'accumulate' under the ice cover at a temperature of 1-2°C.

The electrophoretic survey of 11 freshwater and saline populations from Austria and Italy yielded 31 multilocus genotypes most of them restricted to the saline water bodies in the east of Austria (Neusiedlersee, astatic ponds of Seewinkel). The number of genotypes decreases along an instability gradient from 24 genotypes in the astatic ponds (highest instability, salinity and temperature fluctuate throughout the year, irregular drying), 16 in Neusiedlersee (salinity and water level stable, physical disturbance through wind action) to 3 in Mondsee (freshwater lake littoral). Monoclonal populations were found in man-made water bodies such as gravel pit lakes.

Freshwater and saline clones never co-occur as they cannot tolerate each others' water chemistry. Despite their genetic similarities, the two clonal groups are good candidates to be treated in the future as two separate taxonomic entities.

Two clones of the saline type have been investigated in detail with regard to the influence of temperature (15°, 20° and 25°C) and salinity (2000, 5000 and 8000 $\mu S.cm^{-1}$) on their life history strategies and energy requirements.

Both clones occur in lake Neusiedlersee and in the astatic saline ponds of the nearby Seewinkel. Clone N is one of the two most frequent clones at all seasons (32% in Neusiedlersee and 27% in Seewinkel), whereas clone S is always rare (highest relative abundance: 6% in autumn in Seewinkel) especially in the lake where it is absent during most of the year and was only detected under ice cover (3%).

Both clones differ in all life history parameters summarised as finite rate of population increase and generation time with regard to their response to high temperatures (Fig. 14.2). Although both clones prefer high temperatures, clone S performs best at the highest temperature (25°C) within the range of experimental temperatures. Whilst survival and egg production rates of clone N are decreasing at 25°C, those of clone S keep on increasing. Clone S is only competitively superior at high temperatures (25°C) and performs especially poorly when low temperatures and low salinities are combined. This explains its virtual absence from the lake where salinity is constantly low, favouring the development of clone N which is less susceptible to low salinities. It appears that both clones differ more in life history tactics than in the overall magnitude of fitness. Whilst the clone isolated from the lake (clone N) increases its fitness by increasing its reproductive output, clone S from the temporary pond achieves similar fitness values through faster development.

Metabolic rates, the energetic base of life history strategies, display a similar picture: both clones perform best at high salinities. Whereas clone S shows at both salinities a clear dependence from temperature, clone N is able to maintain its rates constant over a 10°C increase in temperature. These rates representing a snapshot of what is expressed over a much larger time scale in the life history of each clone, we should not wonder that clone S has somewhat higher energetic rates and production values as compared to clone N, especially at high temperatures. These conditions, however, should prevail only for short periods of time like those observed in the ponds after long periods of draught. Therefore, a generalist like clone N, able to

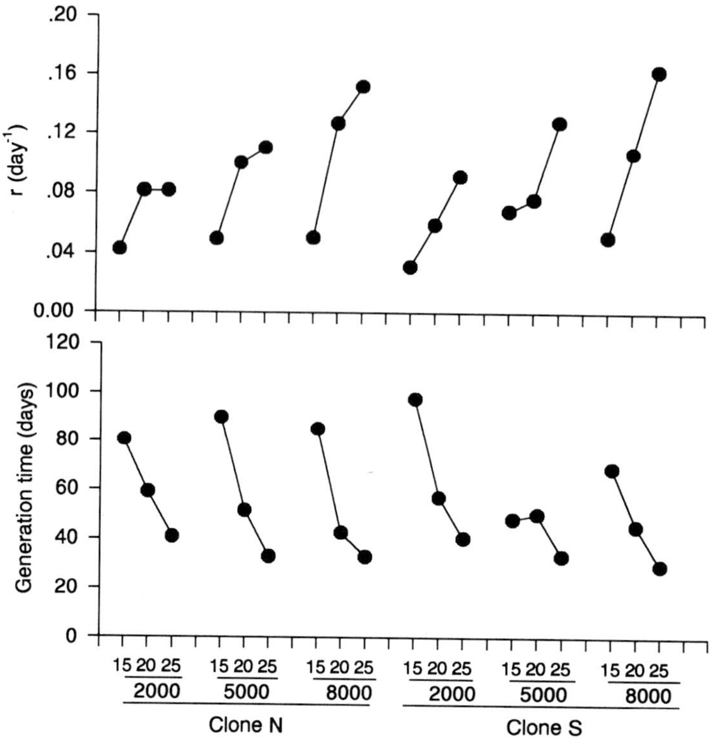

Fig. 14.2 Intrinsic rate of population increase (r) and generation time of two clones of *Limnocythere inopinata* at three temperatures (°C) and three salinities (μS.cm^{-1}).

regulate its metabolic rates with regard to temperature, should be at advantage under the relatively stable conditions typical for the lake. The specialist (clone S), due to its high metabolic efficiencies, can spread rapidly when conditions become favourable, but remains at low numbers for the rest of the year. Both clones are adapted to high temperatures and high salinities, but differ in the fine tuning of their life history strategies and metabolic adaptations.

Darwinula stevensoni

From the three species included in the project, *Darwinula stevensoni* appeared to be the most difficult one to handle. Its long life cycle (up to three years) made experiments of the type described above, where clonal lineages have been kept over several generations under experimental conditions, impossible. Therefore, we are restricted to the interpretation of field data. An ecological comparison of clones from lake and river populations is presently in progress.

The species is described as eurythermal and euryhaline (Hiller 1972, Delorme 1991), occurring over a wide range of habitats from the littoral of lakes, ponds, up to slowly flowing rivers. Reproduction coincides with periods of high temperatures. Life cycle duration depends on water body depth and temperature and is shorter at lower latitudes (McGregor 1969, Ranta 1979). However, even in the tropics it maintains a strictly seasonal abundance pattern (Martens & Tudorancea 1991). Both electrophoretic and DNA data demonstrate low genetic variability and a wide geographic distribution of single genotypes.

In search of the general purpose genotype – does it really exist?

The idea that parthenogenesis goes together with ecological generalism was first stated by Baker (1965). It was further developed for specific animal groups by a variety of authors (Parker 1979, Jaenike *et al.* 1980 amongst others) until it became formalised as a testable scientific hypothesis under the name of 'general purpose genotype' (Box 14.1.) by Lynch (1984).

The theory was developed in order to explain the divergent geographical distribution pattern of parthenogenetic and bisexual populations, which is called by Vandel (1940) 'geographical parthenogenesis'. However, Lynch did not completely exclude the possibility that the widespread distribution of a parthenogen has resulted from the joint distribution of many locally adapted clones rather than from the dispersal of a single, highly generalised clone. At that time, knowledge about the distribution pattern of clones was scarce and, if existing, information about the ecological requirements of clones was lacking.

Hughes (1989) further restricted the hypothesis to parthenogens with a monophyletic origin, where the generation of genetic variation relies entirely on mutations. They should be unable to compete with bisexual relatives for niches in stable environments and therefore occur in marginal habitats where colonising abilities and environmental tolerance are more important than fine-scale tuning and niche specialisation.

Clones with a polyphyletic origin, generated frequently from bisexual populations, with high clonal diversity, should, on the contrary, succeed in their parental habitats, as well as in highly competitive environments. In his taxonomic survey of obligate cloning animals, Hughes (1989) classified 2/3 of the species out of the 18 where data have been available as having a general purpose genotype.

On the contrary, most of the experimental studies on *Daphnia* (Weider & Lampert 1985, Weider & Hebert 1987) proved the existence of highly specialised clones with a high degree of ecological differentiation with regard to their osmoregulatory capacity, temperature preferences or respiratory adaptation. The only study comparing asexual and bisexual lineages (Weider 1993) also failed to demonstrate the existence of a general purpose genotype for the environmental parameters studied (temperature and salinity). However, as these two lineages co-occur in the same pond and are likely to interbreed, the argument that clones with a polyphyletic origin are specialists is supported.

Box 14.1: General purpose genotype

> A genome conferring an ability to exploit a wide range of environments including harsh, peripheral habitats. The existence of such a genotype was proposed in order to explain the observed distribution pattern of parthenogenetic species.

The advantage when working with ostracods is that parthenogenesis developed in different phylogenetic lineages independently. The age of transition from bisexuality to parthenogenesis and the clonal genetic diversity within these lineages differ considerably (see chapters 15 and 16) and therefore allow for a comparative approach to the problem.

From the four species under study, three of them (*Limnocythere inopinata*, *Eucypris virens* and *Heterocypris incongruens*) are known to have bisexual populations mainly confined to the Mediterranean. Nevertheless, at least for *Eucypris virens,* b-pops and p-pops were found to co-occur in the same habitats (chapters 15 and 16). In general, habitats of both types of populations are similar with regard to environmental parameters.

The three species display different degrees of clonal specialisation as a function of the amount of genetic and environmental variability. Whilst the two species with resting eggs (*Heterocypris incongruens* and *Eucypris virens*) modify their resting egg production and egg hatching strategy according to photoperiod and temperature (the latter only in *Heterocypris incongruens*), *Limnocythere inopinata* splits up into two morphologically and ecologically distinct clonal groups: freshwater clones with low clonal diversity and saline clones with up to 24 genotypes in a single pond. Within the saline group, a mixture of 'generalist' and 'specialist' strategies among clones has been found with regard to temperature and salinity. Specialists are rare except under the environmental conditions they have been selected for (e.g. high temperatures). They are not performing badly enough to be outcompeted completely by the generalists during the time the pond exists.

The apparent wide distribution pattern and seemingly high ecological tolerance of the three species are, when analysed at the clonal level, due to a succession of highly specialised clones (*H. incongruens*), a differential colonisation of habitats by specialised clones (*E. virens*) or the existence of distinct clonal groups representing a mixture of generalist and specialist strategies (*L. inopinata*).

Darwinula stevensoni differs from the other three species in that its genetic variability is low and that no bisexual populations are known. Despite the fact that no data about clonal tolerances are available, the general description of its ecology as highly tolerant towards temperature, salinity and oxygen together with the wide distribution range of single clones and a presumptive monophyletic origin makes it a good candidate for an example of a general purpose genotype.

The situation is somewhat more complex in the case of *Cytherissa lacustris*, a species with parthenogenetic reproduction since the Pliocene (Danielopol *et al.* 1990) and a holarctic distribution, living in the profundal zone of moderately deep lakes (Löffler 1975, Delorme 1978). Within-habitat genotypic diversity is restricted (Sywula & Geiger 1990) and genotypes can tolerate under laboratory conditions wider ranges of temperature (Geiger 1990a) and oxygen (Geiger 1990b) than those inferred from their field distribution (Delorme 1982). Similarly to *Darwinula stevensoni*, *Cytherissa lacustris* has, for a meiobenthic species, an extremely long life cycle of up to two years depending mainly on temperature and food conditions but independent from genotype (Geiger 1990a). It is therefore difficult to classify this species in one of the two groups mentioned above, as it combines restricted genetic diversity with the ecology of a specialist.

It should be more than a coincidence that there is a close connection between genetic diversity, ecology and phylogeny within the lineages so far studied (Table 14.2). A further refinement might result once we understand better the genetics and ecology of the Candoninae, the second most diversified group of freshwater ostracods with parthenogenetic representatives in almost any type of water body.

Maintenance of clonal diversity

As described above, high clonal diversity goes together with the existence of bisexual populations and therefore with the potential of generation of genetic diversity. The only species without known bisexual parental populations (*Darwinula stevensoni*) has also the lowest clonal diversity. However, the processes which lead to its maintenance differ in the three species:

Heterocypris incongruens clones produce resting eggs whose hatching is triggered by temperature and photoperiod. Temporal succession of clones adapted to different temperatures allows for existence of up to 36 clones in a single pond. Clones are withdrawn from the effects of competition as resting eggs and coexist only for short time periods. As thermal conditions in these ponds change drastically over the year, each of the clones finds near-optimum conditions for production of subitaneous eggs. If conditions are sub-optimal, they invest reproductive energy in resting eggs. Thermal niche differentiation and resting egg production allow for short term coexistence of a high diversity of genotypes generated most probably by bisexual parental populations.

In *Eucypris virens*, despite the similarity to *H. incongruens* with regard to resting egg production and the existence of bisexual populations, the situation is somewhat different. Living throughout Europe in cold water habitats, the potential for thermal niche differentiation is reduced. The species temperature optimum lies around 15°C, much lower temperatures (5°C) double the juvenile development time (Hounsome 1975), higher temperatures causing high juvenile mortality and low fecundity (Otero *et al.* in prep.). This is in agreement with the fact that no temporal succession of genotypes has been observed.

Table 14.2: Classification of ostracod species according to levels of genetic and ecological diversity and presumptive age of parthenogenetic lineages.

	Cytherideidae	**Limnocytheridae**	**Darwinulidae**	**Cyprididae**
	Cytherissa lacustris	*Limnocythere inopinata*	*Darwinula stevensoni*	*Heterocypris incongruens* *Eucypris virens*
Genetic diversity	low	high	low	high
Ecological diversity	low	high	high	high
Estimated age of parthenogenetic lineages	medium	recent	old	recent

In the absence of data on environmental factors supporting niche differentiation, we propose the following hypothesis to explain clonal coexistence in *Eucypris virens*: the short-term existence of many habitats of this species does not allow for competitive exclusion of the clones which are supposed to be of similar fitness. Resting eggs can survive for a long time in the sediments and therefore serve as a homeostatic mechanism for clonal frequencies.

The fact that the clone from England (Bramhope), investigated with regard to resting egg production, does not change its strategy under long day photoperiod and continues producing subitaneous eggs, can be seen as a reasonable strategy considering the climate conditions on the British Isles in summer. That only one clone could be detected in the English pond can be explained by the absence of bisexual populations at the British Isles and/or the permanency of the pond where resting egg production plays a subordinate role. The fact that monoclonal populations of *E. virens* also exist in areas with bisexual ones (Spain, Italy) rather provides support for the second hypothesis. Given the permanency of the water body, a clone adopting a few-resting egg production strategy is at advantage and could have outcompeted clones which switch to resting egg production at long-day conditions. However, we need more data on the ecology of this species to validate this hypothesis.

The ecological characteristics of *Limnocythere inopinata* clones do not allow for strategies adopted by the two above-mentioned species. They do not produce desiccation-resistant eggs and development only occurs at moderate to high temperatures (15-25°C). Therefore, they are confined to permanent habitats during summer and autumn or, if they live in habitats which frequently fall dry, are rare and have to be colonised from adjacent permanent water bodies. In the case of the Seewinkel ponds in the East of Austria, those which dry out regularly, are lacking *L. inopinata*, whereas in larger ponds, which dry out only occasionally, they can be extremely abundant. In both cases the ponds are located in the vicinity of permanent water bodies with high potentials for frequent colonisation through water fowl.

As this species did not develop mechanisms to overcome the drying periods, these represent strong disturbance events for it. Disturbances of this magnitude, which wipe out the whole population, offer new space for genotypes from the source population (adjacent lakes or other, more permanent ponds). The much higher dynamics of the ponds in terms of salinity and high temperatures allow for physical niche diversification. Clones which are very rare in the lake can reach high relative abundances in the pond as a function of the more unstable water chemistry of the ponds.

Coexistence of clones is further enhanced by the ecological similarity of the clones. The difference between rare and frequent clones in terms of their rate of population increase does not differ by more than 0.05/day for the temperature and salinity ranges investigated, not enough to outcompete them during the timespan available.

The observed decline in clonal diversity from saline ponds (24 clones) via shallow, saline lakes (Neusiedlersee, 16 clones) to freshwater lakes (max. 5 clones) coincides with increasing permanency (10 years, 100 years, 10,000 years), as well as with increasing environmental stability (fluctuating temperatures and salinity, fluctuating temperatures, regularly fluctuating temperatures). Time and the reduced number of available niches might therefore account for the much lower clonal diversity observed in freshwater lakes.

We cannot exclude that the production of rare males observed in all kinds of water bodies has a significant impact on the observed pattern. It would help to explain both the origin of such high diversities in the absence of efficient dispersal mechanisms and the existence of unique clones in the Seewinkel pond. Even if males are extremely rare – from more than 20,000 individuals reared in the laboratory only 2 were males – a single bisexually produced new genotype could, given the high population growth rates which can be achieved under optimum conditions, spread rapidly in the population. The fact that males have only been observed in batch cultures but never in single individual cultures might indicate a density dependence of this process. However, as long as we do not know if there is viable, bisexually-produced offspring (some sort of copulation has been documented, Geiger *et al.* in prep.), this remains pure speculation.

Are ostracods an exception? – a comparison with other crustacean groups

Despite the limited work done on the ecology of parthenogenetic animals at the clonal level, Crustaceans are in so far an exception as there is substantial knowledge on two other groups – the genus *Artemia* within the Anostraca and the genus *Daphnia*. Both served for a long time as excellent organisms for this kind of studies, but their restricted taxonomic range and their type of parthenogenesis (automixis with recombination in some *Artemia*-species and cyclic parthenogenesis in *Daphnia*) reduce their value when studying the long-term effects of parthenogenetic reproduction. Automixis results in genetic diversities comparable to those generated by bisexual reproduction and cyclic parthenogenesis releases large amounts of hidden genetic variation whenever a phase of bisexual reproduction occurs (Gabriel 1988). Furthermore, the obligate parthenogenetic populations of *Daphnia* species are of recent origin and hybridisation with bisexual congeners is a common phenomenon (Lynch *et al.* 1989). Therefore, it is not astonishing that none of the many studies carried out on the search of a general purpose genotype succeeded in detecting it. All studies point towards a high specialisation of clones similar to that observed in bisexual lineages. Even the only study expressely dealing with this topic directed by Weider (1993) that compared obligate parthenogenetic lineages with bisexual ones with regard to niche breadth for temperature and salinity, failed to detect differences which could be attributed to their mode of reproduction. It appears that, as a general pattern, parthenogens with high genetic diversity have also the potential for high ecological differentiation (similar to what has been observed in ostracods with mixed reproduction), even more so when it is combined with cyclic periods of bisexuality (Gabriel 1988).

Maintenance of clonal diversity and coexistence of clones in obligate parthenogenetic *Daphnia* populations is mainly achieved by high ecological differentiation of the clones (Mort 1989, Wilson & Hebert 1992), low fitness differences between clones together with high temporal and spatial heterogeneity of the environment (Loaring & Hebert 1981, Hebert 1987) or intermediate levels of disturbances (Sebens & Thorne 1985, Weider 1992). In contrast, whenever clones exhibit distinct fitness differences and habitats are long-lasting, the fittest clone should rapidly outcompete the others and therefore overall genetic diversity of the assemblage should be reduced (Carvalho 1994).

Conclusions

From what we know from ostracods and other Crustacean groups, it follows that high ecological specialisation always goes together with high genetic diversity. The ways how this genetic diversity is generated and maintained, however, differs among the different lineages and species. Whilst daphnids frequently display cyclic phases of bisexual reproduction, this has been never observed in ostracods. Transition to parthenogenesis seems to be irreversible and of different age in the three phylogenetic groups studied. As a consequence, the oldest group (*Darwinula stevensoni*) has also the lowest genetic diversity and fits best in what has been called 'general purpose genotype'. The other two groups differ from *Darwinula* in that bisexual populations are existing (*Limnocythere inopinata, H. incongruens*) and even coexisting in the same habitats (*Eucypris virens*). This is reflected in high genetic diversity and highly specialised clones. High incidence of generation of genetic diversity and efficient strategies to its maintenance guarantee the short-term and long-term success of these species.

A distinction has to be made between parthenogenetic populations according to their origin (mono- or polyphyletic) and age resulting in differences in genetic diversity and therefore in their potential for ecological differentiation. The general purpose genotype should be restricted to populations with monophyletic origin with low levels of genetic diversity where single genotypes are confronted over large time periods with a wide range of ecological conditions, similar to 'generalists' amongst bisexual species, with no possibility to hybridise with bisexual conspecifics. As a consequence, the general purpose hypothesis, as formulated by Lynch, is impossible to test as it requires to compare parthenogenetic with living bisexual lineages. If bisexual populations exist, we are most likely dealing with parthenogenetic lineages of polyphyletic origin which are, to our knowledge, highly specialised.

With regard to the maintenance of genetic diversity, several of the mechanisms proposed on theoretical grounds or developed for *Daphnia* species proved to be valid also for the ostracod species investigated so far: temporal niche diversification, short-term existence of habitats, differences in resting egg production, low fitness differences among clones, intermediate disturbances and heterogeneity of the environment seem to be the most important factors responsible for the observed high genetic diversities in some of the lineages.

References

Baker, H.G. 1965. Characteristics and modes of origin of weeds. *in*: Baker H.G. & Stebbins, G.L. (eds), *The genetics of colonising species,* Academic Press: 147-168.

Bell, G. 1982. *The masterpiece of nature,* Croom Helm: 625pp.

Carvalho, G.R. 1994. Evolutionary genetics of aquatic clonal invertebrates: Concepts, problems and prospects. *in*: Beaumont, A.R. (ed), *Genetics and evolution of aquatic organisms,* Chapman & Hall: 291-323.

Danielopol, D.L., Olteanu, R., Löffler, H. & Carbonel, P. 1990. Present and past geographical-ecological distribution of *Cytherissa (Ostracoda, Cytherideidae)*. *Bull. Inst. Géol. Bassin d'Aquitaine, Bordeaux*, **47:** 97-118.
Delorme, L.D. 1978. Distribution of freshwater ostracods in lake Erie. *J. Great Lakes Res.*, **4:** 216-220.
Delorme, L.D. 1982. Lake Erie oxygen: the prehistoric record. *Can. J. Fish. Aquat. Sci.*, **39:** 1021-1029.
Delorme, L.D. 1991. Ostracoda. *in*: Thorp, J.H. & Covich, A. (eds), *Ecology and classification of North American freshwater invertebrates,* Academic Press: 691-722.
Dobbin, C.N. 1941. Freshwater Ostracoda from Washington and other western localities. *Univ. Washington Pub. Biol.* **4 (3):** 175-245.
Fuhrmann, R. & Pietrzeniuk, E. 1990. Die Aussage der Ostrakodenfauna zum Sedimentationsablauf im Interglazialbecken, zur klimatischen Entwicklung und zur stratigraphischen Stellung des Interglazials von Neumark-Nord (Geiseltal). *in*: Mania, D., Thomae, M., Litt, T. & Weber, D. (eds), Neumark-Gröbern. Beiträge zur Jagd des mittelpaläolithischen Menschen, *Veröff. Landesmuseum Vorgeschichte, Halle*, **43:** 161-166.
Gabriel, W. 1988. Quantitative genetic models for parthenogenetic species. *in:* de Jong, G. (ed), *Population Genetics and Evolution*, Springer Verlag: 73-82.
Geiger, W. 1990a. Field and laboratory studies on the life cycle of *Cytherissa lacustris* (SARS) (Crustacea, Ostracoda) with special emphasis on the role of temperature. *Bull. Inst. Géol. Bassin d'Aquitaine, Bordeaux*, **47:** 191-208.
Geiger, W. 1990b. The role of oxygen in the disturbance and the recovery of the *Cytherissa lacustris* population of Mondsee (Austria). *Bull. Inst. Géol. Bassin d'Aquitaine, Bordeaux*, **47:** 167-189.
Hebert, P.D.N. 1987. Genotypic characteristics of cyclic parthenogens and their obligately asexual derivatives. *in:* Stearns, S.C. (ed), *The evolution of sex and its consequences,* Birkhäuser Verlag: 175-195.
Hiller, D. 1972. Untersuchungen zur Biologie und zur Ökologie limnischer Ostrakoden aus der Umgebung von Hamburg. *Arch. Hydrobiol. Suppl.*, **40:** 400-497.
Hounsome, M.V. 1975. The effect of water temperature on the growth and allometry of *Eucypris virens* (Jurine) (Crustacea, Ostracoda). PhD thesis, Univ. Manchester.
Hughes, R.N. 1989. *A functional biology of clonal animals,* Chapman & Hall: 331pp.
Jacobs, J. 1990. Microevolution in predominantly clonal populations of pelagic *Daphnia* (Crustacea: Phyllopoda): Selection, exchange, and sex. *J. evol. Biol.*, **3:** 257-282.
Jaenike, J., Parker, E.D. Jr. & Selander, R.K. 1980. Clonal niche structure in the parthenogenetic earthworm *Octolasion tyrtaeum*. *Am. Nat.*, **116:** 196-205.
Loaring, J.M. & Hebert, P.D.N. 1981. Ecological differences among clones of *Daphnia pulex* Leydig. *Oecologia*, **51:** 162-168.
Löffler, H. 1959. Zur Limnologie, Entomostraken- und Rotatorienfauna des Seewinkel-Gebietes. *Sitzungsber. Öster. Akad. Wissensch.*, **168:** 315-362.
Löffler, H. 1975. The evolution of ostracod faunas in alpine and prealpine lakes and their value as indicators. *Bull. Am. Paleont.*, **282:** 434-443.
Löffler, H. 1990. Paleolimnology of Neusiedlersee, Austria. I. Succession of ostracods. *Hydrobiologia*, **214:** 229-238.
Lynch, M. 1983. Ecological genetics of *Daphnia pulex*. *Evolution*, **38:** 465-482.
Lynch, M. 1984. Destabilizing hybridization, general-purpose genotypes and geographic parthenogenesis. *Quart. Rev. Biol.,* **59:** 257-290.
Lynch, M., Spitze, K. & Crease, T.J. 1989. The distribution of life history variation in the *Daphnia pulex* complex. *Evolution*, **43:** 1724-1736.
McGregor, D.L. 1969. The reproductive potential, life history and parasitism of the freshwater ostracod *Darwinula stevensoni*. *in:* Neale, J.W. (ed), *The taxonomy, morphology and ecology of recent Ostracoda,* Oliver & Boyd: 194-221.
Mallwitz, J. 1984. Untersuchungen zur Ökologie litoraler Ostracoden im Schmal- und Lüttauersee (Schleswig-Holstein, BRD). *Arch. Hydrobiol.*, **100:** 311-339.
Martens, K. & Tudorancea, C. 1991. Seasonality and spatial distribution of the ostracods of Lake Zwai, Ethiopia (Crustacea, Ostracoda). *Freshw. Biol.*, **25:** 233-241.
Mertiny, A. 1980. Zur Ökologie und Verbreitung der limnischen Ostrakoden in der Bundesrepublik Deutschland. Dipl. Thesis, Univ. Hamburg.
Mort, M.A. 1989. Ecological genetics of *Daphnia*: response of coexisting genotypes to resource manipulation. *Arch. Hydrobiol.*, **117:** 141-161.

Parker, E.D. Jr. 1979. Ecological implications of clonal diversity in parthenogenetic morphospecies. *Am. Zool.*, **19:** 753-762.

Petkovski, T.K. 1959. Beiträge zur Kenntnis der Ostracodenfauna Jugoslawiens (V). Hydrobiologie, *Publications of the Hydrobiological Research Institute, Faculty of Science, University of Istanbul (ser. B),* **4(4):** 158-165.

Petkovski, T.K. 1977. Ostracodenfauna des Mindelsees (S.W. Deutschland). *Acta Musei Macedonici Scientiarum Naturalium*, **12(1):** 49-94.

Ranta, E. 1979. Population biology of *Darwinula stevensoni* (Crustacea, Ostracoda) in an oligotrophic lake. *Ann. Zool. Fenn.*, **16:** 28-35.

Rossi, V. & Menozzi, P. 1990. The clonal ecology of *Heterocypris incongruens* (Ostracoda). *Oikos*, **57:** 388-398.

Rossi, V. & Menozzi, P. 1993. The clonal ecology of *Heterocypris incongruens* (Ostracoda): Life history traits and photoperiod. *Funct. Ecol.*, **7:** 177-182.

Rossi, V., Gandolfi, A. & Menozzi, P. 1996. Egg diapause and clonal structure in parthenogenetic populations of *Heterocypris incongruens* (Ostracoda). *Hydrobiologia*, **320:** 45-54.

Sebens, K.P. & Thorne, B.L. 1985. Coexistence of clones, clonal diversity and the effects of disturbance. *in:* Jackson, J.B.C., Buss, L.W. & Cook, R.E. (eds), *Population biology and evolution of clonal organisms,* Yale Univ. Press: 357-398.

Sywula, T. & Geiger, W. 1990. Genetic and morphological polymorphism in the freshwater ostracod *Cytherissa lacustris* (Sars) of lakes Mondsee and Attersee (Austria). *Bull. Inst. Géol. Bassin d'Aquitaine, Bordeaux*, **47:** 69-81.

Tétart, J. 1978. Les garnitures chromosomiques des Ostracodes d'eau douce. *Trav. Lab. Hydrobiol. Univ. Grenoble,* **69-70:** 113-140.

Vandel, A. 1940. La parthénogénèse géographique. IV. Polyploidie et distribution géographique. *Bull. Biol. Franc. Belg.*, **74:** 94-100.

Weider, L.J. & Hebert, P.D.N. 1987. Ecological and physiological differentiation of low Arctic clones of *Daphnia pulex. Ecology,* **68:** 188-189.

Weider, L.J. & Lampert, W. 1985. Differential response of *Daphnia* genotypes to oxygen stress: respiration rates, hemoglobin content and low oxygen tolerance. *Oecologia*, **65:** 487-491.

Weider, L.J. 1992. Disturbance, competition and the maintenance of clonal diversity in *Daphnia pulex. J. Evol. Biol.*, **5:** 505-521.

Weider, L.J. 1993. A test of the "general-purpose" genotype hypothesis: Differential tolerance to thermal and salinity stress among *Daphnia* clones. *Evolution*, **47:** 965-969.

Wilson, C.C. & Hebert, P.D.N. 1992. The maintenance of taxon diversity in an asexual assemblage: an experimental analysis. *Ecology*, **73:** 1462-1472.

Yin, Y. 1997. Contribution to the morphology and ecology of Ostracoda Limnocytheridae and Candonidae. Comparative studies between Austrian and Chinese Crustacea. PhD thesis, Univ. Vienna, 163 pp.

CHAPTER 15
CLONAL GENETIC DIVERSITY

V. Rossi, I. Schön, R.K. Butlin and P. Menozzi

Abstract

We present extensive data sets on clonal, genetic diversity of three different ostracod species, which have been derived from allozyme analysis, and compare them with published data in a comprehensive literature review. Our data confirm both the high level of clonal diversity in ostracods with asexual reproduction and its extreme variability among species. *E. virens*, a species with geographical parthenogenesis, showed with more than 200 clones the highest clonal diversity ever reported for ostracods, strongly contrasting with the ancient asexual *D. stevensoni*, which consisted of only few, allozymic clones and *L. inopinata*, a species with mixed reproduction, revealing an intermediate clonal diversity. The majority of clones from *E. virens* and *L. inopinata* were restricted to few populations, while on the contrary, the most common clone of *D. stevensoni* was found in more than 80% of all screened localities.

Allozymic data give evidence for polyploidy in *E. virens* (up to 14% of all samples) and a strong indication for interbreeding of parthenogenetic and sexual lineages, which present an important mechanism for the species to maintain genetic variability despite asexual reproduction. Sympatry of lineages with asexual and sexual reproduction could be demonstrated for at least two different populations in Spain and Sicily.

Introduction

The total amount of genetic variation within a species is the result of complex interactions among evolutionary forces such as mutation, natural selection, drift, migration and geographic distribution, demographic factors, habitat stability, life cycle characteristics and life strategies. As the relative importance of these factors differs among species, the population genetic structure may be quite different, even between closely related taxa.

The mode of reproduction influences genetic variability within and among populations due to the different destiny of mutations that accumulate in asexual lineages (see chapter 1). Freshwater ostracods are of particular interest in this respect. The species *Darwinula stevensoni* is supposed to be an ancient asexual, whereas other groups, such as *Eucypris virens,* show the phenomenon of geographic parthenogenesis, i.e. the existence of both parthenogenetic and bisexual populations of a species in geographically separated areas.

Allelic diversity and heterozygosity are important measures of genetic variation. In parthenogenetic species clonal diversity (number of multilocus genotypes) may be a more useful indicator of genetic variability. A clone consists of genetically

Sex and parthenogenesis: evolutionary ecology of reproductive modes in non-marine ostracods, edited by K. Martens, pp. 257–274

indistinguishable descendants which result from the development of unfertilised eggs produced without recombination (apomictic parthenogenesis) (Hughes 1989). In the present context we will use the term clone as synonymous with multilocus genotype. The genetic markers that allow the identification of a clone are, for the above definition of clone, in linkage with the whole genome. The unique combination of genes that are shared by all individuals that belong to a clone could be an explanation for the different ecological functions and tolerances shown by different clones of the same species (Rossi & Menozzi 1990, 1993, Rossi *et al.* 1996).

Not all animals reproducing by parthenogenesis are clonal. Oogenesis can occur during parthenogenesis through several cytogenetic mechanisms. In automictic parthenogenesis, meiosis and genetic recombination can vary in the mechanisms of restoring the diploidy of the genome and the phase of chromosome doubling (before, during or after meiosis; see chapter 1). In any case, automictic parthenogenesis causes an increase of homozygosity, and will, in the long run, produce completely homozygous clones. In apomictic parthenogenesis, meiosis is suppressed and oogenesis involves only a mitosis without any genetic recombination. In general, this kind of reproduction represents the actual manner in which clones are generated.

Mutation, in general the ultimate source of genetic variation, is the only mechanism that can produce new clones in organisms reproducing by obligate apomictic parthenogenesis, whereas genetically differing lineages in automictic and cyclic parthenogens can also arise through mixis. In organisms reproducing by self-fertilisation or automictic parthenogenesis, the frequency of heterozygotes is reduced by half every generation and eventually all genotypes become homozygous. In apomictic parthenogenesis, we will observe heterozygote excess. All new alleles originated by mutation, will be maintained in heterozygous state unless a new mutation of the non-mutated allele at the same locus generates a homozygote, but this may happen only very rarely.

Apomictic parthenogenesis is typical of the asexual phase of cyclical parthenogens such as *Daphnia* and, as far as we know, is also the way all-female populations of ostracods reproduce. This reproductive mode is usually inferred from the absence of males, from deviations from Hardy-Weinberg equilibrium of genotype frequencies and from non-random association of alleles (linkage disequilibrium). Reproduction by parthenogenesis and its cytological characteristics can be established beyond any doubt by maintaining isofemale lineages (i.e. clonal populations obtained from single mothers), analysing sex ratios and genetic segregation in subsequent generations. Apomictic parthenogenetic reproduction has been demonstrated in this way for *Cypridopsis vidua* (Havel & Hebert 1989), *Candonocypris novaezelandiae* (Chaplin 1992), *Heterocypris incongruens* (Havel & Hebert 1993, Rossi *et al.* 1993), *Potamocypris granulosa* (Havel & Hebert 1993) and *Eucypris virens* (Rossi *et al.* in prep.).

Note that in the literature, often obsolete nomenclatorial combinations are used. We use the correct combinations in the present chapter, i.e. *Heterocypris incongruens* instead of *Cyprinotus incongruens*, *Bradleystrandesia* for the North American species formerly assigned to *Cypricercus*, etc. This facilitates interpretations using correct phylogenies.

Genetic diversity in ostracods

To our knowledge enzyme markers (Box 15.1) were first used in ostracods by Sywula & Lorenc (1982) to study Polish populations of *Heterocypris incongruens*. Later on, Sywula, in collaboration with Suomalainen, Saura & Lokki (1985), compiled detailed methods for enzyme electrophoresis in Ostracoda with special reference to *H. incongruens*, *H. salina* and *Paracyprideis fennica*. In 1989, Hebert & Beaton published a handbook for Cellogel electrophoresis.

Clonal variability of American freshwater ostracods was the subject of several studies. Havel and his co-authors (1989, 1990a, b, 1993) investigated the genetic structure of 33 different bisexual and parthenogenetic ostracod species from North America, including an Arctic site. Their samples represent seven of the ten families described in this region. Other studies on Arctic and temperate localities of North America have been carried out by Turgeon & Hebert (1994, 1995) and Little & Hebert (1997), while Little & Hebert (1994, 1996) described the genetic characteristics of several species from the subtropical island of Jamaica.

Chaplin and her co-author (Chaplin & Ayre 1989, Chaplin 1992, Chaplin, 1993, Chaplin & Ayre 1997) analysed the clonal structure of Australian bisexual and all female populations of *Candonocypris novaezelandiae* (= *Herpetocypris stanleyana* Sars).

In Europe, the genetics of ostracods received relatively little attention with the exception of the studies by Sywula and his collaborators on parthenogenetic and bisexual species (Sywula & Lorenc 1982, Sywula & Geiger 1990, Sywula *et al.* 1995) and Rossi and her colleagues who described the genetic structure of *Heterocypris incongruens* (1990, 1993, 1994).

Although DNA-based techniques such as PCR and sequencing have been in common use for several years, their application to asexually reproducing organisms has been mainly restricted to prokaryotes (Avise 1994). One exception is the study by Foighil & Smith (1996) of the divergence of DNA sequences in different clones of the asexual marine clam species complex *Lasea*. In most other cases, DNA substitution rates have been estimated indirectly with RFLPs as in the work by Quattro *et al.* (1992) where mtDNA of a hybridogenetic clade of fish (*Poeciliopsis monacha-occidentalis*) has been examined or as in the numerous papers on *Daphnia* (Crease *et al.* 1989, Crease *et al.* 1990, Crease & Lynch 1991, Weider & Hobaek 1994, Weider *et al.* 1996). Little & Hebert (1996) and Chaplin & Hebert (1997) used restriction digestions of PCR products to investigate clonality of asexual ostracods. We (Rossi *et al.*, unpublished data) applied random amplified polymorphic DNA (RAPD) techniques in a preliminary analysis of the clonal structure of *Heterocypris incongruens* (see Box 16.2).

In the frame of the European network, nuclear (ITS) and mitochondrial (COI) sequences have been used to identify clones in *Darwinula stevensoni* (Schön *et al.* 1998) and *Eucypris virens* (see chapter 16). To our knowledge, this is the first time that allozyme and DNA sequence data have been used jointly to study the genetic structure of asexual lineages in ostracods. In this chapter, we report the main results of our research on clonal diversity of the three freshwater Ostracoda *Eucypris virens*, *Darwinula stevensoni* and *Limnocythere inopinata*. Populations from different European sites were analysed for all species; data of *D. stevensoni* also include one population from the Middle East (Israel) and one population from Africa (South Africa).

Box 15.1 Methods: the investigation of clonal diversity

The frequencies of different genotypes in a population can be estimated with various techniques such as electrophoresis of enzymatic proteins or different kinds of DNA analysis (Avise 1994).

Allozyme markers are a well established method commonly used: the structure of a protein is determined by one (or more than one) gene locus. At any given locus a different number of alleles may occur coding for enzymes which differ in their primary structure (allozymes) with respect to molecular weight and electric charge. Such a difference causes different migration of different alleles in an electric field. Each individual can be screened for polymorphic loci. Individuals with the same multilocus genotype can be classified as belonging to the same electrophoretic clone. Enzyme electrophoresis is carried out using techniques that differ mainly in the supporting matrix used for the separation: starch gel, acrylamide gel, agarose gel and cellulose acetate gel (commercially know as Cellogel). The separation is performed on tissue extract (fresh or stored at -80°C). Staining of single enzymes is obtained by biochemical reactions of specific substrates and co-factors linked to the reaction catalysed by the protein.

Allozymes are relatively cheap and fast markers for clonal identification. Their Mendelian, codominant mode of transmission allows the identification of heterozygotes and makes allozymes suitable for the analysis of population genetic structure. We have to keep in mind however that allozymes sample a limited and biased part of the genome since they only allow an evaluation of the variability of expressed regions. Moreover, only a relatively small number of polymorphisms can be detected, as synonymous substitutions (and some non-synonymous substitutions in third codon positions) which do not alter the migration of the enzyme remain unresolved. This can generally lead to an underestimation of clonal diversity.

We compared clonal diversity of 6 isofemale lineages of *Heterocypris incongruens* estimated by both PCR and allozyme techniques (see Box 16.1 and chapter 16), by comparing RAPD band-sharing indexes with the same index obtained for allozyme bands (Rossi *et al.* unpubl.). Using 3 polymorphic enzyme loci (MPI, PGM and GPI), we identified 11 different alleles that defined 4 clones (multilocus genotypes). By using 5 random decamers (OPS-03, OPS-09, OPP-03, OPP-09, OPP-16), we identified 72 different amplification products (RAPD loci) which were almost all (96%) polymorphic. The correlation between band sharing indexes obtained by allozymes and RAPD markers was significant ($R=0.945$, $p=0.0001$). Two pairs of clones (multilocus genotypes) defined as identical by allozymes also shared a high percentage of RAPD bands (89%). A much lower proportion of common RAPD bands (44%) was found for clones defined as different from allozyme data. Thus, although DNA-based techniques allow a finer resolution, clonal diversity estimated by allozymes seems to present the same general trend.

Clonal diversity

A review of the literature on ostracods indicates in general a high level of clonal diversity, although great differences in variability are observed among species (Table 15.1 and Table 15.2). Havel & Hebert (1993) identified over 200 clones in 47 populations of *Cypridopsis vidua* from North America and as many as 24 clones in a single population (Havel & Hebert 1989). In a survey of 29 populations from the Laurentian Great Lakes area, Turgeon & Hebert (1995) detected 185 clones of *Bradleystrandesia reticulata* with a maximum of 14 in a single pond. In our studies

Table 15.1 Data from sexual populations of ostracods. Number of populations (Pop), number of specimens scored (N), number of analysed loci (loci) and percentage of polymorphic loci detected (%) are given.

Species	Pop	N	loci	%	Reference
Candona decora	1	75	4	75	Havel & Hebert 1993
C. inopinata	1	44	6	50	Havel & Hebert 1993
Candonocypris novaezelandiae	1	86	3	100	Chaplin & Ayre 1989
Candonopsis spp.	17	1023	8	50	Little & Hebert 1996
C. anitsi	6	141	8	25	Little & Hebert 1996
Cyclocypris ampla	1	24	9	33	Havel *et al.* 1990a
C. serena	1	24	10	30	Havel *et al.* 1990a
Cypria turneri	1	48	4	50	Havel & Hebert 1993
Bradleystrandesia deltoidea	1	24	9	44	Havel & Hebert 1993
B. deltoidea	5	238	10	50	Turgeon & Hebert 1995
B. splendida	2	66	14	21	Havel & Hebert 1993
B. splendida	9	477	10	40	Turgeon & Hebert 1995
B. tincta	7	413	10	40	Turgeon & Hebert 1995
Cypretta brevisaepta	1	72	8	50	Little & Hebert 1994
Cyprideis bensoni	1	48	11	18	Havel & Hebert 1993
C. salebrosa	1	48	7	29	Havel & Hebert 1993
Heterocypris glauca	12	2743	13	31	Havel *et al.* 1990a
H. glauca	11	3932	8	60	Turgeon & Hebert 1994
Cyprois marginata	1	93	11	0	Havel *et al.* 1990a
C. occidentalis	1	78	10	0	Havel & Hebert 1993
Elpidium spp.	22	972	12	50	Little & Hebert 1996
Cypriconcha alba	1	48	13	15	Havel *et al.* 1990a
Notodromas monacha	2	48	15	13	Havel *et al.* 1990a

we have described at least 125 different clones in 49 Italian populations of *Heterocypris incongruens* and 36 different clones (including a polyploid one) in a single Northern Italy rice field (Rossi *et al.*, unpublished data).

The number of clones determined is affected by parameters related to the study design. Factors like the number of populations sampled, the geographic and temporal scale of sampling, the analytical effort, both in terms of number of individuals per site and number of markers used, influence the total number of clones observed. The size of the study area is important in absolute terms and with respect to the geographic distribution of a species. The most reliable estimates of clonal diversity for a species are obtained on the regional scale, as the example of *H. incongruens* shows. In Northern Italy, a large number of ricefields yielded only a few clones, but extending the survey to the whole Po plain revealed a total of 63 clones (Rossi *et al.* 1995) that become 125 by increasing the sampling area in Italy and including populations from temporary ponds.

The results we report here were obtained from studies at the continental scale. *D. stevensoni* was sampled from Finland to Southern Italy and Spain. The study also includes a population from Lake Tiberias in Israel, and a population from Lake Sibaya in South Africa. For *E. virens* we considered populations ranging from Great Britain to Sicily and Spain. *L. inopinata* was sampled from a relatively smaller range from Italian and Austrian sites.

Table 15.2 Data from asexual populations of ostracods. Number of population (Pop), number of specimens scored (N) number of polymorphic loci identified (loci) and total number of clones detected (clones) are given.

Species	Pop	N	loci	clones	Reference
Candocyprinotus ovatus	1	43	1	4	Havel *et al.* 1990b
Candona balatonica	1	42	1	2	Havel & Hebert 1993
C. candida	1	48	2	3	Havel *et al.* 1990b
C. rectangulata	1	48	3	2	Havel *et al.* 1990b
C. rectangulata (II, '89)	3	133	3	18	Little & Hebert 1997
C. rectangulata (BI, '95)	1	65	3	17	Little & Hebert 1997
Candonocypris novaezelandiae	10	676	3	23	Chaplin & Ayre 1989
C. novaezelandiae	42	2255	4	26	Chaplin & Ayre 1997
Chlamydotheca unispinosa	7	260	1	2	Little & Hebert 1994
Cypretta turgida	3	165	3	4	Little & Hebert 1994
Bradleystrandesia fuscata	1	42	2	3	Havel & Hebert 1993
B. fuscata	10	503	7	11	Turgeon & Hebert 1995
B. reticulata	20	1023	6	66	Havel et al. 1990b
B. reticulata	29	1364	10	185	Turgeon & Hebert 1995
Spirocypris horrida	3	96	2	2	Havel *et al.* 1990b
S. horrida	3	50	4	2	Turgeon & Hebert 1995
Cypridopsis sp.	6	**	2	1	Little & Hebert 1997
Sarscypridopsis aculeata	22	799	3	6	Havel *et al.* 1990b
Cypridopsis vidua	23	842	2	77	Havel & Hebert 1989
C. vidua	2	76	2	5	Little & Hebert 1994
Heterocypris incongruens	4	417	3	20	Sywula & Lorenc 1982
H. incongruens	42	1512	3	7	Havel & Hebert 1993
H. incongruens	1	43	4	2	Little & Hebert 1994
H. incongruens	76	2818	7	10	Turgeon & Hebert 1994
H. incongruens	26	2354	3	63	Rossi *et al.* 1995
Cypris decaryi	7	332	4	1	Little & Hebert 1994
C. pubera	2	96	1	4	Havel *et al.* 1990b
C. subglobosa	1	22	2	1	Little & Hebert 1994
Cytherissa lacustris	2	743	1	6	Sywula & Geiger 1990
Darwinula stevensoni	1	48	1	2	Havel & Hebert 1993
Eucypris foveata	16	525	3	4	Havel *et al.* 1990b
Ilyocypris biplicata	1	72	2	1	Havel & Hebert 1993
Potamocypris granulosa	4	109	3	6	Havel *et al.* 1990b
P. illinoisensis	1	72	2	2	Havel & Hebert 1993
Tonnacypris glacialis	1	24	3	7	Havel *et al.* 1990b
T. glacialis	50	2829	2	54	Little & Hebert 1997
Stenocypris major	3	58	2	2	Little & Hebert 1994
Strandesia longula	1	18	2	2	Little & Hebert 1994
Diaphanocypris meridiana	3	56	1	2	Little & Hebert 1994

** number of specimens not reported.

Due to the small body size of two species (*D. stevensoni* and *L. inopinata*), the number of markers which could be used is relatively low. In *E. virens*, a larger species, more enzymes were tested (Rossi *et al.* in prep.). To allow comparisons, we report here the data obtained using GPI (glucose phosphate isomerase), MPI (mannose phosphate isomerase), and PGM (phospho glucose mutase). These markers are usually found to be polymorphic (Havel & Hebert 1993, Rossi & Menozzi 1994). For *D. stevensoni* and *L. inopinata*, only GPI and PGM have been screened, due to the lack of activity of MPI. Results should, nevertheless, represent the clonal diversity of these two species, as, according to data from *Daphnia* (Hebert 1987) and *Heterocypris incongruens* (Rossi & Menozzi 1994), the number of loci analysed is not correlated with the number of detectable multilocus genotypes.

Our current results confirm both the high level of clonal diversity and its variability among species: we found 211 clones in *E. virens* and only 7 in *D. stevensoni*, while the intermediate value of 31 clones was observed in *L. inopinata* (Table 15.3). The relatively low number of clones detected in *L. inopinata*, a species with a reproductive biology with characteristics similar to those of *E. virens,* might be due to the more restricted sampling area that was covered for this species.

In parthenogenetic populations, clonal diversity (number of multilocus genotypes) is correlated with the number of alleles per locus. We found significant differences in the mean number of alleles per locus among the analysed species. As for the number of different clones, the number of different alleles was also the highest in *E. virens* and the lowest in *D. stevensoni* (Table 15.3).

In bisexual populations, genotypic richness is maintained by recombination events. The relationship between the number of markers used and the number of multilocus genotypes detected is obviously different in populations with different reproductive modes. A good example of this relationship is the clonal variability (number of multilocus genotypes) in parthenogenetic and mictic populations of *E. virens* from a small temporary pond in Extremadura (Spain). The two populations are in fact subpopulations inhabiting the same restricted area where both male and female individuals have been found. The analysis of individual genotypes revealed that females belong to two different lineages: one mictic that shares alleles with males and one parthenogenetic that does not (Rossi *et al.* in prep.). The two sub-populations (Table 15.4) have similar sample size. The parthenogenetic sub-population has a higher mean number of alleles per locus and a significantly higher heterozygosity. In the mictic population the number of multilocus genotypes increases exponentially with the number of markers, while in the parthenogenetic population the curve tends to an asymptote (Figure 15.1). The effect of absence of recombination on clonal variation is clear.

Table 15.3 Clonal structure of three different ostracod species.
Numbers of populations and numbers of individuals analysed, mean numbers of alleles per screened locus and numbers of multilocus genotypes per species, based on the variability at 3 * (Mpi, Pgm and Gpi) or 2 ** (Pgm and Gpi) loci per species are given. Percentage of monoclonal populations refers to samples with N>20. Frequencies of the most abundant genotype, of populations where the most widespread genotype has been found, percentages of private genotypes, monoclonal populations and polyploid genotypes are also given.

	E. virens	*D. stevensoni*	*L. inopinata*
Populations	55	36	12
N	1136	1822	806
alleles per locus	*9.3	**2.5	**5.5
multilocus genotypes	211	7	31
most abundant genotype	0.12	0.79	0.52
most widespread genotype	0.24	0.86	0.75
% of private genotypes	74	0	55
% monoclonal populations	5	17	8
% polyploid genotypes	14	0	0

Table 15.4 Mixed reproductive modes in *E. virens*
Number of individuals analysed, mean number of alleles per locus (on the basis of 5 loci Mpi, Pgm, Gpi, Icd1 and Icd2), percentage of polymorphic loci and mean heterozygosity are given in 2 populations, one mictic and one parthenogenetic, from Extremadura (Spain).

Population	N	allele/locus	polymorphic loci	mean heterozygosity
mictic	30	2.2 (0.5)	80	0.260 (0.096)
parthenogenetic	32	2.8 (0.4)	80	0.457 (0.202)

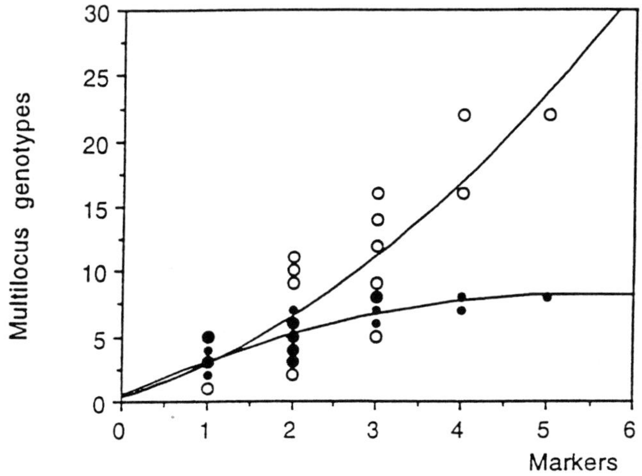

Fig. 15.1 Number of different genotypes detected with respect to the number of enzymatic markers used in parthenogenetic (closed circle) and mictic (open circle) females of *E. virens* from a bisexual population.

Clonal dispersal and geographic distribution of clonal diversity

The genetic structure of few cosmopolitan species has been analysed at a regional scale. The geographic distribution patterns that seem to emerge are to be accepted with caution. Havel & Hebert (1989) surveyed 23 populations of *C. vidua* from Manitoba to Texas and observed that most clones are very rare (62% being found in only a single location) while few (3% of all clones detected) are widespread and found from northern and eastern Canada to the southern United States. Chaplin & Hebert (1997) reported that, in North America, only a few clones of *Heterocypris incongruens* have a broad distribution, while most of them are narrow endemics. We found a similar scenario in a survey of 26 northern Italy populations of *H. incongruens* (the most common species found in ricefield habitats) in an area of over 30,000 km^2. Clones inhabiting only one ricefield were numerous (49%). On the other hand, a few clones (6%) were found in most ricefields and temporary ponds (Rossi *et al.* 1995). Similarly, Chaplin & Ayre (1989, 1997) in a survey of 42 south-eastern Australian water-bodies up to 850 km apart, found a larger but still limited number (23%) of electrophoretic clones of *Candonocypris novaezelandiae* widespread over most of the area.

The continental study of the 3 species reported here seems an important contribution to this kind of investigation. For *E. virens* and *L. inopinata*, we found a pattern similar to that just described from the literature: most clones are present in only one locality (74% in *E. virens* and 55% in *L. inopinata*), while only a few are found in most populations (1% in *E. virens* and 5% in *L. inopinata*). The low clonal variability observed in *D. stevensoni* makes the geographic distribution of this species an exception to the general rule: all clones are found in more than one locality with the most widespread one found almost everywhere.

To compare the clonal structure of the three species we have used rank-abundance diagrams, a tool borrowed from community ecology (Figure 15.2). The proportion of the individuals belonging to a given clone is plotted against the clone's rank (in terms of abundance). *E. virens* shows the highest number of clones and a quite even distribution of specimens among the 211 detected clones. The 2 most common clones make up only 23% of all individuals. The most common clone, found in 6 populations ranging from Spain to Italy, was not the most widespread. The second ranking clone (the most widespread) was present in 13 localities in Italy and Spain. An intermediate situation is represented by *L. inopinata*. The most abundant clone represented by 52% of individuals was present in 75% of Italian and Austrian populations (Tab. 15.3). At the other extreme, for *D. stevensoni* we obtained a typical "geometric series" distribution, an indicator of a high level of dominance. The most common clone represented by 79% of the analysed individuals was found in all but 5 populations (86% of all localities).

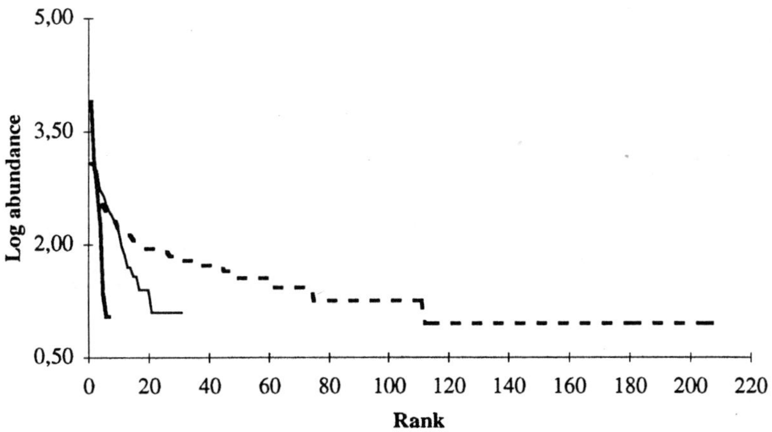

Fig. 15.2 Rank-abundance curves for each species: *E. virens* (dashed line), *L. inopinata*, (thin line) *D. stevensoni* (thick line). Log Pi*10000, where Pi is the frequency of ith clone, plotted against ranks, with Pi for the most abundant clone being plotted first, then the next most common one and so on.

The clonal structure observed in *C. novaezelandiae* (Chaplin & Ayre 1989, 1997) showing a small number of clones represented by a large number of individuals, was justified as the result of the recent and rapid expansion of a single lineage that would leave within each local population only a small fraction of the clonal diversity that was present in an ancestral source population. On the contrary, a relatively high level of variability within all regions may be explained by the stochastic differentiation of populations founded by a small number of individuals and by variation in the abundance of clonal genotypes within local populations.

Chaplin & Ayre (1997) pointed out that the contrast between the distribution of the widespread and local genotypes suggests that some clones may be better colonisers than others and/or that clones differ in their level of habitat specialisation. Such an explanation makes sense when we consider the high colonising ability of ostracods, e.g. the production of desiccation-resistant resting eggs, the most effective way to colonise new habitats even over long distances.

Parthenogenetic populations can be started by a single resting egg. Habitats that are geographically isolated or have been recently colonised should show low clonal diversity. A good example is the case of *Cypridopsis vidua* sampled in Jamaica. This species is highly variable in temperate North America, but shows only 5 different clones out of a total of 76 females screened in the Jamaica study (Havel & Hebert 1989, Little & Hebert 1994).

In our data, if we consider populations with sample size greater than 20, the proportion of populations which consists of only one clone is lower in *E. virens* than in the other two species. The genetic uniformity of the 3 *E. virens* monoclonal populations (Bramhope (GB), Code d'Asino (I) and Morcuera (E)) has been confirmed by DNA sequences for the two latter cases (Schön *et al.* in prep.). The discrepancy for Bramhope is not surprising or inexplicable if we rely on underestimation of genetic variability obtained by allozyme data. Nevertheless, it is interesting to note that these monoclonal populations are geographically isolated: Bramhope lies on an island, while Code d'Asino and Morcuera are at a high altitude. For the latter two populations, characterised by a relatively high density, the lack of clonal variation was confirmed in different years. The genetic uniformity of these 2 populations may be related to the occasional colonisation of these habitats by resting eggs from a single clone and/or the selective adaptation of a specific clone. As the 'Morcuera' clone is also present in other Spanish populations and the only clone described for Code d'Asino is widespread in northern Italian ricefields, a local selective adaptation seems unlikely. In contrast, those results support the idea of colonisation by a single egg. This hypothesis might also explain the intriguing genetic similarity between the *Eucypris* populations from Bramhope and Sicily, based on shared alleles at different allozyme loci, as well as being confirmed by DNA data (Rossi *et al.* in prep., Schön *et al.* in prep.).

Hairston & De Stasio (1988) and Lampert (1995) have stressed the role of egg banks as a time-dispersal system and their function in slowing down the rate of genetic change by sequestering a fraction of the influence of selection generation by generation. *D. stevensoni,* whose most abundant and common clone includes 79% of analysed individuals and is present in 86% of screened populations, does not present any diapausal stage: its spatially and temporally widespread clones are an enigma.

Havel *et al.* (1990b) report that clonal diversity is in general not related to latitude. Our data confirm their results. We also found no relationship with longitude with a fairly constant amount of diversity detected across the geographical range of the species in spite of the highly variable number of clones in different populations. On the other hand, in *H. incongruens*, clonal richness per rice field population decreased according to a west-east gradient in northern Italy; but we might have a special situation here, as this distribution apparently follows the decreasing proportion of land used for rice cultivation (Rossi *et al.* 1995). In *L. inopinata*, the largest genetic differentiation (31%) is found among the 2 main clusters formed on the basis of genetic similarity: one formed by western Italian populations and the other by eastern Austrian ones.

It seems important to remember that different levels of genetic variability are often found in the same species in different geographic regions. *H. incongruens,* according to our data on European populations, shows a high clonal diversity (Rossi *et al.* 1995), while data from North America indicate this species to be only moderately differentiated. Havel *et al.* (1990b) described *H. incongruens* as a low-clonal diversity species. They found only 7 different multilocus genotypes (clones) in 42 populations from low Arctic sites in Canada. From one Canadian Arctic site, Turgeon & Hebert (1994) reported 10 different clones with one clone widespread and with two clones that jointly represent 54% of the individuals. In the temperate USA and Mexico, only two clones have been detected, one unique to a single site and the other widespread. Little & Hebert (1994) reported only 2 different clones in a total of 30 females screened in Jamaica.

Further evidence for geographically related differences can be derived from the data on *E. virens* and *D. stevensoni*. In Europe, *E. virens* shows geographical parthenogenesis, while it is reported as an obligate parthenogen in North America (Bell 1982). *D. stevensoni* is polymorphic for Gpi and monomorphic for Pgm in Europe, the Middle East and South Africa, while in North America, Pgm is variable and Gpi has only one allele (Havel & Hebert 1993). Only general causes can be indicated for these differences. The time and the mechanism that led to the colonisation of different regions are probably major factors to be considered although divergence due to differences in local conditions should not be overlooked.

Polyphyletic origins of different clones: polyploidy, hybridisation and the origin of parthenogenesis by mutation

It is important to investigate whether differences in the level and distribution of clonal diversity are relevant for the different hypotheses about the mechanisms of the origin of parthenogenesis and clonal diversity in parthenogenetic lineages.

Apomixis is generally considered in relation to polyploidy and hybridisation. Suomalainen *et al.* (1987) pointed out that polyploidy could moderate the effect of Müller's ratchet (see also chapter 1) and consider this fact as a possible explanation of the widespread polyploidy in parthenogenetic lineages. In plants, as reported by Suomalainen *et al.* (1987, pp. 20-21), apomixis has a genetic basis, polygenetically controlled with a threshold in the response that may also require a given allele combination. Apomixis is only expressed when the threshold is exceeded. The role of polyploidy and hybridisation in triggering apomixis can be explained in these terms: hybridisation brings the required allelic combination together and polyploidy guarantees that the threshold is exceeded. Loci affecting meiosis are a prerequisite for the expression of apomixis. Gustafsson, cited by Suomalainen and co-authors (1987), hypothesised that the action of these apomixis-inducing loci is stronger in a polyploid genome. There is evidence that polyploidy reinforces the action of loci favouring specific types of apomixis like vivipary or other forms of vegetative reproduction (Suomalainen *et al.* 1987). Moreover, hybrid genotypes may have a wider niche than the parental form: the immediate selective advantage of fixing and multiplying fit genotypes is obvious.

Both cytological (Tétart 1978) and allozyme data (Havel & Hebert 1993, Little & Hebert 1994) indicate a relatively high incidence of polyploidy in ostracods, even if, with the latest technique, typical unbalanced electrophoretic banding patterns allow the detection of polyploid genotypes, but not the assessment of ploidy levels. In *H. incongruens*, we detected 14% of polyploid genotypes (Rossi *et al.*, unpub. data). Data from North America report a higher incidence of polyploidy at high latitudes in ostracods as well as in other taxa (Beaton & Hebert 1988, Weider *et al.* 1987, Havel & Hebert 1993). In *C. vidua*, 30% of clones (Havel & Hebert 1989) are polyploid and their incidence increases in Arctic populations. In the coastline area of Hudson Bay, Turgeon & Hebert (1994) described 10 different clones of *H. incongruens* 7 of which were polyploid, while two clones found in the southern USA were diploid. Data from tropical habitats do not confirm such a pattern. Little & Hebert (1994) in a study of 13 species from Jamaican temporary ponds, reported a mean percentage of polyploid clones of up to 41%, with none in the bisexual species *Cypretta brevisaepta* and 100% in the parthenogenetic species *Stenocypris major*. Our data on European populations do not seem to show a latitudinal gradient in polyploidy. In *E. virens* we detected 14% of polyploid genotypes over the whole continent with 0.5% in Belgian, 3% in Spanish and 4% in Sicilian populations. In Sicily, all-female populations are made up by polyploid clones only.

We found no evidence for polyploidy in *D. stevensoni* and *L. inopinata*, confirming that its occurrence in parthenogenetic lineages varies considerably (from 0% to 100%, according to Havel & Hebert 1993). For *D. stevensoni*, this observation is compatible with the fact that this species is an obligate parthenogen and lacks sexual relatives. It is important to keep in mind that in our study we did not use cytological methods and our evidence of polyploidy was obtained as a by-product of electrophoretic analysis. This most likely leads to an underestimate.

Some studies provide support for the hypothesis that parthenogenetic ostracod species with sexual congeners show a higher clonal diversity than those from all-female genera (Turgeon & Hebert 1994). Low clonal diversity could then be associated with the impossibility of increasing genetic variation via hybridisation or the lack of input of newly asexual clones from the second species. A hybrid origin and the associated variety of polyploid forms is rather common in parthenogenetic animals, i.e. in parthenogenetic fish, salamanders and lizards (Suomalainen *et al.* 1987).

Many authors have pointed out that apomixis is associated with hybridisation and polyploidy. Evidence that diploid asexual species can be fertilised by a sexual species to produce an asexual triploid has been reported by Suomalainen and coauthors (1987). Turgeon & Hebert (1994) suggested that two triploid clones of

Heterocypris incongruens probably originated through hybridisation between a diploid clone of *H. incongruens* and a haploid sperm of *H. glauca*. Other polyploid clones are supposed to be autopolyploids originated by interbreeding between bisexual and parthenogenetic populations of *H. incongruens*. The data from mtDNA analyses on clones of this species are consistent with these hypotheses (Chaplin & Hebert 1997). In *Bradleystrandesia*, Turgeon & Hebert (1995) ascertained a very high incidence of polyploidy by allozyme and cytogenetic analysis. They described a broad array of allopolyploid clones that reach 82% and 44% of the total clonal diversity in *B. fuscata* and *B. reticulata*, respectively. Among parthenogenetic *Bradleystrandesia* species, clonal variability seems related to the hypothesised nature and frequency of hybridisation. The highest diversity described in *B. reticulata* has been related to repeated hybridisation events with the sexual congeners *B. deltoidea*, *B. splendida* and *B. tincta*. In *B. fuscata*, the origin of allopolyploid clones has been attributed to mating between diploid females *of B. fuscata* and males of *B. tincta* (Turgeon & Hebert 1995).

In our survey we found no evidence for hybridisation between parthenogenetic lineages and sexual congeners, but we did observe interbreeding between parthenogenetic and bisexual lineages. One population of *E. virens* in Extremadura (Spain) consists of a mixture: parthenogenetic diploid female lineages, parthenogenetic polyploid female lineages, mictic diploid female lineages and diploid males coexist (Rossi *et al.* in prep.). In this case, the autopolyploid genotypes have probably originated through fertilisation of a diploid parthenogenetic female by a male. This observation seems in accordance with the hypothesis that polyploid apomictic parthenogens evolve from diploid parthenogenetic animals, by the fertilisation of parthenogenetic females by diploid males (Suomalainen *et al.* 1987). For other polyploid clones from the same region a similar origin may be hypothesised. Video-recorded observations of copulatory behaviour, (chapter 10) support the plausibility of such occasional hybridisation. We also described sympatry of lineages with different reproductive modes in Sicilian populations of *E. virens*. Species with mixed breeding systems, such as *E. virens*, would have the opportunity for repeated transition to asexuality and for recruitment of new genotypes from sexual relatives as suggested by Bell (1982) and Havel *et al.* (1990b). Our current results confirm that clonal diversity is higher in species such as *E. virens* with evidence of polyploid lineages due, at least in some cases, to occasional hybridisation between parthenogenetic clones and sexual relatives. For *L. inopinata*, a species with both bisexual populations and sexual congeners, our data do not provide such evidence, but this could be due to the limited number of populations analysed. The low clonal diversity which was observed in *D. stevensoni* may be associated with the absence of a comparable source to retain genetic variability as neither bisexual populations nor sexual congeners are known.

Polyploidy does not seem a general mechanism for the generation of clonal diversity. Several parthenogenetic species exist in which polyploidy has not been observed. Parthenogenesis seems to have arisen as a single mutational event in a bisexual population (e.g. in the case of parthenogenetic weevils), as well as by occasional tychoparthenogenesis (the spontaneous development of unfertilised eggs). This is a natural phenomenon well described in *Drosophila*: in *D. mercatorum* it has been computed that the probability of an unfertilised egg giving rise to an adult female herself capable of parthenogenesis is 10^{-5}. This rate allows the establishment of parthenogenetic lineages directly without selection. In a selection experiment, Stalker, cited by Suomalainen *et al.* (1987) was able to elevate the capacity for initiation of parthenogenesis in *D. parthenogenetica* from 0.91 to 8.3% in 17 generations. In *D. mercatorum*, there is much evidence indicating that natural populations contain variation in this trait and an x-linked recessive allele, abnormal abdomen, shows association with parthenogenesis.

Suomalainen *et al.* (1987), in describing the origin of automixis in *Solenobia triquetrella* (Orthoptera), hypothesised that the decrease of population density due to the new habitats available following the retreating edge of ice, implied that some females remained unfertilised. These tried to lay unfertilised eggs, in which they sometimes succeeded. Sometimes, these unfertilised eggs even gave rise to a new generation through tychoparthenogenesis. The rarity of males, which might have interfered with this development, favoured the genetic fixation of parthenogenesis. Indeed, many populations from temporary ponds, as is the case of *E. virens*, suffer low density and bisexual ones showing a female bias may be sperm-limited. In lab experiments, we have observed that virgin mictic females of *E. virens* from Sicilian bisexual populations do not need males in order to start producing eggs, but do not lay them if copulation fails.

Indeed, parthenogenesis in both *Solenobia* and *Drosophila* is automictic; the transition from mixis does not require any radical change in the cytological mechanism of meiosis. On the contrary, in apomictic parthenogenesis the chromosomes do not pair and their number is not reduced: for this reason, Suomalainen *et al.* (1987) stressed that the adoption of such a mode of reproduction is accompanied by a dramatic change in the mechanism of meiosis. These authors suggest that apomictic parthenogenesis might have been preceded by an automictic transitional period. Even in the silkworm, *Bombyx mori*, artificial parthenogenesis is induced simultaneously with equational division. Alternatively, sperm limitation might have favoured the spread of lineages which, as a result of spontaneous mutation, reproduced by apomictic parthenogenesis.

Mutation and clonal variation

It is assumed that clonal diversity in asexual taxa is generated by the accumulation of mutations. The detrimental nature of mutation-induced variation (the so-called mutation load, see chapter 1) poses a problem for the sustainability of such a mechanism, as clonal diversity should be the result of a balance between the production of variation by mutation and its loss by selection or by genetic drift.

Hardly any data on mutation rates of ostracods exist. Havel & Hebert (1989) reported that lack of variation within the lab cultures of *Cypridopsis vidua* pointed towards a typical mutation rate (not higher than 4.9×10^{-4}).

As mutations accumulate over time, we should expect a significantly higher clonal diversity in *D. stevensoni,* the oldest species we have analysed (Table 15.5) (Birky 1996). On the contrary, *D. stevensoni* shows a very low level of genetic variability and the lowest clonal diversity of the species we investigated. This picture is confirmed by the very low variation of DNA sequences (Schön *et al.* 1998). A complete absence of

Table 15.5 General characteristics of species with respect to habitat (T= temporary, P= permanent), life cycle duration, parthenogenesis characteristics (G= geographical, O= obligate) and presence of diapausal stages.

	E. virens	*D. stevensoni*	*L. inopinata*
Age (Myr)	1	24.6	2
Habitat	T	P	P
Life cycle (months)	1-2	24	2-3
Parthenogenesis	G	O	G
Diapausal stages	+	-	+

substitutions in the ITS1 sequence was found for *D. stevensoni* compared with a genetic distance of up to 10% between European populations of *E. virens*. Schön & Martens (1998) attributed this lack of sequence divergence in *D. stevensoni* to a particularly efficient DNA repair system (see chapter 17 for detailed discussion).

Clonal diversity and species' biological and ecological features

The source of clonal diversity varies among the apomictic species we have analysed and this should be sufficient to explain the different levels of genetic variability among them. The high level of clonal variability observed in *E. virens* might be due to the multiple mechanisms available to the species to generate new clones. In addition to mutation, this species borrows at least a portion of its diversity from sexual ancestors or extant relatives. In this sense, the intermediate level of variability observed in *L. inopinata* would also find an explanation.

In *H. incongruens* from North America, Chaplin & Hebert (1997) obtained evidence for polyphyletic origin of polyploid clones and for repeated transitions to polyploidy. Polyphyletic origin of parthenogenetic taxa has been considered to be a cause of high clonal diversity in different asexual species (Suomalainen *et al.* 1987). A similar origin of asexuality might be hypothesised for *E. virens* and *L. inopinata*, as evidence for interbreeding was found in *E. virens* and sexual populations are known for both species. The lack of sexual relatives and congeners, and the absence of genetic evidence for hybridization, suggest that the only source of genetic variability in *D. stevensoni* is mutation. Nevertheless, ancient clonal lineages would be expected to accumulate mutations that are not under strong selection and, indeed, *D. stevensoni* shows a low level of genetic variability that is incompatible with the age of the species being several million years as indicated by geological data (Butlin & Griffiths 1993, Schön *et al.* 1998).

Clonal diversity is supposed to be increased and maintained by the interaction between genotype and environment or genotype and genotype. As an indirect test of this hypothesis, we can compare the ecological and biological characteristics of different species. We would expect species with high clonal diversity to possess mechanisms to persist in space (for example, habitat segregation) and time (for example, resting eggs and clonal succession). Such a situation has been well described for *H. incongruens* that showed a very high clonal diversity in Europe and in particular in Italy. In this species, seasonal clonal succession was observed, involving the production and hatching of resting eggs in response to temperature and photoperiod (Rossi & Menozzi 1990, Rossi *et al.* 1991, Rossi & Menozzi 1993, Rossi *et al.* 1996).

Although *E. virens* also showed a high clonal diversity, the life history of the species is quite different: it is a typical univoltine, stenotherm species producing resting eggs in early spring which diapause until late autumn or early winter. Neither seasonal succession nor unstable temporal structure have been observed. Nevertheless, a differential response to photoperiod with respect to the production of resting eggs by clones from Great Britain, Italy and Spain has been observed in laboratory experiments. The observed cline in voltinism seems an example of the shaping of clonal spatial structure by the selective action of latitude through photoperiod (Otero *et al.* in press).

L. inopinata shows again an intermediate situation. In this species, the highest clonal diversity has been found in Seewinkel and Neusiedlersee that represent extreme habitats with respect to their high alkalinity (see chapter 14).

On the contrary, *D. stevensoni*, a widely distributed species tolerant of a wide range of temperature and salinity, in which diapausal stages are unknown, seems an ideal material to test the "general purpose genotype" hypothesis (see chapter 14).

References

Avise, J.C. 1994. *Molecular markers, natural history and evolution*, Chapman & Hall.

Beaton, M.J. & Hebert, P.D.N. 1988. Geographic parthenogenesis and polyploidy in *Daphnia pulex*. *Am. Nat.* **132:** 837-845.

Bell, G. 1982. *The masterpiece of nature,* Croom Helm.

Birky, C.W. Jr. 1996. Heterozygosity, heteromorphy, and phylogenetic trees in asexual eukaryotes. *Genetics*, **144:** 427-437.

Butlin, R.K. & Griffiths, H.I. 1993. Ageing without sex. *Nature*, **364:** 680.

Chaplin, J.A. 1992. Variation in the mode of reproduction among individuals of the ostracod *Candonocypris novaezelandiae*. *Heredity*, **68:** 411-424.

Chaplin, J.A. 1993. The local displacement of a sexually reproducing ostracod by a conspecific parthenogen. *Heredity*, **71:** 259-268.

Chaplin, J.A. & Ayre, D.J. 1989. Genetic evidence of variation in the contributions of sexual and asexual reproduction to populations of the freshwater ostracod *Candonocypris novaezelandiae*. *Freshw. Biol.,* **22:** 275-284.

Chaplin, J.A. & Ayre, D.J. 1997. Genetic evidence of widespread dispersal in a parthenogenetic freshwater ostracod. *Heredity*, **78:** 57-67.

Chaplin, J.A. & Hebert, P.D.N. 1997. *Cyprinotus incongruens* (Ostracoda): an ancient asexual? *Mol. Ecol.*, **6:** 155-168.

Crease, T.J., Stanton, D.J. & Hebert, P.D.N. 1989. Polyphyletic origins of asexuality in *Daphnia pulex*. II. mitochondrial DNA variation. *Evolution*, **43:** 1016-1026.

Crease, T.J. & Lynch, M. 1991. Ribosomal DNA variation in *Daphnia pulex*. *Mol. Biol. Evol.*, **8:** 620-640.

Crease, T.J., Lynch, M. & Spitze, K. 1990. Hierarchical analysis of population genetic variation in mitochondrial and nuclear genes of *Daphnia pulex*. *Mol. Biol. Evol.*, **7:** 444-458.

Foighil, D.O. & Smith, M.J. 1996. Phylogeography of an asexual marine clam complex, *Lasaea*, in the Northeastern Pacific based on Cytochrome Oxidase III sequence variation. *Mol. Phylogen. Evol.*, **6:** 134-142.

Hairston, N.G. Jr. & De Stasio, B.T. Jr. 1988. Rate of evolution slowed by dormant propagule pool. *Nature*, **336:** 239-242.

Havel, J.E. & Hebert, P.D.N. 1989. Apomictic parthenogenesis and genotypic diversity in *Cypridopsis vidua* (Ostracoda: Cyprididae). *Heredity*, **62:** 383-392.

Havel, J.E. & Hebert, P.D.N. 1993. Clonal diversity in parthenogenetic ostracods. *in*: McKenzie, K. G. & Jones, P. J. (eds), *Ostracoda in the earth and life sciences*, A. A. Balkema: 353-368.
Havel, J.E., Hebert, P.D.N. & Delorme, L.D. 1990a. Genetics of sexual Ostracoda from a low arctic site. *J. Evol. Biol.*, **3:** 65-84.
Havel, J.E., Hebert, P.D.N. & Delorme, L.D. 1990b. Genotypic diversity of asexual Ostracoda from a low arctic site. *J. Evol. Biol.*, **3:** 391-410.
Hebert, P.D.N. 1987. Genotypic characteristics of the Cladocera. *Hydrobiologia*, **145:** 183-193.
Hebert, P.D.N. & Beaton, M.J. 1989. *Methodologies for allozyme analysis using Cellulose Acetate Electrophoresis,* Beaumont.
Hughes, R.N. 1989. *A functional biology of clonal animals,* Chapman and Hall.
Lampert,W. 1995. Egg bank investment. *Nature*, **377:** 479.
Little, T.J. & Hebert, P.D.N. 1994. Abundant asexuality in tropical freshwater ostracodes. *Heredity*, **73:** 549-555.
Little, T.J. & Hebert, P.D.N. 1996. Endemism and ecological islands: the ostracods from Jamaican bromeliads. *Freshw. Biol.*, **36:** 327-338.
Little, T.J. & Hebert, P.D.N. 1997. Clonal diversity in high arctic ostracodes. *J. Evol. Biol.*, **10:** 233-252.
Otero, M., Rossi, V., Baltanás, A. & Menozzi, P. (in press). Effect of genotype and photoperiod on diapause strategies in *Eucypris virens* (Jurine, 1820) (Crustacea: Ostracoda). *Arch. Hydrobiol.*
Quattro, J.M., Avise, J.C. & Vrijenhoek, R.C. 1992. An ancient clonal lineage in the fish genus *Poeciliopsis* (Atheriniformes: Poeciliidae). *Proc. Natl. Acad. Sci. USA*, **89:** 348-352.
Rossi, V., Bucci, G., Giordano, P. & Menozzi, P. 1995. Variabilità genetica, modalità riproduttive ed ecologia in popolazioni di ostracodi d'acqua dolce. *S. It. E. Atti*, **16:** 479-481.
Rossi, V., Gandolfi, A. & Menozzi, P. 1996. Egg diapause and clonal structure in parthenogenetic populations of *Heterocypris incongruens* (Ostracoda). *Hydrobiologia*, **320:** 45-54.
Rossi, V., Giordano, P. & Menozzi, P. 1993. Genetic variability in parthenogenetic populations of *Heterocypris incongruens* (Crustacea, Ostracoda). *in*: McKenzie, K.G. & Jones, P.J. (eds), *Ostracoda in the earth and life sciences*, Balkema: 369-383.
Rossi, V. & Menozzi, P. 1990. The clonal ecology of *Heterocypris incongruens* (Crustacea: Ostracoda). *Oikos*, **57:** 388-398.
Rossi, V. & Menozzi, P. 1993. The clonal ecology of *Heterocypris incongruens* (Ostracoda): life-history traits and photoperiod. *Functional Ecol.*, **7:** 177-182.
Rossi, V. & Menozzi, P. 1994. Enzyme and DNA polymorphism in ostracod evolutionary ecology. *in*: Horne D.J. & Martens K. (eds), *The evolutionary ecology of reproductive modes in non-marine Ostracoda,* Greenwich Univ. Press: 43-54.
Rossi, V., Rozzi, C. & Menozzi, P. 1991. Life strategies among electrophoretic clones of *Heterocypris incongruens* (Crustacea, Ostracoda). *Verh. Int. Verein. Limnol.*, **24:** 2816-2819.
Schön, I., Butlin, R.K., Griffiths, H.I. & Martens K. 1998. Slow molecular evolution in an ancient asexual ostracod. *Proc. R. Soc. Lond. B*, **265:** 235-242.
Schön, I. & Martens, K. 1998. Opinion: DNA repair in ancient asexuals – a new solution for an old problem? *J. Nat. Hist.*, **32:** 943-948.
Suomalainen, E., Saura, A. & Lokki, J. 1987. *Cytology and evolution in parthenogenesis*, CRC Press.
Sywula, T. & Geiger, W. 1990. Genetic and morphological polymorphism in the freshwater ostracod *Cytherissa lacustris* (Sars) of lakes Mondsee and Attersee (Austria). *Bull. Inst. Géol. Bass. Aquit.*, **47:** 69-81.
Sywula, T., Glazewska, I., Whatley, R.C. & Moguilewsky, A. 1995. Genetic differentiation in the brackish-water ostracod *Cyprideis torosa*. *Marine Biol.*, **121:** 647-653.
Sywula, T. & Lorenc, R. 1982. Esterase polymorphism as a marker of interpopulation differentiation in *Cyprinotus incongruens* (Ramd.) (Ostracoda, Crustacea). *Bull. Acad. Polonaise Sci.*, **29:** 317-322.
Sywula, T., Suomalainen, E., Saura, A. & Lokki, J. 1985. Genetic structure in natural populations of Ostracoda. I. Electrophoretic methods. *Z. Syst. Evolutionsf.*, **23:** 194-198.
Tétart, J. 1978. Les garnitures chromosomiques des Ostracodes d'eau douce. *Trav. Lab. Hydrobiol. Univ. Grenoble*, **69-70:** 113-140.
Turgeon, J. & Hebert, P.D.N. 1994. Evolutionary interactions between sexual and all-female taxa of *Cyprinotus* (Ostracoda: Cyprididae). *Evolution*, **75:** 561-570.
Turgeon, J. & Hebert, P.D.N. 1995. Genetic characterization of breeding systems, ploidy levels and species boundaries in *Cypricercus* (Ostracoda). *Heredity*, **75:** 561-570.

Weider, L.J., Beaton, M.J. & Hebert, P.D.N. 1987. Clonal diversity in high-Arctic populations of *Daphnia pulex*, a polyploid apomictic complex. *Evolution*, **41:** 1335-1346.

Weider, L.J. & Hobaek, A. 1994. Molecular biogeography of clonal lineages in a high-Arctic apomictic *Daphnia* complex. *Mol. Ecol.*, **3:** 497-506.

Weider, L.J., Hobaek, A., Crease, T.J. & Stibor, H. 1996. Molecular characterization of clonal population structure and biogeography of Arctic apomictic *Daphnia* from Greenland and Iceland. *Mol. Ecol.*, **5:** 107-118.

CHAPTER 16

GENETIC DIVERSITY AND MOLECULAR PHYLOGENY

I. Schön and R.K. Butlin

Abstract

DNA sequence data can be used to estimate the ages of parthenogenetic lineages and their phylogenetic relationships to one another and to sexual lineages. Thus, the number of independent origins of parthenogenesis can be inferred and hybrid origins can be detected. Furthermore, aspects of the evolution of parthenogenetic lineages, such as mutation accumulation, can also be detected. Here we consider the significance of DNA sequence data from two different, genomic regions, the nuclear ITS1 and the mitochondrial COI in three species of non-marine ostracods.

Intraspecific, genetic divergence of both nuclear and mitochondrial DNA was exceptionally high in *Eucypris virens*, an ostracod species with geographical parthenogenesis, both in sexual and parthenogenetic lineages. This could indicate an ancient origin for some parthenogenetic lineages with subsequent accumulation of mutations or a faster rate of molecular evolution, or multiple origins from diverse sexual ancestors. Phylogenetic reconstructions revealed that parthenogenesis has originated several times within this species, and that some origins are recent. Furthermore, we found some evidence for hybridisation between parthenogenetic females and males. The continuous generation of new clones could help to explain the persistence of parthenogenesis, while occasional hybridisation between lineages with different reproductive modes could increase genetic diversity. Although some clones of *E. virens* might have persisted for several million years, we cannot say at which point of their history they abandoned sexual reproduction.

DNA sequence data from *Darwinula stevensoni*, a supposedly ancient asexual ostracod, revealed surprising results. Instead of the expected accumulation of mutations, we found a low substitution rate. This could be the result of a generally low evolutionary rate. However, a significant difference in the substitution rate between the mitochondrial and the nuclear genome indicates that other processes like gene conversion or a more efficient DNA repair are more likely to be responsible.

What can molecular data tell us?

Within any group of organisms, mutations (permanent changes of the DNA sequence) accumulate over time. As a result, the DNA sequences of two lineages become progressively more different as the time since they shared a common ancestor progresses. The comparison of the nucleotide sequence of the same gene (or DNA region) between species, populations or individuals, therefore, allows the reconstruction of their phylogenetic relationships. If one assumes that divergence accumulates at a steady rate, inferences about the time-scales involved in speciation and radiation can also be made: the principle commonly known as the 'molecular clock' (Avise 1994). The 'ideal' concept of a constant, universal clock does not exist

Sex and parthenogenesis: evolutionary ecology of reproductive modes in non-marine ostracods, edited by K. Martens, pp. 275–293

in reality (Avise 1994) but, nevertheless, sequence comparison is an extremely powerful technique capable of providing strong conclusions about the relative timing of evolutionary events and at least a guide to absolute timings.

One of the most critical questions concerning the evolutionary balance between sexual and parthenogenetic reproduction is the time-scale involved (see chapter 1). Assuming that asexually reproducing populations have a short-term advantage but are destined to extinction in the long term (Judson & Normark 1996, Kondrashov 1993, Maynard Smith 1989), we have to get an idea what 'long' or 'short' periods could mean in evolutionary terms. Griffiths & Butlin (1995) suggested that sexual lineages might already gain an evolutionary advantage in a period of 5,000 years, whereas Maynard Smith (1992) stated that in evolutionary terms, 100,000 years is 'but an evening gone'. Molecular clock estimates from unisexual vertebrates indicate that most have an age of 500,000 years or less (Avise 1994, Quattro *et al.* 1992), with some exceptions (Hedges *et al.* 1992). In invertebrates, mitochondrial data for asexual clam lineages suggest an age of up to 3 Myr (Foighil & Smith 1996). Based on restriction digestions, Chaplin & Hebert (1997) estimated that some asexual lineages of the ostracod *Heterocypris incongruens* could have existed for several million years. The assumption of a constant clock rate is not the only problem where molecular data are used for age estimates in asexuals (see below) but, still, the 'molecular clock' represents the only alternative beside the fossil record.

In addition to information about the age of asexual lineages, molecular data can be used to resolve phylogenetic relationships. This allows inferences about the origin of parthenogenetic lineages and the sources of diversity among clones. One can determine whether extant, asexual clones in a certain species have originated separately from a sexual ancestor, or whether they are the result of diversification within a monophyletic parthenogenetic lineage (see chapter 11). If the sexual ancestor is not known, it can be identified. Using DNA-based techniques, it has been demonstrated that all 70 unisexual vertebrate taxa investigated so far (e.g. *Poeciliopsis* fish, *Cnemidophorus* lizards and *Ambystoma* salamanders) are hybrids originating from at least two parental sexual species (Avise 1994). Such a hybridisation event has also been suggested for asexual lineages of the ostracod species *Heterocypris incongruens* (Turgeon & Hebert 1994, Chaplin & Hebert 1997). The high heterozygosity that results from a hybrid origin may provide the parthenogenetic lineage with an evolutionary advantage, at least in the short term. Heterozygosity may be replenished, and clonal diversity increased, where parthenogenetic lineages are able to interbreed occasionally with closely related sexual species (Chaplin and Hebert 1997; see chapter 1). This too can easily be detected with molecular analyses, especially by comparing mitochondrial and nuclear DNA-based phylogenies (Crease *et al.* 1989).

Other predictions concerning the evolutionary history of asexual clones, such as the effects of 'Müller's ratchet' (Müller 1964) or the mutational load (Crow 1994), can only be verified with DNA sequence data. Meselson (http://golgi.harvard.edu/meselson; in Judson & Normark 1996) has predicted that heterozygosity in ancient apomictic parthenogenes will not only be preserved, since recombination is absent, but will increase indefinitely. The consequences of this process for a parthenogenetic, diploid lineage will be that the two alleles of the same gene accumulate (different) mutations over the time, becoming progressively more different from each other. Moreover, the two alleles within an individual share a common ancestor in the original sexual population, while alleles in different individuals, clones or even species that have diverged since the transition to asexuality, have more recent common ancestors. Only molecular sequence data, in this case the comparison of the two (or more in a polyploid lineage) copies of the same gene within the same individual, can verify this theory, and use this theoretical prediction to test the cases of 'ancient asexuals'.

In this chapter, we describe measurements of DNA sequence divergence among clones, and in comparison with outgroups, for each of the three focal taxa in this study: *Limnocythere inopinata*, *Eucypris virens* and *Darwinula stevensoni*. The data complement the allozyme data reported in chapter 15 in terms of assessing clonal diversity in Europe and go further in that they can be used to estimate how divergent clones are, and thus how long ago their separate evolution began. In the case of *Eucypris*, our data also provide information on the separate origins of clones, or groups of clones, from their sexual ancestor and suggest occasional gene exchange between sexual and asexual lineages. For *Eucypris* and *Darwinula*, comparisons between nuclear and mitochondrial DNA sequences have been made, with surprising results in both cases.

Box 16.1 DNA-related techniques:

Most techniques of modern molecular biology require first the extraction of DNA from tissue (or whole organisms), a process in which the nucleic acids are separated from cell proteins, lipids and other organic and inorganic contents. A broad collection of different protocols is available (for review see e.g. Palumbi 1996), which have been adapted for various taxonomic groups. Not only fresh material, but also dried samples or samples fixed with ethanol or even formalin might still be suitable for DNA extractions.

Especially for small animals such as ostracods, but also for a directed molecular analysis, the application of the polymerase chain reaction (PCR, see Fig. 16.1) was the indispensable break-through in molecular techniques. PCR comprises the following three steps at different temperatures: the separation of the DNA double strand (denaturation), the binding of a short DNA fragment, the 'primer', to a specific region of the genomic DNA (annealing), and 'extension', when the thermostable *Taq* polymerase enzyme synthesises a complementary strand, starting with the primer. After 1 PCR cycle, the number of copies of the target region has doubled, after 2 cycles, it has increased four times and so on, until after 30 or more cycles, the PCR region has been amplified several million times.

The design of the primers determines which genomic region will be amplified. If no sequence data are available, either 'universal' primers can be used, which anneal in conserved regions, or RAPD (Random Amplified Polymorphic DNA) can be used as an alternative approach where short primers bind not to a specific region, but randomly to several loci in the DNA.

The amplified DNA ('PCR product') can then be further analysed, either by restriction fragment length polymorphisms (RFLPs) or DNA sequencing. Restriction enzymes, which are used for RFLPs, cut nucleic acids at specific sites, mostly comprising 2 to 4 consecutive nucleotides (e.g. AATT). If a mutation has occurred at a site being specific for this restriction enzyme, the banding pattern of the RFLP changes, and this can be made visible by gel electrophoresis and subsequent staining of the DNA fragments. However, if the mutation has occurred somewhere else, it will remain undetected. DNA sequencing, nowadays mostly by the Sanger 'chain termination method' on automatic sequencing machines, reveals the exact nucleotide sequence of the whole PCR product, thus providing the finest tool available for genetic analysis. It requires specially marked primers or nucleotides and highly sophisticated equipment and software but these are becoming widely available.

How was genetic diversity measured?

DNA was extracted from single ostracods with different geographical origin for each of the three species: *Limnocythere inopinata*, *Eucypris virens* and *Darwinula stevensoni* (Tab. 16.1 to 16.3). DNA was amplified by the polymerase chain reaction (PCR, see Fig. 16.1) and sequenced automatically. Two different genomic regions were analysed: the nuclear Internal Transcribed Spacer 1 region (ITS1) and part of the mitochondrial Cytochrome Oxidase I gene (COI). Details concerning the techniques and primers used are described elsewhere (Box 16.1, Schön *et al.* 1998, Schön *et al.* in prep.).

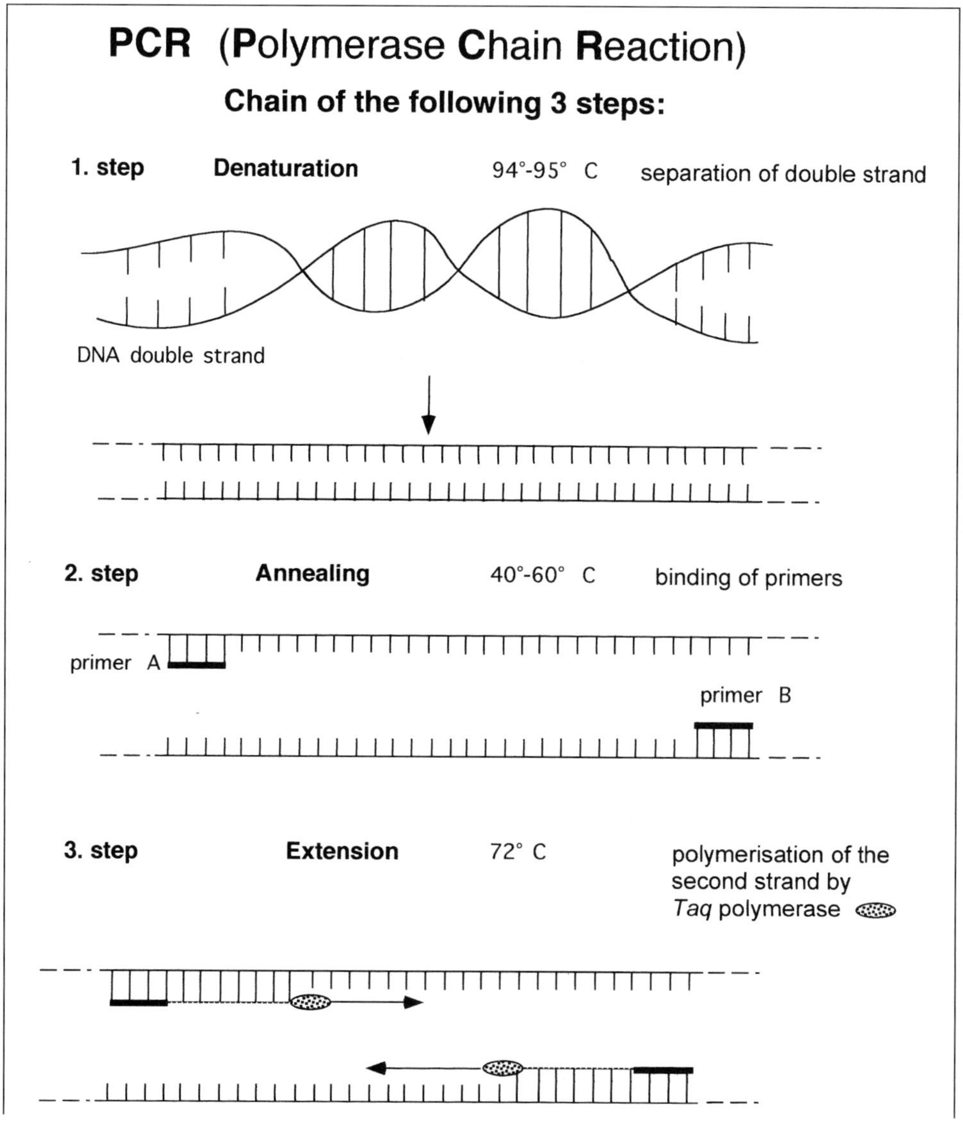

Fig. 16.1 Different steps of the Polymerase Chain Reaction (PCR).

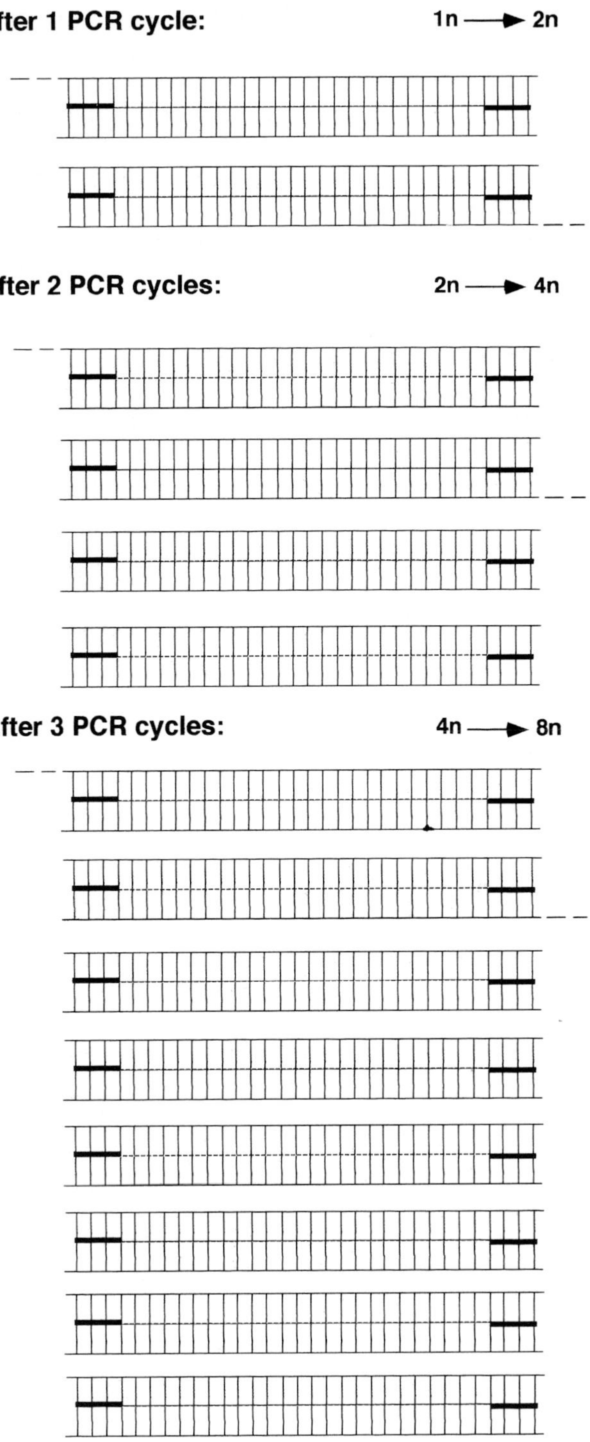
after 1 PCR cycle:
1n → 2n
after 2 PCR cycles:
2n → 4n
after 3 PCR cycles:
4n → 8n

It is quite important to analyse both the nuclear and mitochondrial genomes, as the evolutionary rate in mitochondrial DNA is often higher (Avise 1994) and might, therefore, provide a finer resolution. Since mitochondrial DNA is only inherited maternally, it enables us to compare maternal and paternal lineages. Such an approach is particularly interesting for species with a mixed reproductive system like *Eucypris virens*, because it might either help to detect hybridisation or to determine the female parent where hybridisation is discovered.

The nuclear ITS1 is an interspersed 'spacer' region, which is transcribed into RNA, but not translated into a protein. It separates the 18S and 5.8S genes, both coding for ribosomal RNA (Fig. 16.2), and is present in many copies per genome. These two RNA genes represent highly conserved regions, facilitating the use of '-universal' primers for PCR, whereas ITS1 itself evolves quite rapidly (Hillis & Dixon 1991), and can be used to estimate genetic divergence at the intraspecific level. The region of the mitochondrial genome investigated, Cytochrome Oxidase I, codes for a mitochondrial enzyme.

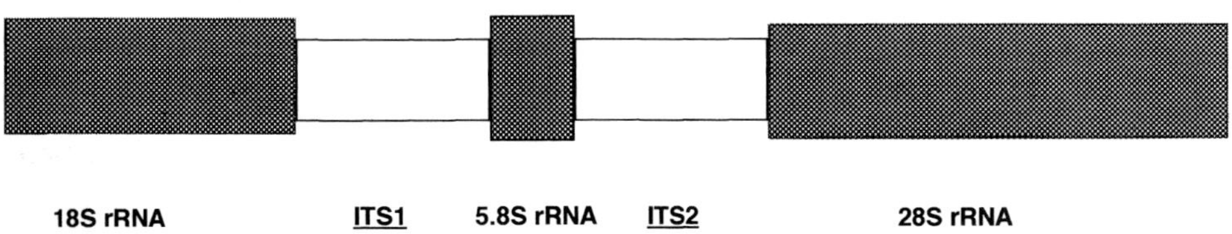

Fig. 16.2 Grouping of the ribosomal RNA genes, showing the position of ITS1

For each species, genetic variation was estimated at different levels: within populations, between European populations and, in the case of *Darwinula stevensoni*, also among African and European populations. Except for *Limnocythere inopinata*, where only a small number of samples could be analysed, genetic distances were not estimated by the simple 'p-ratio' (total number of changes, relative to the total number of sequenced nucleotides). Instead, the 'Kimura 2-parameter' distance method (Kimura 1980) was used, which accounts for different transition/transversion ratios (s. Box 16.3), as the estimated values of genetic diversity could otherwise be biased.

Box 16.2 Molecular evolution

Selection pressure on DNA sequence variation in regions that code for functioning proteins or enzymes is higher than on non-coding regions like ITS, but selection acts mostly at the amino acid level. When the nucleotide sequence is translated into the protein sequence, each amino acid is coded by three consecutive DNA nucleotides (a 'triplet'). However, the genetic code is degenerated, which means that the majority of amino acids can be translated from several triplet combinations, which mostly differ in the third position. The logical consequence is that at the DNA level, third codon positions often have the least selective constraint and change rather frequently, whereas second positions are most conserved.

Box 16.3 Transitions and transversions

DNA consists of four nucleotides, which fall into two structural classes, purines and pyrimidines. The observed frequency of substitutions within one structural class (called transitions; A would change to G, C to T, or vice versa) is higher than the frequency of substitutions between the two classes (called transversions; A would change to C or T, C to A or G and so on). If substitutions are randomly distributed, transversions will outnumber transitions by the ratio 2:1, but in most cases, the opposite is observed: transitions occur more frequently in evolution than transversions (Wakeley 1996). Therefore, since transversions are rarer than transitions, they can be considered to be more informative about evolutionary history.

Limnocythere inopinata

Although only a limited number of samples (4) from 2 populations could be sequenced in the frame of this survey (Tab. 16.1), the preliminary results give a first impression about the genetic variation in *L. inopinata*. Furthermore, these are the first DNA sequences, which have ever been published for the genus and the species. The number of differences between populations of *L. inopinata* is low, at least compared to *Eucypris virens*, but reaches a maximum of 1.4% for nuclear DNA in the ITS1 region, for which 429 bp could be sequenced. The differences within the mitochondrial genome are even smaller (0.6% in 655 bp). While ITS1 variation could only be observed between populations, some mitochondrial variation exists, both within and between the two populations compared.

Table 16.1 Samples analysed for *Limnocythere inopinata*.
The total number of individuals which have been sequenced for ITS1 and COI, is given.

Name of site	Country	Number of individuals for ITS/COI
Semerwater	Great Britain	1
Seewinkl	Austria	2/1
Monate	Italy	3/2

Eucypris virens

342 bp from the internal transcribed spacer region (ITS1) were sequenced for 38 specimens from 13 localities (Tab. 16.2). A considerable amount of genetic variation occurred within *E. virens*, with divergence among sequences varying from zero to 4.7% within localities and reaching a maximum of 10.3% for individuals from different geographical sites. Asexual populations turned out to be as variable as sexual ones. Geographically close populations were more similar genetically than distant ones (Mantel correlation: $Z = 0.56$, $p < 0.01$).

The analyses of mitochondrial DNA included a 423 bp long strand, which was sequenced for 18 individuals from 7 sites. Genetic divergence was also very high in this part of the genome, with a maximum of 20.9% for individuals from different areas. As for ITS, the amount of variation in the mitochondrial region did not depend on the reproductive mode.

Table 16.2 Samples analysed for *Eucypris virens*.
The total number of individuals which have been sequenced for ITS1 and COI, is given. Numbers in brackets represent the numbers of analysed males and females.
parth = parthenogenetic; sex = sexual.

Name of site	Abbreviation	Country	Number of individuals for ITS/CO I (males/females)	Reproductive mode
Extremadura	Extr	Spain	8/6 (3/3)	parth/sex
Pico	Pico	Spain	4/3	parth
Berzosa	Berz	Spain	3/0	parth
Morcuera	Morc	Spain	1/0	parth
Code d'Asino	Asin	Italy	1/0	parth
Biglinia	Bigli	Italy	1/0	parth
Sanguinaro	Sang	Italy	3/3	parth
Sicily	Sicili	Italy	7/1 (1/1)	sex
Colorno	Color	Italy	0/1	parth
Ventina	Vent	Italy	3/3	parth
Leeds	Leed	Great Britain	1/1	parth
Bramhope	Bram	Great Britain	3/0	parth
Ketton	Kett	Great Britain	1/0	parth
Haantjen	Haan	Belgium	2/0	parth

Darwinula stevensoni

A 320 bp region of the nuclear ITS1 region was sequenced for 56 individuals from 20 different sites (Tab. 16.3). No genetic differences could be detected at all, not even between samples from Africa and Europe. Within the mitochondrial region, some differentiation was observed, rising from 2.1% (average) between different European populations to 4.0% (maximum) for the comparison between Europe and South Africa.

In order to judge the magnitude of the substitution rate in *Darwinula stevensoni*, DNA from *Microdarwinula zimmeri* and *Penthesilenula brasiliensis* was also sequenced. By using fossil records and the observed genetic distances between these two species and *D. stevensoni*, we were able to estimate an evolutionary rate for mitochondrial DNA within the Darwinulidae of 0.19-0.32% per Myr (Schön *et al.* 1998). If we use this rate, the COI variability suggests that the split between African and European populations of *D. stevensoni* probably occurred at least 4.4 Myr ago (loc. cit.). This makes the complete lack of ITS1 variation among *D. stevensoni* samples all the more surprising.

Molecular phylogenies

Eucypris virens

Individuals from the two species *Prionocypris zenkeri* and *Prionocypris aragonica* were sequenced as outgroups to allow rooting of the *E. virens* phylogeny. A 'neighbour-joining tree' with its bootstrap-values is shown (Fig. 16.3); the length of the horizontal branches correlates with the genetic distance between the taxa and the bootstrap values (of 1000 replicates) indicate the robustness of the hypothesised relationships.

Table 16.3 Samples analysed for *Darwinula stevensoni*.
The total number of individuals which have been sequenced for ITS1 and COI, is given.

Name of site	Country	Number of individuals for ITS/CO I
Pääjärvi	Finland	3/0
Semerwater	Great Britain	2/1
Blikkreek	Belgium	2/1
Hollanders Gatkreek	Belgium	3/0
Wörthersee	Austria	1/1
Keutschachersee	Austria	2/2
Mondsee	Austria	2/0
Ossiachersee	Austria	2/1
Faaker See	Austria	2/1
Afritzer See	Austria	2/1
Lago di Endine	Italy	2/1
Lago di Monate	Italy	2/2
Lago di Garda	Italy	3/1
Comabbio	Italy	1/1
Montorfano	Italy	5/1
Mantova	Italy	4/0
Annone	Italy	2/1
Clou de la Fou	France	4/0
Lake Bled	Slovenia	7/0
Sibaya	South Africa	5/2

Box 16.4 Phylogenetic reconstructions

There are three major methods for reconstructing phylogenies from DNA or other data matrix, and all require the extensive use of specific computer packages such as PAUP (Swofford 1993), PHYLIP (Felsenstein 1993) or MEGA (Kumar *et al.* 1993).

1. Distance matrix methods, such as neighbour-joining or UGMA use the evolutionary distances (usually the number of nucleotide or amino acid substitutions between sequences), which are computed for all pairs of taxa. A phylogenetic tree is then constructed with an algorithm, on the basis of functional relationships among the distance values.
2. In parsimonious methods 'character states' are compared, which are usually the nucleotide or amino acid (or morphological character) at a certain site. The best tree chosen is the one which requires the smallest number of evolutionary changes to explain the differences among the studied taxa. Often, more than one maximal, parsimonious tree with the smallest number of changes can be found.
3. Maximum likelihood methods use 'character state configurations', which are patterns of nucleotide differences amongst the investigated sequences at a certain site, to form the maximum likelihood function. They then search for branch lengths that maximise the likelihood for a given tree.

For more details, see e.g. Li (1997).

The *Prionocypris* species are quite distant from *Eucypris virens*, as expected for another genus. The *E. virens* populations form two major branches, one including three Spanish asexual populations (Pico, Berzosa and Morcuera), and all other populations belonging to the second group. Thus, lineages are not separated according to their reproductive mode, and the two sexual populations (Sicily and Extremadura)

Eucypris virens

```
AAATCGGAAG GACCTCGGGG GTTCCAGAGA TGAGAGTGAG ACTTTGGTGA
....T..... .......... .......... ...A...... .........T
....T..... .......... .....G.... ...A...... .......C.T
....T..... .......... .....G.... ...A...... .......C.T
....T..... .......... .......... ...A...... .........T
....T...G. .......... .......... ...A...... .......C.T
....T..... .......... .......... ...A...... .......C.T
....T..... .......... .......... ...A...... .......C.T
....T...G. .......... .......... ...A...... .......C.T
....T..... .......... .......... ...A...... .........T
....T...G. .......... .......... ...A...... .........T
....T..... .......... ......A... ...A...... .......C.T
....T..... .......... ......A... ...A...... .......C.T
....T..... .......... .......... ...A...... .......C.T
....T..... .......... .......... ...A...... .......C.T
....T..... .......... .......... ...A...... .......C.T
....T..... .......... .......... ...A...... .........T
....T..... .......... .......... ...A...... .......C.T
```

Darwinula stevensoni

```
CACCCGCGCT CCTTGGGGCT TCTTCCCAGG GGAGAAGAGC CTTCGTCCCC
.......... .......... .......... .......... ..........
.......... .......... .......... .......... ..........
.......... .......... .......... .......... ..........
.......... .......... .......... .......... ..........
.......... .......... .......... .......... ..........
.......... .......... .......... .......... ..........
.......... .......... .......... .......... ..........
.......... .......... .......... .......... ..........
.......... .......... .......... .......... ..........
.......... .......... .......... .......... ..........
.......... .......... .......... .......... ..........
.......... .......... .......... .......... ..........
.......... .......... .......... .......... ..........
.......... .......... .......... .......... ..........
.......... .......... .......... .......... ..........
```

Fig. 16.3 Intraspecific varation in *E. virens* and *D. stevensoni*, shown for 50 bp of ITS1 and COI. Identical nucleotides are indicated by '.', otherwise, the different nucleotide is shown.

Eucypris virens

```
CTCATTGGTA ATGACCAAAT TTATAACACA ATTGTCACCG CACATGCTTT
.......... .C........ .......... .....A..T. .T........
.......... .......... .......... .....G.... ..........
..T..C..A. .......... .......... .....G..T. .T........
..T..C..A. .......... .......... .....G..T. .T........
..T..C..A. .......... .......... T....G..T. .T........
..T..C..A. ....T..... .......... .....A..T. .T........
..T..C..A. ....T..... .......... .....A..T. .T........
..T..C..A. ....T..... .......... .....A..T. .T........
..T..C..A. .C........ .......... .....A..T. .T........
..T.....G. .......... .......... .....T..T. .T........
..T.....G. .......... .......... .....T..T. .T........
..T.....G. .......... .......... .....T..T. .T........
..T.....G. .C..T..... ...C..T... .....G..T. .T.....C..
..T.....G. .C..T..... ...C..T... .....G..T. .T.....C..
..T.....G. .C..T..... ...C..T... .....G..T. .T.....C..
..T.....G. .C..T..... ...C..T... .....G..T. .T.....C..
..T.....G. .C..T..... ...C..T... .....G..T. .T.T.ATCC.
```

Darwinula stevensoni

```
TTGAGAGTTT TAATTCGATC CGAACTAGGA CTTCCAGGAA CCCTCATCGG
.......... .......... .......... .......... ..........
.......... .......... .......... .......... ....T.....
.......... .......... .......... .......... ....T.....
.......... .......... .......... .......... ....T.....
.......... .......... .......... .......... ....T.....
.......... .......... .......... .......... ....T.....
.......... .......... .......... .......... ....T.....
.......... .......... ....T..... .......... ....T.....
.......... .......... ....T..... .......... ....T.....
.......... .......... ....T..... .......... ....T.....
.......... .......... ....T..... .......... ....T.....
.......... .......... G...T..... .......... ....T.....
.......... .......... G...T..... .......... ....T.....
.......... .......... G...T..... .......... ....T.....
.......... .......... ....T..... .......... ....T.....
.......... .......... G...T..... .......... ....T.....
.......... .......... ....T..... .......... ....T.....
```

Fig. 16.3 (Cont.)

each form separate branches within their group. Samples from the same site usually cluster together, with one exception: two samples from Extremadura form a group with Italian asexual populations (Asino and Bigliana), and are thus separated from the other samples from Extremadura. These two individuals are probably asexual females which are known to coexist with sexual females at this site (chapter 15).

In fully apomictic lineages, nuclear and mitochondrial phylogenies should correspond perfectly. We therefore compared nuclear and mitochondrial divergence for those samples which were sequenced for both regions. Irrespective of the divergence at ITS1, the majority of the mitochondrial sequences differ from each other by 15-20% base substitutions. This indicates a saturation of the more rapidly evolving mitochondrial sequence, especially at third positions (see Box 16.2). Actually, the mtDNA data contain more phylogenetic information (more informative sites) than the ITS data.

However, in *E. virens*, there is one exception to the expectation of concruence: the phylogenetic position of the two putative asexual females from Extremadura. Although they cluster in one group with Italian asexuals in the ITS-tree, their mitochondrial sequences are more similar to those from the asexual, Spanish populations such as Pico (Fig. 16.5). We suggest that these individuals represent a parthenogenetic lineage, living sympatrically with the Extremadura sexual population, and being influenced by occasional hybridisation with sexual males. Hybridisation between a parthenogenetic female and a sexual male would allow transfer of nuclear alleles without influencing the mitochondrial lineage. Thus, the parthenogenetic ancestor of the present lineage was most likely a clone related to the other Spanish clones at Pico, Berzosa and Morcuera, as indicated by the mitochondrial DNA, while the sexual parent was from Extremadura. The substantial divergence in ITS1 between the 'atypical' Extremadura individuals and the Extremadura males implies either that this hybridisation occurred a long time ago, or that the rate of evolution of ITS1 has increased since the transfer. The latter explanation might be favoured given the unequal branch lengths in this part of the tree.

Darwinula stevensoni

No tree was constructed with the ITS data, as all specimens of *D. stevensoni* had identical sequences. The following description only reflects the topology of the tree constructed with combined mitochondrial COI and nuclear ITS sequence data (Fig. 16.6).

The two outgroups, *Penthesilenula brasiliensis* and *Microdarwinula zimmeri*, have a position far from the *D. stevensoni* branches, as was expected. African *Darwinula stevensoni* form a distinct cluster outside all European populations. Within Europe, two groups can be distinguished, which do not follow any obvious biogeographical pattern: some Italian samples cluster together with the Belgian and British populations, whereas Garda, Montorfano and Annone (also from Italy) form a second group with samples from Austria and France.

Conclusions

The genetic variability of the three ostracod species investigated is strikingly different (Tab. 16.4). *Eucypris virens* showed the highest variation in both nuclear and mitochondrial DNA. A similar level of variability in ITS1 has only been reported for one beetle species (Vogler & De Salle 1994), but in that case, it could be explained by divergent ITS1 copies within single individuals. In all other studies, any variability, comparable to the one in *E. virens*, has only been reported at higher taxonomic levels, e.g. between species (McLain *et al.* 1995, Miller *et al.* 1996). Literature data from COI reveal a comparable result (Satta & Takahata 1990,

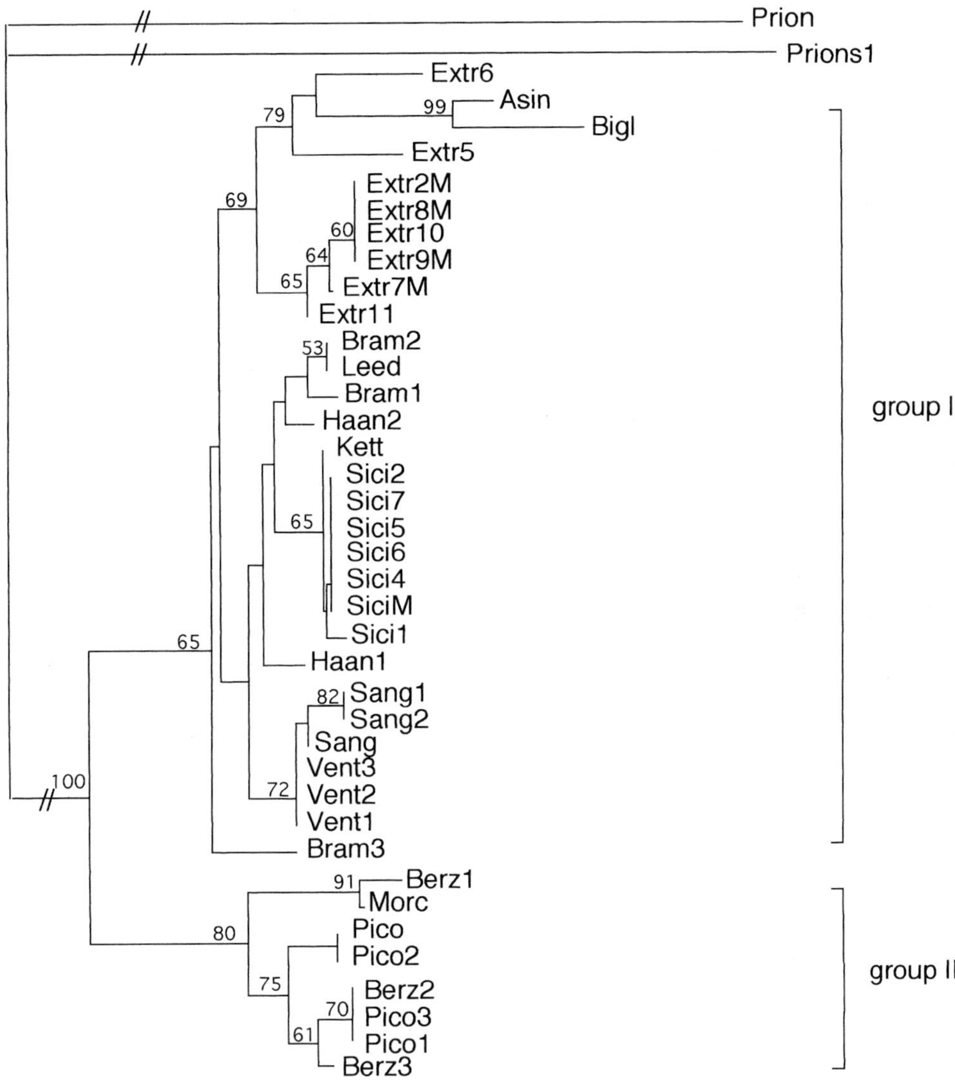

Fig. 16.4 Neighbour-joining tree, constructed from ITS1 data for *E. virens,* using Kimura two-parameter distances and including all samples. Branch lengths are proportional to genetic distances, except for the outgroup branches, which were shortened manually after tree construction. Bootstrap percentages, based on 1000 replicates, are given above branches for values > 50%. Prion = *Prionocypris aragonica*, Prions1 = *Prionocypris zenkeri*, M = male.

Okamoto *et al.* 1995, Howland & Hewitt 1995) with two exceptions. In the tiger copepod, divergence reached 18% between different populations (Burton & Lee 1994), and Perez *et al.* (1994) detected a high variability of COI among different lineages of the brine shrimp *Artemia parthenogenetica,* on which basis they suggested to regard the lineages as separate species. Thus, *E. virens* shows a remarkable variability in both the nuclear and the mitochondrial genome.

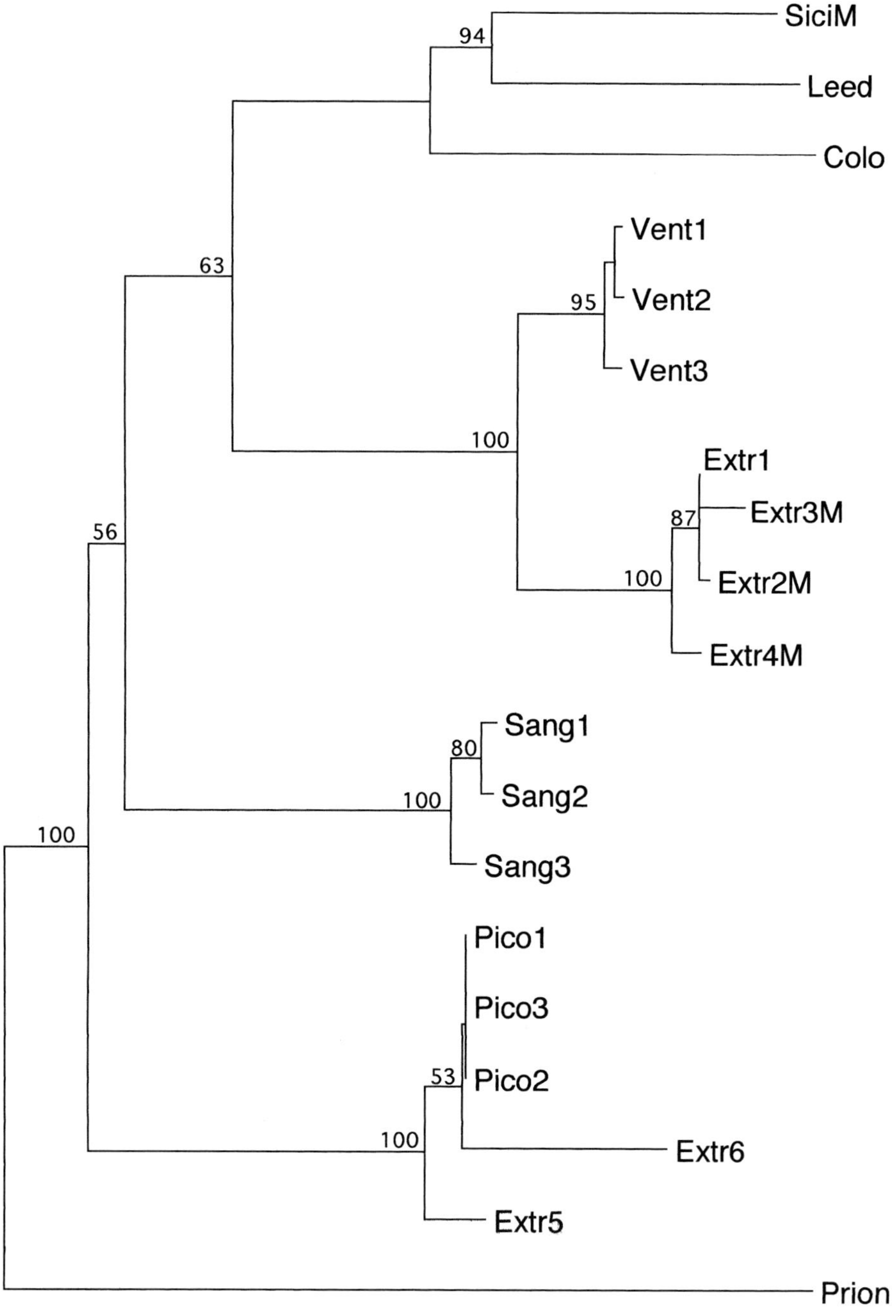

Fig. 16.5 Neighbour-joining tree, constructed from COI data for *E. virens*, using Kimura 2-parameter distances and including all samples. Branch lengths are proportional to genetic distances, including the outgroup branch. Bootstrap precentages, based on 1000 replicates, are given above branches for values > 50%.
Prion = *Prionocypris aragonica*, M = male.

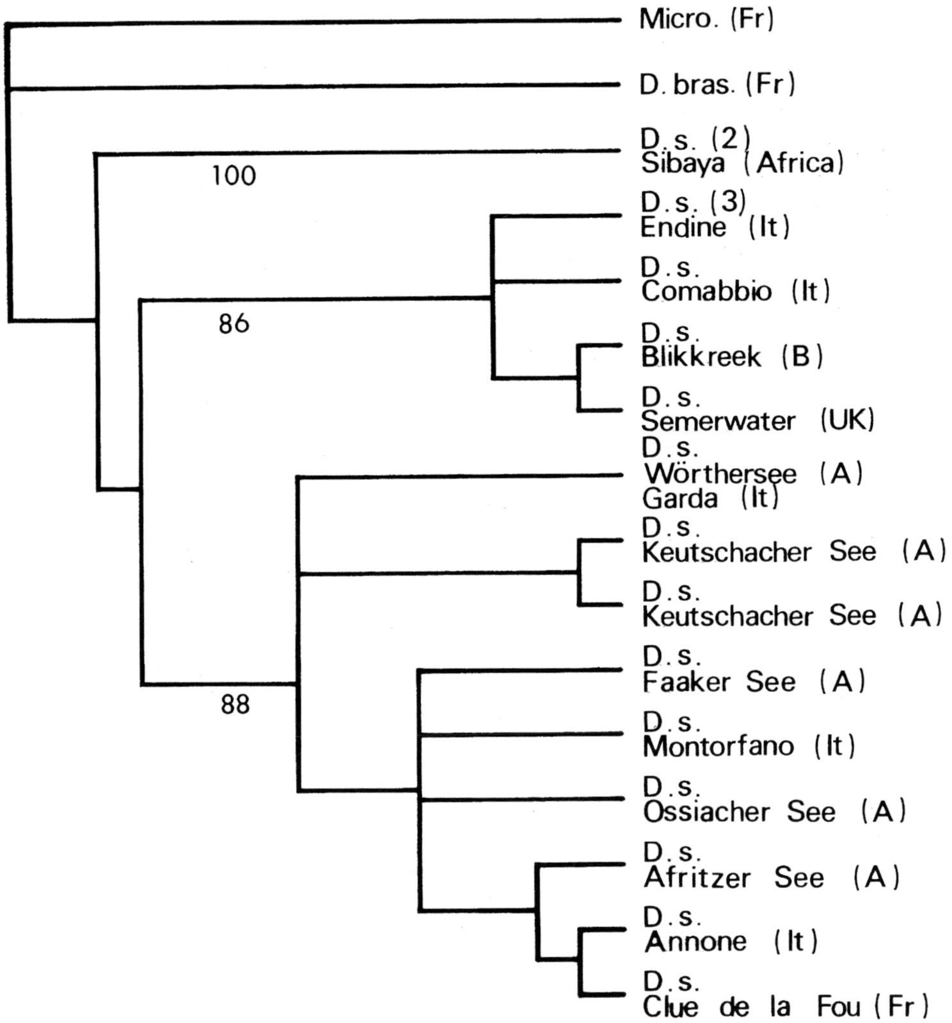

Fig. 16.6 Cladogram of *D. stevensoni*, based on 447 bp of the COI gene and 320bp of the ITS1 region for a set of congruent samples. Bootstrap values in percentage are shown in parentheses for major branches. *Micro. = Microdarwinula zimmeri*, *D. bras. = Penthesilenula brasiliensis*, *D. s. = Darwinula stevensoni*. Fr = France, It = Italy, B = Belgium, UK = United Kingdom, A = Austria. Redrawn from Schön *et al.* 1998.

We can use this variation for estimating the time-scale of divergence among the different lineages of *E. virens*. It is difficult to use the ITS1 data for a molecular clock, however, as no good calibration exists. So far, comparable studies have only been conducted on *Drosophila* (Schlötterer *et al.* 1994), where no variation within species could be detected at all and a divergence of 1.2% was estimated to be equivalent to 1 My. If we apply this estimation to our data on *E. virens*, either the asexual clones must be extremely old (about 3.6 to a maximum of 8.6 Myr within Europe)

Table 16.4 Maximal genetic divergence
For each species, the length of the sequenced fragment in basepairs (bp), the total number of changes and the maximal genetic divergence in percent is given for different comparisons.

species	nuclear ITS	mitochondrial CO
Limnocythere inopinata	429 bp	655 bp
within populations	0	4 (0.61%)
within species	6 (1.39%)	4 (0.61%)
Eucypris virens	298 bp	423 bp
within populations	14 (4.7%)	2 (0.47%)
within Europe	42 (14.09%)	97 (18.59%)
Darwinula stevensoni	320 bp	451 bp
within Europe	0	12 (3.2%)
within species	0	19 (4.0%)

or evolution is faster in this group. However, a comparable age is estimated with the mitochondrial data: for crustaceans, COI is supposed to evolve with a rate of 2.2 to 2.6% per My (Knowlton *et al.* 1993). This rate would lead to a maximum age of 5.8 My for the divergence of the different European populations, which is in line with the ITS1 based estimate. If this estimate is reliable, the divergence of *E. virens* lineages began before the ice ages, in the Pliocene! A faster evolutionary rate in *E. virens* may be a more likely explanation, especially if we consider the observed saturation in COI, but this would have to influence both nuclear and mitochondrial loci.

It could be possible that the substitution rate of ITS in *E. virens* is increased by the effect of Müller's ratchet (see chapter 1). If the separation between parthenogenetic and sexual lineages occurred a long time ago, one would expect that only the asexual lineages would have suffered from the ratchet. However, a difference in the variation between lineages with different reproductive modes did not occur. If we assume on the other hand that parthenogenetic lineages originated recently from sexual populations, no varying effects of the ratchet would be detectable between populations with different reproductive modes. At this point, we approach a major problem, when using DNA data to estimate the age of asexual clones. We cannot distinguish between two alternatives: a high, genetic divergence in asexual lineages could either be the result of a recent spin-off of asexuals from sexual lineages, in which case the variability of sexual populations would still be preserved. Alternatively, asexuals could have separated from sexuals a long time ago, and the observed genetic variability would then be the result of accumulating mutations since this event has happened. For *Eucypris virens*, both alternatives most likely are valid: some parthenogenetic clones might have been able to survive for a long time, like the three Spanish parthenogenetic clones that form a separate group, although we cannot tell at what point in their history they abandoned sexual reproduction. Our data imply, further, that parthenogenesis within *Eucypris virens* originated several times and the spin-off of new asexual clones from sexual populations probably still continues. The different branches of the tree (Fig. 16.4) suggest that this happened at least seven times within Europe. Griffiths & Butlin (1995) also suggested that the high proportion of parthenogens in fossil assemblages might be due to high origination rates; and a multiple origin of asexual clones has been described for other ostracod species, too (Chaplin *et al.* 1994, Chaplin & Hebert 1997).

The two samples from Extremadura, which cluster outside the rest of this population that includes both males and females, most likely represent parthenogenetic clones coexisting with the sexual population. Allozyme data (see chapter 15) have also shown that some females from the Extremadura site are genetically closer to parthenogenetic females from other sites than to males from the same site. This finding supports the hypothesis that the different reproductive modes in *E. virens* are genetically controlled and that lineages with different modes can co-exist, even in the same pond.

The different levels of divergence of these individuals for the nuclear and mitochondrial markers suggest introgression of nuclear genes into a parthenogenetic lineage in the past through hybridisation. Males are known to court and attempt to copulate with parthenogenetic females (chapter 10). Hybridisation would produce triploid offspring (as have been observed in some localities, chapter 15) unless meiosis still occurs occasionally in parthenogenetic females. The initial hybridisation would clearly produce offspring heterozygous for alleles derived from the sexual and parthenogenetic parent. A complete replacement of the ITS1, as is apparently observed, would require either repeated backcrossing or replacement of the parthenogenetically derived allele by homogenising evolution, which has been reported from asexual *Daphnia* clones (Crease & Lynch 1991). Hybridisation between parthenogenetic and sexual lineages may be an important mechanism aiding the persistence of parthenogenetic lineages by providing them with a genome that has not suffered from mutation accumulation. It has been suggested to occur in other ostracod taxa (Chaplin & Hebert 1997).

A most surprising result is the total identity in the ITS1 sequence of all *Darwinula stevensoni* populations, even those from Europe and Africa. The fact that at least some variation occurs in the mitochondrial genome of *Darwinula* weakens the explanation that the Recent species *Darwinula stevensoni* consists of only one genetically identical clone. This could have happened, if one extremely successful clone outcompeted all other clones and colonised the whole of Europe and Africa, a scenario called a 'selective sweep'. Such an event would have to have taken place recently, as otherwise mutations would be expected to have accumulated in ITS1, too. But if we use fossil data and the observed substitution rates between the outgroups to calculate the time of the split between the African and European population, this has not occurred recently, but 4 to 5 Myr ago (Schön *et al.* 1998).

Meselson's test for ancient asexuals did not work in *Darwinula stevensoni*. Cloning experiments, as well as the unambiguous nature of our sequencing results, indicate that the multiple copies of ITS1 within one individual are identical and have not diverged (Schön *et al.* 1998). This result is not necessarily an argument against *D. stevensoni* being an ancient asexual. ITS1 is not only an unusually organised sequence, being part of a tandemly repeated complex, but has also a special mode of evolution that homogenises the copies within one lineage (Birky 1996, Hillis & Dixon 1991). The 'concerted evolution' of ITS1 can explain the absence of the Meselson effect but it cannot explain why ITS copies are identical between different individuals from widely spaced localities. Predictions for an ancient asexual would be just the opposite, as one would expect effects of Müller's ratchet to lead to an accumulation of substitutions. Explanations for this apparently zero substitution rate in ITS1 could be a generally reduced mutation rate, some sort of 'molecular drive' like gene conversion or unequal crossing over, or rare automixis. However, further discussion of these hypotheses, especially that *Darwinula* has reduced its mutation rate by efficient DNA repair (Schön & Martens 1998), will be reserved for chapter 17.

References

Avise, J.C. 1994. *Molecular markers, natural history and evolution*, Chapman & Hall.

Birky, C.W. 1996. Heterozygosity, heteromorphy, and phylogenetic trees in asexual eukaryotes. *Genetics*, **144:** 427-437.

Burton, R.S. & Lee, B.-N. 1994. Nuclear and mitochondrial gene genealogies and allozyme polymorphism across a major phylogeographic break in the copepod *Tigriopus californicus*. *Proc. Natl. Acad. Sci. USA*, **91:** 5197-5201.

Chaplin, J.A., Havel, H.E. & Hebert, P.D.N. 1994. Sex and ostracods. *Trends Ecol. Evol.*, **9:** 435-439.

Chaplin, J.A. & Hebert, P.D.N. 1997. *Cyprinotus incongruens* (Ostracoda): an ancient asexual? *Mol. Ecol.*, **6:** 155-168.

Crease, T.J., Stanton, D.J. & Hebert, P.D.N. 1989. Polyphyletic origins of asexuality in *Daphnia pulex*. II. Mitochondrial DNA variation. *Evolution*, **43:** 1016-1026.

Crease, T.J. & Lynch, M. 1991. Ribosomal DNA variation in *Daphnia pulex*. *Mol. Biol. Evol.*, **8:** 620-640.

Crow, J.F. 1994. Advantages of sexual reproduction. *Dev. Genet.*, **15:** 205-213.

Felsenstein, J. 1993. *PHYLIP* – phylogeny inference package, version 3.5. *Dept. Genetics*, Univ. Washington.

Foighil, D.O. & Smith, M.J. 1996. Phylogeography of an asexual marine clam complex, *Lasea*, in the Northeastern Pacific based on Cytochrome Oxidase III sequence variation. *Mol. Phylogen. Evol.*, **6:** 134-142.

Griffiths, H.I. & Butlin, R.K. 1995. A timescale for sex vs parthenogenesis: Evidence from Holocene ostracods. *Proc. R. Soc. Lond. B.*, **260:** 65-71.

Hedges, S.B., Bezy, R.L. & Maxson, L.R. 1992. Ancestry of unisexual salamanders. *Nature*, **356:** 708-710.

Hillis, D.M. & Dixon, M.T. 1991. Ribosomal DNA: molecular evolution and phylogenetic inference. *Quart. Rev. Biol.*, **66:** 411-453.

Howland, D.E. & Hewitt, G.M. 1995. Phylogeny of the Coleoptera based on mitochondrial cytochrome oxidase I sequence data. *Insect Mol. Biol.*, **4:** 203-215.

Judson, P.O. & Normark, B.B. 1996. Ancient asexual scandals. *Trends Ecol. Evol.*, **11:** 1-46.

Kimura, M. 1980. A simple method for estimating evolutionary rates of base substitutions through comparative studies of nucleotide sequences. *J. Mol. Evol.*, **16:** 111-120.

Kondrashov, A.S. 1993. Classification of hypotheses on the advantage of amphimixis. *J. Hered.*, **84:** 372-387.

Knowlton, N., Weight, L.A., Solorzano, L.A., Mills, D.E.K. & Bermingham, E. 1993. Divergence of proteins, mitochondrial DNA, and reproductive compatibility across the Isthmus of Panama. *Science*, **260:** 1629-1632.

Kumar, S., Tamura, K. & Nei, M. 1993. Molecular evolutionary genetics analysis (*MEGA*, version 1.01). *Pennsylvania State Univ.*

Li, W.-H., 1997. *Molecular evolution*, Sinauer: 487 pp.

Maynard Smith, J. 1989. *Evolutionary genetics*, Oxford Univ. Press.

Maynard Smith, J. 1992. Age and the unisexual lineage. *Nature*, **356:** 661-662.

McLain, D., Wessen, D.M., Collins, F.H. & Oliver, J.H. JR 1995. Evolution of the rDNA spacer ITS2, in the ticks *Ixodes scapularis* and *I. pacificus* (Acari: Ixididae). *Heredity*, **75:** 303-319.

Miller, B.R., Crabtree, M.B. & Savage, H.M. 1996. Phylogeny of fourteen *Culex* mosquito species, including the *Culex pipiens* complex, inferred from the internal transcribed spacers of ribosomal DNA. *Insect Mol. Biol.*, **5:** 93-107.

Müller, H.J. 1964. The relation of recombination to mutational advance. *Mut. Res.*, **1:** 2-9.

Okamoto, M., Bessho, Y., Kamiya, M., Kurosawa, T. & Hori, T. 1995. Phylogenetic relationships within *Taenia taeniaformis* variants and other taenied cestodes inferred from the nucleotide sequence of the cytochrome c oxidase subunit I gene. *Parasitol. Res.*, **81:** 451-458.

Palumbi, S.R. 1996. Nucleic Acids II: The Polymerase Chain Reaction. *in*: Hillis, D.M., Moritz, C. & Mable, B.K. (eds), *Molecular systematics*, Sinauer Ass.: 205-248.

Perez, M. L., Valverde, J. R., Batuecas, B., Amat, F., Marco, R. & Garesse, R. 1994. Speciation in the *Artemia* genus: mitochondrial DNA analysis of bisexual and parthenogenetic brine shrimps. *J. Mol. Evol.,* **38:** 156-168.

Quattro, J.M., Avise, J.C. & Vrijenhoek, R.C. 1992. An ancient clonal lineage in the fish genus *Poeciliopsis* (Atheriniformes: Poeciliidae). *Proc. Natl. Acad. Sci. USA*, **89:** 348-352.

Satta, Y. & Takahata, N. 1990. Evolution of *Drosophila* mitochondrial DNA and the history of the *melanogaster* group. *Proc. Natl. Acad. Sci., USA*, **87:** 9558-9562.

Schlötterer, C., Hauser, M.-T., Van Häseler, A. & Tautz, D. 1994. Comparative evolutionary analysis of rDNA ITS regions in *Drosophila. Mol. Biol. Evol.,* **11:** 513-522.

Schön, I.,Butlin, R.K., Griffiths, H.I. & Martens, K. 1998. Slow molecular evolution in an ancient asexual ostracod. *Proc. R. Soc. Lond. B*, **265:** 235-242.

Schön, I. & Martens, K. 1998. Opinion: DNA repair in ancient asexuals – a new solution for an old problem? *J. nat. Hist.*, **32:** 943-948.

Swofford, D.L. 1993. *PAUP*. Phylogenetic analysis using parsimony, version 3.1.1. *Illinois Nat. Hist. Survey*.

Turgeon, J. & Hebert, P.D.N. 1994. Evolutionary interactions between sexual and all-female taxa of *Cyprinotus* (Ostracoda, Cyprididae). *Evolution*, **48:** 1855-1865.

Vogler, A.P. & De Salle, R. 1994. Evolution and phylogenetic information content of the ITS1-region in the tiger beetle *Cicindela dorsalis. Mol. Biol. Evol.*, **11:** 393-405.

Wakeley, J. 1996. The excess of transitions among nucleotide substitutions: new methods of estimating transition bias underscore its significance. *Trends Ecol. Evol.*, **11:** 158-163.

CHAPTER 17

SEX AND OSTRACODS: A NEW SYNTHESIS

K. Martens

Abstract

The present chapter integrates taxonomic, genetic, ecological and palaeontological information on non-marine ostracod lineages, in order to assess the impact of the reproductive mode on evolutionary ecology. At least three fundamentally different strategies are distinguished within non-marine ostracod groups: fully sexual (in some Candonidae and Cytheroidea), mixed reproduction (Cyprididae, Candonidae and Limnocytheridae) and ancient asexuals (Darwinulidae). The reproductive modes have a large impact on genetic, morphological and ecological diversities (extant diversities, as well as the dynamic aspects) and as a result, species with different reproductive modes have different competitive abilities with regard to colonisation of habitats, habitat predictability, habitat stability and ecological specialisation. This improved understanding of reproductive strategies is applied to geographical parthenogenesis, clonal taxonomy, palaeoecological and –climatic interpretations and conservation biology.

Introduction

The present chapter integrates the data and reviews provided in the preceding chapters and attempts to present a synthesis of the various genetic, phenotypic and ecological strategies associated with the different reproductive modes in non-marine ostracods. This is a personal interpretation of the different patterns and processes, to which other authors of the present volume might not necessarily subscribe. For example, I distinguish and characterise different evolutionary reproductive strategies in non-marine ostracods; a view which is at least partly inspired by my own background in systematics, as the ecologists in our team tend to see these different strategies more as a continuum between fully sexual and fully asexual reproduction. Secondly, many people presently consider *group selection* as an amply refuted concept. However, I use the term 'strategy' with clear reference to this concept, as I feel both sexual and mixed reproduction offer arguments to at least reconsider this idea.

As the canvas of this review is very wide, I have attempted to make its structure as transparent as possible. After a brief overview of sex determination systems in ostracods, some important concepts are (re-) defined and the three reproductive strategies described. A brief discussion on rare male occurrences next assesses their functionality. The history of reproductive modes, assessed by both the fossil record and age estimates derived from molecular divergences in the three non-marine lineages is subsequently discussed. Genetic, morphological and ecological diversities

Sex and parthenogenesis: evolutionary ecology of reproductive modes in non-marine ostracods, edited by K. Martens, pp. 295–321

(and their interactions) are described in the next three sections on sexuality, mixed reproduction and ancient asexuals. The fitness of the three reproductive modes in view of four ecological capacities is next summarised; these new concepts are then applied to four different spin-offs: geographical parthenogenesis, clonal taxonomy, palaeoecology and conservation biology.

In the introduction to this book, we claimed that non-marine ostracods constitute an excellent model-group for the study of asexual reproduction and its effect on evolution and persistence of such lineages, because information from several disciplines is available. The present chapter aims to validate this claim.

Sex determination and parthenogenesis in non-marine ostracods

The ability to reproduce asexually via unfertilised eggs in ostracods requires several parallel mechanisms, amongst which are: (1) the suppression of meiosis; (2) the suppression of the production of males and (3) the ability to auto-initiate cleavage of the diploid (unfertilised) eggs. All of these separate processes have to act together in order to allow clonal reproduction. It is presently still uncertain why parthenogenesis is so unusually common in (non-marine) ostracods, but it seems likely that this group either has the above processes better developed or more tightly integrated.

Sex determination in ostracods in general, and in the non-marine groups in particular, is assumed to be genetic (chapter 2), and this already considerably facilitates the suppression of the production of males (process 3) as compared to taxa where the presence of phenotypical males (as in Decapoda) or cyclic sexuality (as in daphnids) can be environmentally cued. A possible exception, discussed in chapter 7, is the suppression of the male phenotype by cytoplasmic, symbiont bacteria. The presence of such bacteria in *Heterocypris incongruens* has been established (chapter 7), but the extent to which these bacteria indeed suppress males in non-marine ostracods in general (and in *H. incongruens* in particular) remains yet to be determined.

Males in ostracods are the heterogametic sex and both XO and XY systems have been found. These two systems are not taxonomically segregated, and two species in the same genus (e.g. *Cyclocypris*) can have different systems. Rather peculiar in non-marine ostracods is the occurrence of multiple sex-chromosomes, as several X and Y chromosomes can occur. Chaplin *et al.* (1994) cite the case of *Heterocypris incongruens*, where females have 6 pairs of X chromosomes and males have one set of 6 X and 1 Y chromosome. With four pairs of autosomes, males produce two types of sperm: one with five chromosomes (giving rise to males) and the other with ten chromosomes (for female offspring). In general, chromosomal patterns are very variable amongst and especially within species (chapter 2).

Females either reproduce sexually, in which case haploid eggs are produced through meiosis, or parthenogenetically, in which case there is no meiosis and eggs are at least diploid. In all parthenogenetic populations of non-marine ostracod species thus far investigated, parthenogenesis appears to be apomictic, although doubts seem justified at least for the darwinulids. Individual females are not known to switch between either mode (although both types can occur within one species and even population), neither cyclically nor erratically, although intraspecific hybridisation between parthenogenetic females and sympatric males has been reported for a number of species; it is believed that the resulting offspring consist of parthenogenetic polyploid females with mixed genotype (Dowling & Secor 1997).

Following Butlin *et al.* (in press), a PARTHENOGENETIC (asexual) species can be defined as: *a set of all-female lineages that form a morphological and genetic cluster that is not known to include any sexually reproducing lineages*. Species with MIXED REPRODUCTION form *a genetical and morphological cluster of clonal and sexual lineages*. FULLY SEXUAL SPECIES *are defined by the biological species concepts, they only comprise sexual populations.*

Note that a species can for a long time be considered fully asexual, until a sexual population is discovered, or *vice versa* (see below). It is also important to make a clear distinction between parthenogenetic populations and species. All-female populations are parthenogenetic populations, populations with both males and females are commonly called bisexual populations. However, the sexual populations of Cyprididae thus far investigated mostly turned out to consist of males, bisexual and parthenogenetic females. Such populations are therefore mixtures of parthenogenetic and sexual populations. Also, a population can generally not be considered sexual, or even mixed, simply because of the occurrence of a single male.

Reproductive strategies in non-marine ostracods

On the basis of the evidence presented in the previous chapters, I distinguish three different reproductive strategies within the non-marine ostracod lineages.

(1) Fully sexual reproduction
A number of lineages appear unable to reproduce parthenogenetically, this in spite of the fact that their species occupy wide geographical and ecological ranges. Good examples of this are species in the genus *Cyclocypris*. Of course, these might represent the opposite situation of the so-called fully asexual species in Cyprididae (which may in fact be mixed, but await discovery of populations with sexual reproduction), but as at this stage no all-female populations of these species are known, we must accept that these species either have lost, or never had, the ability to reproduce asexually. Other potential candidates of fully sexual taxa are the Cytherideinae from Lakes Tanganyika (*Cyprideis s.l.*) and Baikal (*Cytherissa*), although extra-lacustrine relatives of these species-flocks are known from all-female populations (Martens 1994), and groundwater candonids (Danielopol 1978). An intriguing example from the former habitats is that of *Cytherissa lacustris,* which occurs in parthenogenetic populations in European and North American lakes and lives in waters associated with Lake Baikal in sexual populations. Recent molecular screening indicates that the European populations are recent spin-offs of the Baikalian flock (c 55 species), in which case the ability to reproduce by parthenogenesis was either not lost from the genome of the otherwise exclusively sexually reproducing Baikalian endemics (Schön *et al.* 1998b) or (re-) developing the ability requires a small genetic change only (switching on silent genes, for example).

The evolutionary strategy of these sexuals is congruent with that amply described for other fully sexual taxa and not much attention will be devoted to this strategy here.

(2) Mixed reproduction
Most (all?) Cyprididae and significant numbers of species of candonid and cytheroid non-marine groups have mixed reproduction. Species with mixed reproduction occur in both all-female populations and in populations with sexual and parthenogenetic females and males. Populations with sexual reproduction are indeed generally of mixed reproduction in these groups. A number of species in the Cyprididae are deemed fully asexual, as males have never been found. Future research will show whether they are indeed fully asexual and if so, if such lineages are evolutionary short-lived as theory predicts, or if they behave like persistent asexual spin-offs of clusters with mixed reproduction. Most research on reproductive strategies in non-marine ostracods has thus far focused on the Cyprididae, for the obvious reason that the majority of non-marine ostracod species (up to 73% of all species in Africa) belong to this family.

Species in Candonidae and Limnocytheridae are either fully sexual or occur in both parthenogenetic and bisexual (mixed?) populations. For reasons more fully outlined below, their reproductive strategy is suspected to be different from that of the Cyprididae. For example, what little information exists, indicates that Limnocytheridae do not have hybridisation (as there is no polyploidy), neither of these groups has geographical parthenogenesis, while both Candonidae and Limnocytheridae seem better preadapted for radiations in long-lived habitats such as ancient lakes and groundwater.

(3) Ancient asexuals
The Darwinulidae, the only family of Darwinulocopina to have survived to date, all reproduce exclusively asexually and are believed to have done so for at least 100 Myr, maybe longer (Butlin & Griffiths 1993). Because they are one of the few examples that seem to defy the many theoretical considerations predicting early extinction of fully asexual lineages (chapter 1), they were called *ancient asexual scandals* by Judson & Normark (1996). The only other ancient asexual group is the bdelloid rotifers (Poinar & Ricci 1992), as the status of the putative third group cited in Judson & Normark (*loc.cit.*), *Artemia parthenogenetica,* has been doubted by Butlin *et al.* (in press).

The distribution of these different reproductive modes is not entirely congruent with the phylogeny of non-marine ostracods (Table 3.1), although the Darwinulidae are most likely the only ancient asexuals and fully sexual species are far more common in Cytheroidea and Candonidae than in Cyprididae. As will be outlined below, however, almost no information is available on reproductive systems in Cytheroidea and Candonidae, and this makes comparisons with the better-known reproductive strategy in Cyprididae (mixed, with apomictic parthenogenesis) difficult.

Rare males

Sex ratios in extant non-marine ostracod populations can show significant female excess, on average a 2:1 female bias (Chaplin *et al.* 1994). Several possible explanations have been suggested for this: sympatric occurrence of sexual and parthenogenetic females, segregation distorters on multiple sex-chromosomes, life cycle incongruence of the genders, metapopulations structure and patchiness of ostracod populations (reviewed in chapter 2) and/or partial infection of a population by cytoplasmic symbionts (chapter 7).

The occurrence of single (rare) males, e. g. one male for 10,000 females (chapter 14) can be seen as an extreme case of skewed sex ratios. Possible underlying genetic causes are discussed in chapter 2, but I am more concerned here with the functionality of such rare males. Rare males are either atavistic, non-functional individuals (triggered through the activation of otherwise 'silent genes' – Fryer in press), possibly reflecting very ancient affinities, or they can be at least partly functional and a response to environmental factors, either abiotic (e.g. alkalinity – Löffler 1990, chapter 6) or biotic (crowding effect – chapter 14). In both cases, these males can result directly from diploid eggs, so that meiosis is not necessary at this stage (chapter 2). A third possible explanation holds that even in (seemingly) all-female populations of species with mixed reproduction, there are still some sexual females and males present, albeit at very low ratios. Such males are fully functional, and can fertilise both sexual females (giving rise to bisexual offspring) or produce intraspecific hybrids (mostly polyploid) with parthenogenetic females. Finally, rare males can also be individuals that have escaped the male-suppressing cytoplasmic symbiont bacteria. Some environmental factors may influence this and the effect of, for example, alkalinity (see above) on the activity and even survival of such cytoplasmic bacteria is certainly worth investigating.

I argue that single males have different origins and functions in different lineages. In *Limnocythere inopinata,* for example, such rare male occurrences are not uncommon. They have been found in several Recent and fossil populations and have even been produced in clonal laboratory cultures (see review in chapter 14). The males from these lab-culture experiments must have been produced by an otherwise parthenogenetic female, so that the 'mixed population' hypothesis can be excluded, at least in this case.

Video-recordings of the behaviour of such males amongst thousands of females shows that the male certainly attempts to mate (see chapter 10), but no successful copulation has been observed with certainty. This, together with the absence of polyploidy in *L. inopinata* (chapter 15), which would be the expected result of an intraspecific hybridisation between a male and a parthenogenetic female, suggests that single males in this lineage are non-functional, atavistic occurrences.

Single males in cypridinid lineages with mixed reproduction have been reported numerous times (e.g. in *Heterocypris*), and in such taxa, high incidences of polyploidy have been reported (up to 100% – chapter 15). However, it is not certain whether such polyploid strains result from the activities of rare males, or are exclusively the result of (repeated) intraspecific hybridisations in populations with mixed reproduction. In short, even in groups with mixed reproduction, where genuine sexual reproduction is known at least in certain populations, the functionality of single males has thus far not been demonstrated.

The case of the single male of *Darwinula stevensoni* reported by Brady & Robertson (1870) provoked Little & Hebert (1996) into arguing that the status of darwinulids as ancient asexuals is untenable. Schoen *et al.* (1996), Rossetti & Martens (1996) and Schön *et al.* (1998a) (see also chapter 3) offered several strong arguments to show that the single male in that unique case certainly was a non-functional, atavistic specimen: the very low incidence in general (one male amongst hundreds of thousands of investigated individuals), no known sexuality in the entire (super-)family, chromosomes which have seemingly lost all potential to experience meiosis, a high proportion of homozygosity and absence of polyploidy in all screened specimens (chapters 15, 16), etc.

In conclusion, rare males are most likely not functional, even in groups with mixed reproduction (although they are relatively frequently found there) and certainly not in ancient asexual darwinulids. In spite of the fact that the mystery of their occurrence thus remains largely unsolved, it seems logical to consider populations with single males as parthenogenetic populations.

History of reproductive modes: age of lineages

Methodological difficulties with the fossil record

The history of reproductive modes in non-marine ostracods (based on fossil evidence) has extensively been described in chapters 3 and 6. Several general observations are here summarised, as they contribute to our understanding of the age and origin of the different lineages and their reproductive strategies.

The methodological difficulties of recognising sex in the fossil record must be cited again. Although ostracods indeed offer unique opportunities to compare high extant global diversities with an excellent long-term fossil record, the problems related to the interpretation of the latter must not be underestimated. Only taxa with sexual dimorphism in valves can be recognised as sexual. Bisexual species without this valve dimorphism may thus not be considered as such. Valve dimorphism in a fossil assemblage can be sexual, but also eco-morphological, caused by two different species or even by differences between adults and last-instar juveniles. These possible confusions can lead to erroneous interpretations. As was clearly outlined in chapter 6, the problems are limited for lineages where comparisons with extant relatives are possible, but for completely extinct groups (*Carbonita, Cypridea*) such interpretations are difficult.

Ancient asexual darwinulids: 100 or 200 Myr without sex ?

The Darwinuloidea originated about 360 Myr ago in the Palaeozoic (Lower Carboniferous) (Sohn 1987, 1988). When the early darwinulids invaded non-marine habitats, they almost certainly had an (at least partly) sexual reproduction, certainly during the Permian, when they were furthermore at the height of their specific and supra-specific diversity. However, already during the Permian some all-female populations of darwinulids have been reported (Abushik 1990), so that asexuality in darwinulids might have coincided with their invasion of non-marine habitats. After the Permian-Triassic mass extinction, the diversity of the group declined slowly but drastically, until at the beginning of the Jurassic (c 200 Myr) only a few dozen species in one genus (*Darwinula s.l.*) were left (chapter 3). The last reported darwinulid assemblages with several putative males date from the Lower Cretaceous (c 100 Myr – Zheng 1995), but these records can be refuted on the basis of the above arguments; most likely the so-called sexual dimorphism actually refers to the sympatric occurrence of two different species. After the Lower Cretaceous, no populations with sexual reproduction have been found. It seems thus firmly established that the darwinulids are indeed ancient asexuals, having abandoned sex for at least 100 Myr. If the putative Cretaceous males can be refuted, then asexuality has been the sole reproductive mode in this group for 200 Myr.

The age of darwinulid species is more difficult to ascertain as conspecificity between Recent and fossil populations is often difficult: morphological differences between darwinulid species are small and features are not visible on badly preserved specimens. Nevertheless, both *Microdarwinula zimmeri* and *Darwinula stevensoni* are at least 20 Myr old (Straub 1952, Martens *et al.* 1997). The molecular data (chapter 16, Schön *et al.* 1998a) of interspecific differences were actually calibrated using these fossil records and thus contribute little to the present discussion. The age of (intraspecific) divergence of the different screened populations of *D. stevensoni* (Finland to South Africa) was estimated (using mitochondrial DNA sequences) at c 4.8 Myr of evolution of the branches, i.e. a divergence c 2.4 Myr ago, during which no changes in the screened nuclear sequence had occurred. The relevance of this slow molecular evolution will be discussed below.

Mesozoic Cypridoidea: 150 Myr of mixed reproductive strategies

The first genuine Cypridoidea have been found in the Middle Jurassic (c 150 Myr – Anderson 1985, Whatley 1990a,b, 1992). Most of the expansion of this group in non-marine habitats in the Upper Jurassic and the Lower Cretaceous is due to the radiation of the (now extinct) *Cypridea*-lineage, which was deemed fully parthenogenetic by Sylvester-Bradley (1976). Colin & Danielopol (1980) and Horne & Martens (in press a), however, argue that a rapid specific and generic diversification is unlikely for a fully asexual group, so that a mixed reproductive strategy seems more likely. The *Cypridea*-lineage disappeared completely during and shortly after the Cretaceous-Tertiary mass extinction (Moore 1961), after which the radiation of the present-day Cyprididae started, according to all available evidence, also with mixed reproductive strategies. Nevertheless, the ancestors of the present-day Cyprididae were already present during the *Cypridea*-reign, lurking in the background and awaiting their chance, much like the first mammals were biding their time when dinosaurs roamed the Earth. Some cypridinid (and non-*Cypridea*) taxa found then actually strongly resemble present day species and genera (Horne & Martens in press a), and although the preservation of the Jurassic & Cretaceous material is insufficient to establish conspecificity with modern species, we assume that reproductive strategies have been much the same in both the earlier and present-day cypridinid lineages, so that Cyprididae could have had mixed reproduction for about 150 Myr.

Chaplin & Hebert (1997) estimated that some of the clonal lineages of *Heterocypris incongruens* diverged about 5 Myr ago, a figure corroborated for *Eucypris virens* by Schön *et al.* (in prep.). The apparent persistence of clonal lineages for such a long time caused Chaplin & Hebert (1997) to suggest that also *H. incongruens* could be considered an ancient asexual. This claim is certainly not valid, as several sexual populations of this species are known to occur at least in

America and Europe (Fryer 1997), while the problems of estimating the age of clonal lineages in clusters with repeated origin from sexual roots are discussed by Butlin *et al.* (in press). *Heterocypris incongruens* is clearly a species with mixed reproduction, and although these species could have clonal lineages that have survived for millions of years, it will be very difficult to actually demonstrate this, as age of divergence between clones could just as easily reflect the age of divergence between (unknown) sexual root lineages, as it is impossible to say when asexual clones actually originated from sexual roots.

Non-marine Cytheroidea

Although there are non-marine representatives of at least seven different cytheroid lineages (chapter 3), the most important one is that of the Limnocytheridae. Whatley & Moguilevsky (in press) situate the origin of this family, evidently from marine sexual ancestors, in the Early to Middle Devonian (380 Myr). The actual origin of the modern subfamilies Timiriaseviinae and Limnocytherinae is situated somewhat later, in the case of the former even around the Permian-Triassic boundary, but these two groups are believed to have persisted in continental habitats for at least 250 Myr (but see chapter 6). Whereas the Timiriaseviinae dominated most of the Mesozoic (Colin & Danielopol 1980), the Limnocytherinae became more abundant and diverse from the Tertiary onwards (c 65 Myr). As nearly all Timiriaseviinae have externally visible brooding pouches, sexual dimorphism is generally apparent. In Limnocytherinae, males are generally larger and more elongate and mostly have different carapace shapes. Exceptions exist for extant members of both groups (Martens 1993, 1996), but are rare.

Sexuality in all cytheroid lineages is known from the ancestral taxa onwards. Asexuality, or at least mixed reproduction, is known in the Timiriaseviinae at least since the Tertiary: species in the genus *Frambocythere* were sexual before the Cretaceous-Tertiary boundary (known as K-T boundary, c 65 Myr), but asexual afterwards (Colin & Danielopol 1980, chapter 6), but Colin & Danielopol (*loc. cit.*) assume that asexual reproduction in Timiriaseviinae became more generally introduced from the Oligocene (c 30 Myr) or Neogene onwards (c 20 Myr), e.g. in the genus *Kovalevskiella* (see also Carbonel *et al.* 1986). In Limnocytherinae, some of the earliest records of populations in which sexual dimorphism was not apparent, are from the Bathonian (Middle Jurassic – c 160 Myr) (Wakefield, 1994). In general, however, sexuality in Limnocytheridae is more widespread than in Cyprididae and parthenogenesis can be assumed to have been of less evolutionary significance in the former than in the latter group (see chapter 5). No molecular information on the age of clonal lineages in Limnocytheridae is available.

Conclusion

All three lineages of non-marine ostracods started off from sexual marine ancestors. In Darwinuloidea and Cypridoidea, the invasion of continental waters was accompanied by an increasing importance of asexual reproduction; in Cytheroidea, parthenogenesis might have originated later. Parthenogenesis thus seems to have existed in darwinulids for at least 300-350 Myr, and in exclusivity in this group for anything between 100 and 200 Myr, and for at least 150 Myr in both Cypridoidea and Limnocytheridae. Data on persistence of species and of clonal lineages are scarce and confusing, but in darwinulids, fully asexual species have persisted for at least 20 Myr, maybe longer, while there are indications that clonal lineages in Cyprididae can survive for several million years.

Reproductive modes and diversity

Genetic, morphological and ecological diversity in the three different reproductive strategies will now be discussed in more detail.

Sexual and natural selection and stochasticity

The advantages of sexual reproduction for the persistence and diversification of lineages are manifold (chapter 1, Kondrashov 1993). Furthermore, evolutionary rates (both micro-evolution and cladogenesis) are higher in lineages with sexual reproduction (Bell 1982). The low extant diversity of the Darwinulidae (Rossetti & Martens 1998) is an excellent example of how asexuality (possibly in combination with several other factors – Schön *et al.* 1998a) leads to slow evolution and low diversity.

The higher evolutionary tempo of sexual reproduction is due to recombination, which provides a much faster changing gene pool, with constant creation of new gene combinations, upon which selection can act, but also because an additional type of selection (sexual selection) can act upon the formation of recognition systems and thus create reproduction isolation and hence species formation. The latter mechanism is deemed to have had a major impact on the ultra-fast (adaptive) radiation of cichlid fish in the great African Lakes (Turner 1994 and others). Sexual selection most likely also has been an important factor in the evolution and speciation of non-marine ostracods (chapter 10), although the example of the Limnocytherinae shows that sexual and natural selection and stochastic effects interplay very closely in the evolutionary history of the diversification of a lineage (Martens in press). The importance of sexual selection in taxa with mixed reproduction has not yet been discussed.

Survival through diversification

Origin and maintenance of clonal diversity

Application of molecular methods on cypridinid populations with asexual reproduction showed that all-female populations of such species held a surprisingly high clonal diversity (Tab. 15.3), for example, not less than 211 different clones of *E. virens* (1136 individuals, 55 populations) and 31 in *L. inopinata* (806 individuals, 12 populations) were found.

Potential and demonstrated mechanisms contributing to the origin and maintenance of clonal diversity are cited in chapter 1 and in Butlin *et al.* (in press) and are recapitulated in Table 17.1. Most important for the present discussion is that taxa with mixed reproduction, i.e. those in which a sexual root has survived (either con-

Table 17.1 Mechanisms of clonal origin in species with different reproductive strategies. S= fully sexual species. M= species with mixed reproduction; M_1= Cyprididae and Candonidae; M_2 Limnocytheridae. AA= Ancient asexuals.

	S	M_1	M_2	AA
mutations	-	+	+	+
multiple origin from sexuals	-	+	+	-
Intraspecific hybridisation				
parthenogenetic female	-	+	-	-
Interspecific hybridisation				
parthenogenetic female	-	+	-	-
sexual female	+	+	?	-

specific or at least congeneric) have a powerful additional engine to increase clonal diversity through intra- and/or interspecific hybridisation, as well as through the multiple origin of asexual lineages directly from the sexual root. These mechanisms are obviously unavailable to lineages which have lost their sexual root, like the darwinulids (see below). At least circumstancial, and sometimes direct, evidence has been found for all three types of hybridisation in cypridinid ostracods (Butlin *et al.*, in press, chapters 15,16). Three main mechanisms allow clonal diversity to be maintained at high levels in groups with mixed reproduction (Box 17.1). The importance of egg banks in the sediment for the maintenance of clonal diversity in aquatic micro-Crustacea in general has been discussed extensively in the literature (Hairston & De Stasio 1988, Rossi *et al.* 1996). The other two mechanisms will be further discussed below.

Box 17.1 Mechanisms maintaining clonal diversity in non-marine Ostracoda

1. *Ecological (niche) diversification and specialisation,* including:
   spatial (geographical) segregation
   temporal segregation
   degrees of specialisation (highly specialised to generalist)
2. *Metapopulation structure:* high diversities depend on non-equilibrium interactions between patchily distributed species, i.e. stochastic sequence of local extinctions and recolonisations. This implies a degree of redundancy in each waterbody.
3. Removal of genotypes from the selective arena by storage in *egg banks in sediment.* This facilitates short-term persistence through environmental instability.

METAPOPULATIONS AND REDUNDANCY

The recognition of non-equilibrium maintenance of diversity through a metapopulation structure, either through patchiness or in discrete water bodies (but with some interchange of genetic material), is relatively new. The concept has been discussed at the species level (see for example several papers in Schulze & Mooney 1994), but was thus far not yet elaborated at a clonal level. Chapter 11 strongly argues against the recognition of all clonal lineages as separate species; nevertheless, clonal lineages can behave in several aspects as independent entities which are (due to ecological specialisation – see below) in direct competition with each other. This can be illustrated by applying a number of hypotheses dealing with specific diversities to clonal diversities. For example, Little & Hebert (1994) found low clonal diversities in ponds on Caribbean islands and postulated that these resulted from colonisation bottlenecks, but their data fit the surface area versus species diversity 'island' model (MacArthur & Wilson 1967) equally well. A similar test for this hypothesis would be to compare clonal diversity across water bodies of different size and habitat complexity. Such comparisons have yet to be conducted in a systematic way (Butlin *et al.* in press), although Rossi *et al.* (1995) have explained clonal diversity of *Heterocypris incongruens* in northern Italian ricefields using this hypothesis. If ecologically differentiated clones are indeed independent and competing entities, then all models related to dynamics of (specific) diversity can be applied to clones; i.e. niche diversification, metapopulation dynamics and, resulting from the latter, redundancy hypotheses (Gitay *et al.* 1996, Tilman *et al.* 1997). The concept of *redundancy* refers to the presence of several species belonging to the same functional guild in one ecosystem, in other words 'there are too many species'. It has been put forward that redundancy in fact adds to ecosystem resilience and provides a buffering capacity against environmental fluctuations (Solbrig 1994). The redundancy concept provides causality, in addition to the classical niche diversification and egg bank explanations, for sympatry of closely related clonal lineages of micro-crustaceans and in general for maintenance of high clonal diversity.

Genetic diversity and ecological specialisation

Lynch (1984) suggested that asexual lineages are competitively inferior and can therefore only survive if they develop a very wide tolerance range for a large set of variables. This *general purpose genotype* hypothesis was tested and found invalid for the cyclic parthenogens *Daphnia* by Weider (1993). Also the evidence presently available for non-marine cypridinids and cytherids with mixed reproduction does not corroborate this hypothesis. Strong indications of high ecological specialisation in ostracod clones have recently been found: seasonal adaptation in *Heterocypris incongruens* (Rossi & Menozzi 1990, 1993, Rossi *et al.* 1996); strong salinity–specialisation in different sets of clonal lineages of *Limnocythere inopinata* in Austrian ponds and lakes (Yin 1997 and chapter 14), as well as more general data on ecological differentiation amongst clones of *Eucypris virens* (Otero *et al.* 1996 and in press). Note that not all clones identified by allozyme screening are necessarily occupying different ecological niches; as was pointed out by Butlin *et al.* (in press), some clones differ in a few neutral mutations only. The redundancy concept deals with such situations.

Clonal specialisation does not necessarily exclude the presence of one or more clonal lineages which are more generally adapted, and indeed the fact that certain clones are found in nearly every screened population, sometimes over large geographical distances (chapter 15), seems to corroborate the existence of such clones. Nevertheless, high clonal diversity appears to coincide with high ecological diversity (chapters 13, 14) and this is not according to the original general purpose genotype concept. Species with mixed reproduction, i.e. those which have retained a sexual root, increase their fitness through diversification.

Phenotypic plasticity and genetic diversity

Phenotypic variability (here exemplified by morphology) can have several underlying causes and can either be genetically induced, or be ecophenotypic plasticity or be the result of an interaction of both genetic and ecological effects. Chapters 8 and 9 (for valve outlines in *Eucypris virens* and limb morphology in *Limnocythere inopinata*, resp.), amply illustrate that phenotypic variability in species with mixed reproduction is generally a consequence of all three factors. Cladograms of different populations of *Eucypris virens* based on both valve outlines and on molecular divergence are remarkably congruent (Schön *et al.* 1997), further strengthening the linkage between morphological and genetic variability. Both types of cladograms furthermore have the sexual lineages nested amongst the parthenogenetic ones, which illustrates that both types of reproductive modes belong to one and the same morphological and genetic cluster. This corroborates our initial definition of species with mixed reproduction.

High levels of genetic, phenotypic and ecological diversity

The results summarised here indicate that species with mixed reproduction strongly bet on diversification at all three levels as a means to increase the fitness of the cluster as a whole. This means that both sexual and asexual lineages have high levels of genetic variability (respectively as haplotypes and clonal lineages), of which the latter are to a large degree ecologically specialised. Some of these clonal lineages, furthermore, have typical morphologies, and can even be identified morphologically.

For such species it is most important to maintain a high genetic variability, and several (ecological) mechanisms exist which allow high levels of clonal diversity to persist sympatrically. Diversity levels are furthermore maintained by increased rates of clonal origin through hybridisation with conspecific or congeneric sexual roots and through multiple origin of clones directly from the sexual root, which apparently manages to balance levels of extinction. No information is as yet available on clonal extinction rates (chapter 13) and this constitutes an excellent topic for future research.

The faithful ones: Limnocytheridae

The above strategy implies a high tempo of origin of clonal diversity through 3 types of hybridisation. Either sexual or parthenogenetic females of one species can mate with a male of another species (interspecific), or males can successfully mate with conspecific parthenogenetic females (intraspecific). The latter two types of hybridisation, those involving parthenogenetic females, most likely give rise to polyploid clones, as haploid spermatozoa have to fuse with (at least) diploid eggs. The occurrence of these mechanisms in Cyprididae is corroborated by high levels of polyploidy in cypridinid species (review in chapter 15).

In the one asexual limnocytherid species screened extensively, *Limnocythere inopinata* (chapter 15), however, no trace of polyploidy was found. This would effectively exclude two out of three types of hybridisation (inter- and intraspecific with parthenogenetic females) as mechanisms for origin of clonal diversity. Species with such a strategy thus only have direct mutations in existing lineages and multiple spin-off from sexual roots as engines for the production of new clones.

If this situation is valid for the entire family, then the reasons behind the inability of the present group to hybridise may well be related to a higher effect of sexual selection in the Limnocytheridae (chapter 10) or have other morphological causes, preventing parthenogenetic females in this group to successfully mate with males. Although speculative at this stage, these areas constitute excellent avenues for future research.

The unknown: Candonidae

Polyploidy in species of the Candonidae has been reported: up to 25% in *Candona ovata* (Havel *et al.* 1990) and up to 56% in *Candona rectangulata* (in Havel & Hebert 1993), although no polyploidy was found in *Candona candida* and *Candona balatonica* (in Havel & Hebert 1993). The link between hybridisation and polyploidy is not necessarily absolute, as polyploids can theoretically originate without hybridisation, i.e. through chromosome non-disjunction. However, no indications of this exist at present in the Ostracoda (chapter 2). At least a number of Candonidae thus seem to be able to hybridise, although this group differs in other aspects of its reproductive evolutionary strategy from the Cyprididae. Geographical parthenogenesis, or at least a form of it, occurs only in the Cyprididae and not in the Limnocytheridae, nor in the Candonidae as a group (see chapter 8 and below). Finally, Martens (1994, 1997) already mentioned that the overwhelming presence of Cytheroidea and Candonidae in long-lived lakes, paralleled by the nearly complete absence of Cyprididae from such habitats, might be due to different reproductive factors. This is further corroborated by the higher diversity of Candonidae in even older aquatic habitats, namely groundwater, as compared to Cyprididae.

Although Limnocytheridae, Candonidae and Cyprididae all have mixed reproductive strategies, clear differences exist. In Cyprididae, parthenogenesis is the predominant reproductive mode, with sexuality mainly persisting to generate high clonal diversity. In Limnocytheridae and Candonidae, sexuality is more prevalent, with parthenogenesis being more an optional and less important mode. To establish a clearer image of the differences in strategies between these groups with mixed reproduction is a research priority. Combined morphological, ecological and molecular research on a number of candonid species is most urgent.

A life of chastity: Darwinulidae

Genetic and phenotypic uniformity

Schön *et al.* (1998a – see also chapter 16) found very low intraspecific genetic variability (as compared to other ostracod groups – Schön *et al.* 1998b) in the mitochondrial genome of *D. stevensoni* (about 3.8%, reflecting almost 4 Myr of molecular evolution), and absolutely identical sequences in 320 bp of ITS1, part of the nuclear genome, in the full set of individuals analysed (N= 56), covering 20 populations in a range from Finland to Southern Africa. This low variability in mitochondrial DNA and uniformity in nuclear DNA is matched by a very low clonal variability in the same species as detected by allozyme analysis: only 7 multilocus genotypes were found in *D. stevensoni* (1822 individuals, 36 populations – compare to higher diversity in *E. virens* and *L. inopinata* in Tab. 15.3). Rossetti & Martens (1996 – see also chapter 9), finally, found a remarkable morphological uniformity, congruent with the absence of significant genetic variability.

Only few data on genetic distances between darwinulid species and genera are available (Schön *et al.* 1998 and unpubl., chapter 16), but also here variability is relatively low. For example, genetic distances (for the COI sequence) between *Darwinula stevensoni* on the one hand and *Penthesilenula brasiliensis* and *Microdarwinula zimmeri* on the other, is 22% and 27.7%, respectively, which is only slightly more than the total intraspecific diversity recorded for the same gene-sequence in, for example, *Eucypris virens* (chapter 16, Schön *et al.* in prep.). This low supraspecific genetic diversity is, again, matched by low morphological diversity in the Darwinulidae. In a recent taxonomic revision of the Holocene and Recent representatives of this family (sole surviving member of a whole infraorder), Rossetti & Martens (1998) retained less than 30 species in 5 genera, while they specifically noted that morphological differences between species are small (low supraspecific diversity) but remarkably constant (low intraspecific diversity). As fossil evidence shows that most lineages (species and genera) have nevertheless persisted for several millions of years (more than 20 Myr for both *D. stevensoni* and *M. zimmeri* – Straub 1952), this ancient asexual group experiences slow molecular and morphological evolution.

Slow evolution

Whereas the slow evolution in the mitochondrial DNA of *D. stevensoni* is already significant, the total absence of genetic variability in a sequence of nuclear DNA of this species is intriguing. The latter is indeed an anomaly, as in the time of evolution deduced from the COI divergence, c 5 substitutions in 320 bp of ITS1 would have been expected. A recent selective sweep of a super clone, replacing all other genotypes, is impossible, as it would also have affected the mitochondrial sequences and the same logic applies to background selection (Charlesworth 1994). Most likely a selective sweep did take place, but not recently (Schön *et al.* 1998). There are no reasons to believe that this species is in fact not a genuine apomictic asexual, as no sexual populations have been discovered in the entire (super-) family for the last 100-200 Myr, there is to date no evidence for automixis, wide-ranged allozyme screening (chapter 15) did not find any evidence for polyploidy (weakening the hypothesis on occasional sex in this group, Little & Hebert 1996), and the chromosomal structure is deteriorated following the prediction of Judson & Normark (1996) for ancient asexuals (Tétart 1978, chapter 2). Finally, even for an ancient asexual with several biological adaptations reducing the rate of molecular evolution (low fecundity combined with brooding – allowing the abortion of defective embryos, Lively & Johnston 1994 – and long life cycles), the nuclear mutation rate in ITS1 is aberrant. This leads to the formulation of a completely different reproductive strategy than those summarised above for lineages with mixed reproduction.

General purpose genotype and DNA repair

Of the 28 Recent and Holocene species presently recognised in the Darwinulidae (Rossetti & Martens 1998), only one (*D. stevensoni*) is truly cosmopolitan and ubiquitous. All other species have more restricted distributions (two further species have intercontinental distributions, but are very rare, all others are known from 3 localities or less) and mostly occur in specialised habitats (interstitial, mosses in splash zones, dolomitic springs, etc.). I will therefore concentrate on the strategy of *D. stevensoni* first, and afterwards discuss the potential relevance of this hypothesis for the darwinulids as a group.

Darwinula stevensoni mostly occurs in the shallow littoral of postglacial lakes, but can also be found interstitially, in springs, rivers and other types of habitats, while the species furthermore occurs from subglacial conditions (Finland) to the (sub-) tropics in Africa, and in salinity ranges from completely fresh water to even highly carbonated environments (Griffiths & Butlin 1994, Martens *et al.* 1997). This asexual species could thus have a *general purpose genotype*, as postulated by Lynch (1984). It was explained in chapter 14 that the presence of a general purpose genotype cannot be tested experimentally, as the asexual lineage would have to be compared to a matching sexual lineage and such a sexual root does not exist in the Darwinulidae. Nevertheless, field data suggest that *D. stevensoni* is a likely candidate for this type of ecological strategy. Assuming that this is so, then an ancient asexual would do well to ensure that its genotype is well protected. To abandon sex altogether already goes a long way, as recombination rapidly forms new genetic combinations, but equally rapidly destroys the existing ones (chapter 1). Nevertheless, over long periods of time, mutation would still affect an increasing part of the populations, while stochastic events would cause irreversible loss of unaffected genotypes (Müller's ratchet – chapter 1). Other biological strategies such as those cited above (brooding, etc.) can further decrease the rate of molecular evolution, but to explain long-term intact preservation of a genome, none of these mechanisms is sufficient.

Schön *et al.* (1998a) and Schön & Martens (1998) therefore assumed that at least this ancient asexual species has developed ways to increase the efficiency of its DNA repair mechanism(s). This hypothesis is rather controversial, as Gabriel *et al.* (1993, see also Judson & Normark 1996) actually suggested the completely opposite strategy, namely that ancient asexuals could reduce the risk of mutational load by having no DNA repair at all and simply losing a larger fraction of each generation through direct mortality. However, Schoen *et al.* (1996) already deemed this an unlikely possibility, based upon the fact that at present no eukaryotic organisms are known to exist without any DNA repair mechanisms. The presence of one or more highly efficient nuclear DNA repair mechanism in this species still requires experimental confirmation, but it is as yet the most plausible explanation for the absence of nuclear molecular variability, while from an evolutionary point of view, a general purpose genotype would be worth preserving. Finally, in most individuals of *D. stevensoni* thus far screened, either for allozyme diversity or for DNA diversity, no evidence for heterozygosity is found; whereas, according to Meselson's rule (in Judson & Normark 1996), the accumulation of mutations occurs randomly in the two DNA strands, so that ancient asexuals are expected to be generally heterozygous. Efficient DNA repair will cause and maintain homozygosity.

The strategy of this asexual organism is thus, contrary to the one summarised above for taxa with mixed reproduction, to avoid diversification.

There are a number of potential counter arguments. Genetic diversity in *D. stevensoni* is low, but not completely absent (Tables 15.2 & 15.3). However, complete genetic uniformity is neither required by the above hypothesis, nor entirely possible. One of these rare aberrant genotypes seems to be more common in lotic habitats than the normal clone (V. Rossi, pers.comm.). This only shows that a general purpose genotype does not necessarily allow a 'super clone' to be competitively superior in all habitats, otherwise the species would indeed occur in all habitats and this is not the case.

How valid is this hypothesis for the other darwinulids, equally fully asexual, but obviously with a far less general purpose genotype (as judged from their ecological and geographical distribution) ? This question certainly needs further elaboration, as almost no data on intraspecific variability (genetic or morphological) of other darwinulid species are available (Schön *et al.* 1998a). Some, like *Penthesilenula brasiliensis,* also seem to have relatively wide ecological tolerances (occurring in lakes and bogs in both high latitude Scottish lochs and subtropical Brazilian water bodies – Rossetti & Martens 1998), but in spite of its intercontinental distribution, it is by no means as common as *D. stevensoni* (Martens *et al.* 1997). Whereas all extant darwinulids must have several adaptations allowing them long-term survival without sex, most likely all involving limiting diversity, only *D. stevensoni* seems to have perfected this strategy. Further corroboration of the present hypothesis can be obtained by testing the extent of the general purpose genotype of several darwinulids (experimentally determined tolerance ranges in the absence of competition), screening for genetic variability in darwinulid species other than *D. stevensoni*, corroborating the absence of nuclear genetic variability by screening other nuclear genes, also in *D. stevensoni,* and by experimentally demonstrating the presence of a highly efficient DNA repair mechanism acting on nuclear DNA. All of these tests are presently in progress.

Reproductive modes and group selection

These strategies should of course not be seen from a humanistic point of view, i.e. as pre-considered and planned goals. The reproductive modes are the results of the evolutionary processes of selection working on variation. Nevertheless, group selection has been cited as a mechanism to maintain sexual reproduction. Similar arguments can be invoked for the maintenance of high clonal diversification in the mixed reproductive strategy, as the specialisation of different clones will increase the probability of survival and the longevity of the species as a whole (seen as the inclusive cluster of all sexual and asexual lineages), but in doing so reduce their own long-term fitness. Clonal specialisation might turn out to be one of the best examples of group selection, but this will be discussed elsewhere.

Reproductive modes and ecological fitness

Habitat type

Much speculation about possible relations between reproductive mode and clonal diversity on the one hand and habitat type on the other has been put forward, but few of these were backed by significant correlations. Attempts to show associations between clonal diversity and permanence of water body (Chaplin & Ayre 1989) and latitude (Little & Hebert 1994) failed (see also Butlin *et al.* in press). Claims have been made that asexuality would be more common in temporary water bodies (Chaplin *et al.* 1994) and in the tropics (Little & Hebert 1994). Both tests have a limited validity, because the databases used were not fully representative of the environmental factor assessed. For example, Chaplin *et al.* (1994) based their data on obsolete generic taxonomies, while Little & Hebert (1994) failed to look at more ancient habitats in the tropics. However, the major flaw in such analyses up to now was their coarseness, as they insufficiently distinguished between parthenogenetic, bisexual and mixed populations, and between populations and species, and generally did not distinguish the fundamentally different evolutionary strategies outlined above. I will briefly examine four ecological properties of species with the different reproductive strategies, which might lead to differential fitness and competitive abilities. Relevant data are mostly reviewed in chapter 14, so the present discussion only provides a summary.

Colonisation ability

The ability to colonise a new habitat fast depends more on the presence of drought-resistant stages than on reproductive mode. Cyprididae and some Limnocytheridae have desiccation-resistant eggs (cysts), while both Cyprididae and Candonidae, as well as Darwinulidae, can survive some time (up to a few months) without water in a torpid stage (Delorme & Donald 1969). All groups thus seem to have some degree of resistance against desiccation, although the cysts are of course by far the most efficient mechanism. Parthenogens are superior colonisers to sexuals as, theoretically, one egg is sufficient to found a new population in a free habitat, whereas sexuals need at least two eggs, of which the resulting individuals have to locate one another in a spatially and temporally very diluted environment. Much of the latter drawback for sexuals can be overcome by brooding, like in most Timiriaseviinae. Darwinulids are also brooders, which might partly compensate for the absence of drought-resistant eggs in this group. However, making abstraction of brooding and desiccation-resistant mechanisms, in principle sexuals are at a disadvantage compared to parthenogens with regard to colonisation. This hypothesis is not corroborated by fossil data from the beginning of the Holocene (Griffiths & Butlin 1995), nor by experimental results (chapter 14), but maybe time scales used were not appropriate to distinguish between colonising efficiency of conspecific sexual and asexual lineages. The Darwinulidae also seem to contradict this general hypothesis, as they appear unable to produce desiccation-resistant cysts, although this would seem to be fully in line with their broad ecological tolerance. Most likely, in their case it was a trade-off between the advantages of brooding and of producing desiccation-resistant, diapausing eggs. Drought resistance and diapause are indeed often co-occurring strategies, and a brooding organism is not well served by obligately diapausing eggs in its brooding cavity! At the specific level, theory dictates that species with mixed reproduction would take an intermediate position between fully sexual and fully asexual populations.

Habitat longevity

As outlined by Hughes (1989), sexuals should be better competitors, as they can adapt to changing ecological circumstances with a fine scale tuning unavailable to parthenogens. This can be partly overcome by animals with high diversity of clones with ecological specialisation (i.e. with mixed reproduction – called polyphyletic clones by Hughes 1989), as well as by a general purpose genotype (Lynch 1984), but neither of these solutions can reach the high levels of competitivity provided by fully sexual reproduction. Therefore, in long-lived habitats (with geological stability), sexuals will always be at an advantage. This explains the high incidence of fully sexual reproduction in long-lived habitats such as ancient lakes (Martens *et al.* 1994, Martens 1994, 1997) and groundwater (Danielopol 1978 a.o.) and is in accordance with the 'Tangled-bank' hypothesis (Bell 1982, chapter 1).

Habitat predictability

The relation between reproductive strategy and habitat predictability, here including climatic stability, is primarily relevant in unstable, short-lived habitats. If such habitats are predictably unstable, then parthenogens will have an advantage over sexuals. If the habitat is unpredictably unstable, sexuality will give the greater genetic plasticity to adapt to the new situation, in which case sexual populations will have the advantage. Up to a certain measure of unpredictable instability can be absorbed by parthenogenetic populations by having a general purpose genotype and by mixed reproduction by having high clonal variability, with differently adapted clones. However, in highly unpredictable situations, the probability that clonal lineages can persist becomes low.

Specialisation and adaptation

The degree of specialisation and genetic diversity related to the different reproductive strategies has been discussed extensively in previous sections. The ancient asexuals have a general purpose genotype with low genetic variability; the taxa with mixed reproduction increase genetic and ecological diversity, with several clones showing degrees of ecological specialisation and the sexual taxa are the best equipped to provide dynamic genetic diversity and plasticity. Especially the dynamic aspect in adaptation is an important difference between sexuals and parthenogenetic taxa and taxa with mixed reproduction.

The shamrock diagram

When these different adaptations of reproductive strategies are plotted in a hypothetical diagram of ecological space (Fig. 17.1), it becomes apparent that the differ-

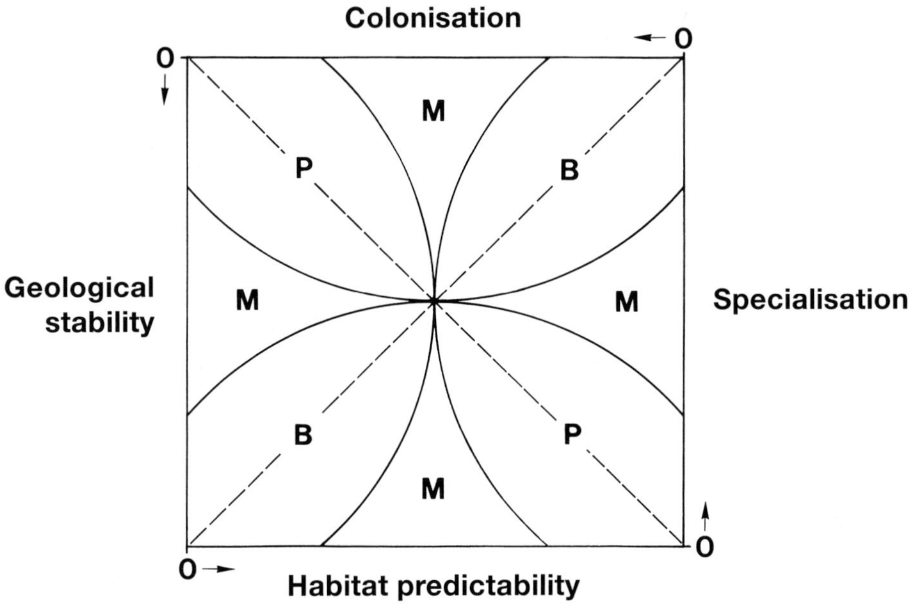

Fig. 17.1 Shamrock diagram, showing theoretical scores of species with different reproductive modes for four ecological characteristics: the ability to colonise new habitats, habitat longevity (as geological stability of the basin and area), habitat predictability (including predictability of climatic fluctuations) and ability to rapidly adapt and specialise. P= fully parthenogenetic, B= fully sexual, M= mixed reproduction.

ent reproductive strategies have advantages and disadvantages, depending on which biological factor one is dealing with. This explains why non-marine ostracods, which basically all have similar ecological functions (they are nearly all benthic detritus feeders), can coexist with different reproductive strategies without outcompeting one another.

It also shows that the different factors have significant interactions, which means that questions such as 'which reproductive mode occurs most in temporary pools ?' are not precise enough. Superior colonisation ability should better equip parthenogens to occur in short-lived, erratic habitats, but in temporary pools with long-term geological stability, sexuals are better adapted, as the longevity of the habitat means that fine-tuned adaptations provide superior competitive abilities. It is important to distinguish between geological and ecological stability: the latter favours asexuals, but habitat longevity combined with ecological fluctuations will favour sexuals. This is corroborated by the data from temporary pools in a geologically stable, long-lived arid area in South Western Africa, where most temporary pool ostracods reproduce sexually (Martens 1998). It also explains why predominantly sexual lineages of Limnocytheridae managed to persist and adapt to such long-lived temporary pools in South Africa (Martens 1996).

Latin lovers: geographical parthenogenesis

The distribution patterns

Vandel (1928) coined the term *geographical parthenogenesis* for taxa with segregated distributions of different reproductive modes (of course, only taxa with mixed reproduction can show these distributions), with sexual populations mostly occurring in the southern part of Europe. Several authors (McKenzie 1971 a.o.) have pointed out that reproductive modes in non-marine ostracods are distributed following this pattern. Recent GIS analyses of geographical distributions of non-marine ostracods compiled in the NODE database (chapter 5), however, have shown that only the family Cyprididae has sexual populations restricted to the Mediterranean area, whereas other groups such as Candonidae and Limnocytheridae have both modes of reproduction scattered throughout Europe, although certain isolated species, such as *Limnocythere inopinata*, might display patterns similar to those found in the Cyprididae.

However, even the Cyprididae do not have genuine geographical parthenogenesis. Firstly, there are several exceptions to the pattern, as *Scottia pseudobrowniana* and two species in *Heterocypris* have sexual populations in Central Europe (see chapter 5). Secondly, and most importantly, the two reproductive modes do not have segregated distributions, as sexual populations indeed are restricted to the Mediterranean area, but parthenogenesis occurs in both northern and southern Europe. Taking into account the importance of hybridisation (including intraspecific) for the generation of clonal diversity in the Cyprididae, the fact that parthenogenetic females co-occur with males is not surprising. Thirdly, of the one hundred or so cypridinid species presently in NODE, sexual populations have been found in Europe of 20 species only (chapter 5). This means that of 80% of the cypridinid species, no males have thus far been described from Europe. This is of course an underestimation, as of these 80%, sexual (or mixed) populations have been described from outside of Europe (North America, Siberia, etc.), while the probability that many sexual populations around the Mediteranean simply have not been found yet is quite high.

Existing explanatory hypotheses

There are two main groups of hypotheses explaining these distribution patterns, namely the historical and ecological scenarios.

The *historical* explanation invokes the Pleistocene glaciations in Europe, which would have wiped out all aquatic faunas in Europe, leaving only refugia in southern Europe (Vandel 1928 and later papers). After the retreat of the glaciers, parthenogenetic populations managed to recolonise the newly-available habitats first and thus far the sexuals have not (yet) managed to displace them competitively. To explain the latter failure of the sexuals, several e*cological* explanations are offered, which can go either way. Some authors claim that parthenogenetic populations are less well adapted to suboptimal conditions, so that only sexual populations can survive in the fringes of a species' areal; parthenogens are then either superior competitors in the optimal parts of the areal (northern parts of Europe), or the historical scenario is invoked again to explain the absence of sexuals in Western and Central Europe. Other authors (chapter 13) claim that parthenogens are better adapted to suboptimal conditions, whereas sexuals outcompete asexuals in more stable environments. Peck *et al.* (1998) provide a new model, based on longer reproductive periods (up to 3 times longer in the south!), hence more offspring, in southern areas. Löffler (1990), finally, cites alkalinity as a factor promoting the presence of males in populations of *L. inopinata*, but this is most likely to account for the presence of rare male occurrences and less for the presence of sexual populations (see also chapter 6).

Holocene stability

Horne & Martens (in press b) provide a new hypothesis, which is based on new evidence on Holocene climates and also takes the differential competitiveness of reproductive strategies, in relation to different environmental factors, into account.

The hypothesis partly adopts the historical scenario: during glaciations, non-marine faunas are indeed wiped out, to be followed by recolonisation from southern refugia. The ecological scenarios are refuted, however, as they all concentrate on one ecological factor, which is moreover often not corroborated by the data:

(1) Analyses of cores from different parts of Europe for the first 5000 years of Holocene history (Griffiths & Butlin 1995), for example, do not show that parthenogenetic species are better colonisers than sexual species, rather the contrary.

(2) It is furthermore unlikely that sexual populations would have failed to outcompete parthenogens after 10,000 years of Holocene history, as sexual populations of several species have been found soon (a few thousand years) after the retreat of the glaciers during previous interglacial periods (Diebel & Pietrzeniuk 1975).

(3) The difference in the length of reproductive period in northern and southern parts of Europe (Peck *et al.* 1998) is absolutely not founded on hard data (certainly no studies support a threefold increase in offspring in the south for ostracods) and, furthermore, cannot account for the sympatric presence of both parthenogenetic and sexual lineages there.

New geological evidence indicates that the Holocene is climatically unusually stable, when compared to other interglacials (see also chapter 6). Horne & Martens (in press b) argue that this more stable (predictable) climatic situation has allowed and still allows parthenogens to out-compete the sexuals in the present-day habitats of western and central Europe. Sympatric persistence of both reproductive modes in southern parts of the continent, during glacial and interglacial (at least the Holocene) periods, is either due to higher ecological (or climatic) instability in circum-Mediterranean habitats, for which there is some evidence (Horne & Martens in press b, chapter 6), or more likely to the greater geological stability of the area, leading to habitat longevity in which sexuality is competitively superior (see above).

This attractive scenario, owing its higher robustness to a combination of evidence from several disciplines, can nevertheless still not account for the puzzling persistence of both modes of reproduction in several candonid lineages in most of Europe (chapter 5). I suggest that an improved understanding of the reproductive strategy of Candonid lineages will contribute to solve this remaining enigma.

Clonal taxonomy

The ability to distinguish between different types of asexuals (see Box 1.1) also allows a better understanding of a long-standing problem, that of clonal taxonomy. Mayr (1970) admitted to resorting to the morphological species concept when dealing with clonal organisms. At least in the ostracods, the clonal lineages within the different reproductive strategies can now be better included into taxonomies, using morphological, phylogenetic and ecological arguments. The full discussions are conducted in chapter 11, the results are summarised here.

Ancient asexuals can be divided into good and discrete morphological units, which are furthermore mostly monophyletic. Although morphological differences between darwinulid species are small, they are mostly constant (Rossetti & Martens 1998). For clonal lineages belonging to species with *mixed reproduction,* Mallet's (1995) cluster genotype species concept is used. This is an extrapolation of Mallet's original concept, which is amply corroborated by the facts presented in previous chapters: sexual and asexual strains of species with mixed reproduction form an inclusive cluster, using morphological (chapters 8,9), ecological (chapter 14) and genetic (chapters 15,16) data. This conclusion effectively puts a stop to the allocation of taxonomic identities (specific or subspecific name) to populations with different reproductive modes. *Limnocythere sappaensis* thus becomes a junior synonym of *L. inopinata*; *Cytherissa lacustris baicalensis* simply becomes *Cytherissa lacustris*; etc. *Interspecific hybrids* are treated as proposed for the extensively discussed vertebrate examples (Echelle 1990), i.e. they can be treated as different species (Griffiths *et al.* 1998).

The only remaining problematic cases are the fully asexual spin-offs of clusters of mixed reproduction, as for these lineages still only morphological species concepts can be used. Whether or not certain populations are morphologically sufficiently different (and discrete) from the majority of the mixed cluster, will always be a subjective decision, based on insufficient material. The revision of European *Herpetocypris* was conducted with this problem in mind (Gonzalez Mozo *et al.* 1996), and gives an example of how morphological variability in such higher taxa with mixed reproduction can be assessed and translated into a taxonomy. Differences between fully asexual spin-offs of mixed clusters (in Cyprididae) and ancient asexuals (Darwinulidae) are discussed in chapter 11.

Palaeoecological reconstructions

There are at present three main methods which use fossil assemblages of ostracod valves for the interpretation of palaeoecologies and palaeoclimates (Carbonel *et al.* 1980). Roughly summarised, these are: (1) the use of indicator species with specific ecologies (e.g. salinity tolerance); (2) valve morphologies, implying causality between ecological factors and reticulation, nodosity, spinosity, etc. (Peypouquet *et al.* 1988); and (3) analysis of valve chemistry (based upon experimental results with living populations of the same species) (Holmes 1996). The improved understanding of the reproductive strategies of non-marine ostracods can refine at least the first method. Firstly, because clones with ecological specialisation which can be identified on valve shape, provide refined information on the palaeoecology of the habitat, thus improving the indicator method. Yin *et al.* (in press) give an example of different sets of clones of *L. inopinata*, which they deem typical of either fresh or saline waters. Nevertheless, they also advise to apply these extrapolations to fossil situations only in localised areas from which clonal ecologies of recent populations are known.

Secondly, intrinsic differences between sexual species and those with mixed reproduction are now better understood, so only the presence of sexual (or mixed) populations of the latter type of species are informative for palaeontological interpretations. For example, the presence of a sexual population of *Cyclocypris ovum* is not informative, only the presence of the species as such is, as it always occurs in sexual populations. But the presence of sexual populations of species with mixed reproduction at northern latitudes may refer to unstable (less predictable) climates (Horne & Martens in press b).

It is now clear that populations with genuine sexual reproduction and rare male occurrence are different situations which cannot be treated as the same. The (palaeo-) ecological significance of these rare male occurrences still eludes us and this is a topic for future research.

Conservation biology

A significant part of research on biodiversity relates to its conservation (Heywood 1995). Different conservation strategies exist, but most methods concentrate on prioritising regions, habitats and populations for conservation. The data summarised in the previous chapters can lead to the following recommendations:

1. localities with localised ancient asexual species must receive priority, as loss of these taxa is irreversible;
2. certain areas have higher clonal diversity than others; other areas hold more rare genotypes. Ways must be found to integrate the conservation of both types;
3. the importance of sexual roots for the persistence of species with mixed reproduction now being clear, areas where sexual (or mixed) populations of species with mixed reproduction occur must receive high priority for conservation. Their disappearance will not lead to short-term impact, but will reduce the long-term resilience and eventually cause the disappearance of species.

Conclusions

Non-marine ostracod lineages comprise at least three, maybe more, very different reproductive modes, with different evolutionary and ecological strategies. *Fully sexual taxa* occur at least in a number of candonids and cytherids and these taxa all have the known advantages and disadvantages of sexual reproduction outlined in chapter 1. *Taxa with mixed reproduction* combine advantages of both types of reproductive modes and have an evolutionary strategy which aims at high diversity (genetic, phenotypic and ecological) amongst clonal lineages. This is especially true for the Cyprididae, in which the function of the sexual root in most species is effectively reduced to an additional engine for the generation of clonal diversity, either through hybridisation or through a multiple origin of clonal lineages directly from sexual populations. Candonidae and Limnocytheridae also have mixed reproduction, but with sex still being their prime mode of reproduction. At least the latter group seems unable to increase clonal diversity through hybridisation. Differences between reproductive modes in Cyprididae, Candonidae and Limnocytheridae constitute an excellent field for future research. *Ancient asexual taxa*, in this case the darwinulid ostracods, have found ways to defy several theoretical problems, which normally should lead to early extinction of fully asexual lineages, amongst others: general purpose genotype and possibly increased efficiency of DNA repair in at least one species. Future research on these groups will have bearing on general evolutionary theory, but possibly also in applied spin-offs, such as cancer research.

References

Abushik, A.F (ed) 1990. *Manual of the microfauna of the USSR. Vol. 4.* Palaeozoic Ostracoda. *VSEGEI, NEDRA*: 356 pp. (in Russian).

Anderson, F.W. 1985. Ostracod faunas in the Purbeck and Wealden of England. *J. micropalaeontol.*, **4:** 1-68.

Bell, G. 1982. *The Masterpiece of Evolution, the Evolution and Genetics of Sexuality.* Croom Helm: 635 pp.

Brady, G.S. & Robertson, D. 1870. The Ostracoda and Foraminifera of Tidal Rivers. *Ann. Magazine Nat. Hist., ser.* **4, 6:** 1-33, 307-309.

Butlin, R.K. & Griffiths, H.I. 1993. Ageing without sex? *Nature,* **364:** 680.

Butlin, R.K., Schön, I. & Martens, K. (in press). Asexual reproduction in non-marine ostracods. *Heredity.*

Carbonel, P., Colin, J.-P., Danielopol, D.L. & Londeix, L. 1986. *Kovalevskiella* (Ostracoda, Timiriaseviinae), genre a mode de vie benthique depuis l'Oligocène, son adaptation a la vie interstitielle. *Géobios,* **19:** 677-687.

Carbonel P., Colin, J.-P., Danielopol, D.L., Löffler, H. & Neustrueva, I. 1980. Paleoecology of limnic ostracodes: a review of some major topics. *Palaeogeogr., Palaeoclimatol., Palaeoecol.*, **62:** 413-461.

Chaplin, J.A. & Hebert, P.D.N. 1997. *Cyprinotus incongruens* (Ostracoda): an ancient asexual? *Mol. Ecol.,* **6:** 155-168.

Chaplin, J.A. & Ayre, D.J. 1989. Genetic evidence of variation in the contributions of sexual and asexual reproduction to populations of the freshwater ostracod *Candonocypris novaezelandiae. Freshw. Biol.,* **22:** 275-284.

Chaplin, J. A., Havel, J. E. & Hebert, P. D. N. 1994. Sex and ostracods. *Trends Ecol. Evol.,* **9:** 435-439.

Charlesworth, B. 1994. The effect of background selection against deleterious mutations on weakly selected, linked variants. *Genet. Res.*, **63:** 213-227.

Colin, J.P. & Danielopol, D.L. 1980. Sur la morphologie, la systématique, la biogéographie et l'évolution des ostracodes Timiriaseviinae (Limnocytheridae). *Paléobiol. continent.,* **9:** 1-51 + 16 plts.

Danielopol, D.L. 1978. Über Herkunft und Morphologie der Süsswasser-hypogäischen Candoninae (Crustacea, Ostracoda). *Sitzungsber. Österr. Akad. Wissensch. Mathem.-naturwissensch. Kl., Abt. I,* **187(1-5):** 162 pp.

Delorme, L.D. & Donald, D. 1969. Torpidity of freshwater ostracods. *Can. J. Zool.*, **47:** 997-999.

Diebel, K. & Pietrzeniuk, E. 1975. Mittel- und jungpleistozäne Ostracodenfaunen des Raums Potsdam-Brandenburg-Rathenow in stratigraphischer und ökologischer Sicht. *Z. geol. Wissensch.,* **3:** 1197-1233.

Dowling, T.E. & Secor, C.L. 1997. The role of hybridisation and introgression in the diversification of animals. *Annu. Rev. Ecol. Syst.,* **28**: 593-620.

Echelle, A.A. 1990. Nomenclature and non-Mendelian (clonal) vertebrates. *Syst. Zool.*, **39:** 70-78.

Fryer, G. 1997. The horse-trough ostracod *Heterocypris incongruens. Naturalist,* **122:** 121-135.

Fryer, G. (in press). The case of the one-eyed brine shrimp: are ancient atavisms possible? *Trans. R. Soc. Lond.*

Gabriel, W., Lynch, M. & Burger, R. 1993. Müller's ratchet and mutational meltdown. *Evolution,* **12:** 627-640.

Gitay, H., Wilson, J.B. & Lee, W.G. 1996. Species redundancy: a redundant concept? *J. Ecol.,* **84:** 121-124.

Gonzalez Mozo, M., Martens, K. & Baltanás, A. 1996. A taxonomic revision of European *Herpetocypris* Brady and Norman, 1889 (Crustacea, Ostracoda). *Bull. k. belg. Inst. Natuurw. Biol.,* **66:** 93-132

Griffiths, H.I. & Butlin, R.K. 1994. *Darwinula stevensoni*: a brief review of the biology of a persistent parthenogen. *in*: Horne, D.J. & Martens, K. (eds), *The evolutionary ecology of reproductive modes in non-marine ostracods.* Greenwich. Univ. Press: 27-36.

Griffiths, H.I. & Butlin, R.K. 1995. A timescale for sex vs parthenogenesis: Evidence from Holocene ostracods. *Proc. R. Soc. Lond. B.,* **260:** 65-71.

Griffiths, H.I., Pietrzeniuk, E., Fuhrmann, R., Lennon, J.L., Martens, K. & Evans, J.G. 1998. *Tonnacypris glacialis* comb. nov. (Ostracoda, Cyprididae): taxonomic position, (palaeo-) ecology and palaeodistribution. *J. Biogeogr.* **25:** 515-526.

Hairston, N. G. & De Stasio, B. T. 1988. Rate of evolution slowed by a dormant propagule pool. *Nature,* **336:** 239-242.

Havel, J. E. & Hebert, P. D. N. 1993. Clonal diversity in parthenogenetic ostracods. *in*: McKenzie, K. G. & Jones, J. P. (eds), *Ostracoda in the Earth and Life Sciences,* Balkema: 353-368.

Havel, J. E., Hebert, P. D. N. & Delorme, L. D. 1990. Genotypic diversity of asexual Ostracoda from a low arctic site. *J. evol. Biol.,* **3:** 391-410.

Heywood, V.H. (ed) 1995. *Global Biodiversity Assessment,* Cambridge Univ. Press: 1140 pp.

Holmes, J.A. 1996. Trace-elements and stable-isotope geochemistry of non-marine ostracod shells in Quaternary palaeoenvironmental reconstructions. *J. Palaeolimnol.,* **15:** 223-235.

Horne, D.J. & Martens, K. (in press a). The importance of resting eggs for the evolutionary success of non-marine Ostracoda (Crustacea). *in*: Brendonck, L., De Meester, L. & Hairston, N.(eds), Diapause in the crustacea. *Adv. Limnol.*

Horne, D.J. & Martens, K. (in press b). Geographical parthenogenesis in European non-marine ostracods (Crustacea, Ostracoda); post-glacial invasion or Holocene stability? *Hydrobiologia.*

Hughes, R.N. 1989. *A functional biology of clonal animals,* Chapman & Hall: 331pp.

Judson, P.O. & Normark, B. B. 1996. Ancient asexual scandals. *Trends Ecol. Evol.,* **11:** 41-46.

Kondrashov, A.S. 1993. Classification of hypotheses on the advantage of amphimixis. *J. Heredity,* **84:** 372-387

Little, T. & Hebert, P.D.N. 1994. Abundant asexuality in tropical freshwater ostracodes. *Heredity,* **73:** 549-555.

Little, T. & Hebert, P.D.N. 1996. Ancient asexuals: scandal or artefact? *Trends Ecol. Evol.,* **11:** 296.

Lively, C.M. & Johnston, S.G. 1994. Brooding and the evolution of parthenogenesis: strategy models and evidence from aquatic invertebrates. *Proc. R. Soc. Lond. B*, **256:** 89-95.

Löffler, H. 1990. Paleolimnology of Neusiedlersee, Austria. I. Succession of ostracods. *Hydrobiologia*, **214:** 229-238.

Lynch, M. 1984. Destabilizing hybridization, general-purpose genotypes and geographic parthenogenesis. *Quart. Rev. Biol.,* **59:** 257-290.

MacArthur, R.H. & Wilson, E.O. 1967. *The Theory of Island Biogeography,* Princeton Univ. Press.

Mallet, J. 1995. A species definition for the modern synthesis. *Trends Ecol. Evol.,* **10:** 294-299.

Martens, K. 1993. On the taxonomy and zoogeography of the genus *Gomphocythere* Sars, 1924 (Crustacea, Ostracoda), with the description of a new species from the Nahal Dan (Israel). *Belg. J. Zool.,* **123:** 39-54.

Martens, K. 1994. Ostracod speciation in ancient lakes: a review. *in*: Martens K., Goddeeris B. & Coulter G. (eds), *Speciation in ancient lakes, Adv. Limnol.,* **44:** 203-222.

Martens, K. 1996. On *Korannacythere* nov.gen. (Crustacea, Ostracoda) a new genus of temporary pool limnocytherids from southern Africa, with the description of three new species and a generic reassessment of the Limnocytherinae. *Bull. k. belg. Inst. Natuurw. Biol.,* **66:** 51-72.

Martens, K.1997. Speciation in ancient lakes. *Trends Ecol. Evol.,* **12:** 177-182.

Martens, K. 1998. Diversity and endemicity of recent non-marine ostracods (Crustacea, Ostracoda) from Africa and South America: a faunal comparison. *Verhandl. int. Verein. Limnol.,* **26:** 2093-2097.

Martens, K. (in press). On the origin and development of different recognition systems in the Limnocytherinae (Crustacea, Ostracoda, Cytherocopina): chance or selection? *in*: Horne, D.J. & Martens, K. (eds), Proceedings of the XIII international symposium for ostracodologists. *Hydrobiologia.*

Martens, K., Coulter, G. & Goddeeris, B. 1994. Speciation in ancient lakes – 40 years after J.L. Brooks. *in*: Martens K., Goddeeris, B. & Coulter, G. (eds), *Speciation in ancient lakes, Adv. Limnol.*, **44:** 75-96.

Martens, K., Rossetti, G. & Fuhrmann, R. 1997. Pleistocene and recent species of the family Darwinulidae Brady & Norman, 1889 (Crustacea, Ostracoda) in Europe. *Hydrobiologia,* **357:** 99-116.

McKenzie, K.G. 1971. Palaeozoogeography of freshwater Ostracoda. *In*: Oertli, H.J. (ed), *Paléoécologie des ostracodes., Bull. Centre Rech. Pau-SNPA, Suppl.,* **5:** 207-237.

Mayr, E. 1970. *Populations, species and evolution*, Belknap Press: 453 pp.

Moore, R.C. (ed) 1961. *Treatise on Invertebrate Palaeontology. Part Q, Arthropoda 3, Crustacea, Ostracoda. Geol. Soc. Am.,* Univ. Kansas Press. 442 pp.

Otero, M., Valenti, M., Gentile, G., Rossi, V., Baltanás, A. & Menozzi, P. 1996. Effect of photoperiod on life-history traits of electrophoretic clones of *Eucypris virens* (Crustacea, Ostracoda). *Abstracts of the 3rd European Ostracodologists Meeting*, Paris-Bierville: 63.

Otero, M., Rossi, V., Baltanás, A. & Menozzi, P. (in press): Effect of genotype and environment on diapause strategies in *Eucypris virens* (Jurine, 1820) (Crustacea, Ostracoda). *in*: Brendonck, L., De Meester, L. & Hairston, N. (eds), Diapause in the Crustacea. *Adv. Limnol.*

Peck, J.R., Yearsley, J.M. & Waxman, D. 1998. Explaining the geographic distributions of sexual and asexual populations. *Nature,* **391:** 889-892.

Peypouquet, J.P., Carbonel, P., Ducasse, O., Tölderer-Farmer, M. & Lete, C. 1988. Environmentally cued polymorphism ostracods. A theoretical and practical approach. A contribution to geology and to the understanding of ostracod evolution. *in*: Hanai, T., Ikeya, N. & Ishizaki, K. (eds), *Evolutionary Biology of Ostracoda, Developm. Palaeontol. Stratigr.*, **11:** 1003-1019.

Poinar, G.O. Jr & Ricci, C. 1992. Bdelloid rotifers in Dominican amber: evidence for parthenogenetic continuity. *Experientia,* **48:** 408-410.

Rossetti, G. & Martens, K. 1996. Redescription and morphological variability of *Darwinula stevensoni* (Brady & Robertson, 1870) (Crustacea, Ostracoda). *Bull. k. belg. Inst. Natuurw. Biol.,* **66:** 73-92.

Rossetti, G. & Martens, K. 1998. Taxonomic revision of the recent and holocene representatives of the family Darwinulidae (Crustacea, Ostracoda), with a description of three new genera. *Bull. k. belg. Inst. Natuurw., Biol.,* **68.**

Rossi, V. & Menozzi, P. 1990. The clonal ecology of *Heterocypris incongruens* (Ostracoda). *Oikos,* **57:** 388-398.

Rossi, V. & Menozzi, P. 1993. The clonal ecology of *Heterocypris incongruens* (Ostracoda): life-history traits and photoperiod. *Funct. Ecol.*, **7:** 177-182.

Rossi, V., Bucci, G., Giordano, P. & Menozzi, P. 1995. Variabilità genetica, modalità riproduttive ed ecologia in popolazioni di ostracodi d'acqua dolce. *S. It. E. Atti*, **16:** 479-481.

Rossi, V., Gandolfi, A. & Menozzi, P. 1996. Egg diapause and clonal structure in parthenogenetic populations of *Heterocypris incongruens* (Ostracoda). *Hydrobiologia*, **320:** 45-54.

Schoen, I., Martens, K. & Rossi, V. 1996. Ancient asexual scandals. *Trends Ecol. Evol.,* **11:** 296-297.

Schön, I., Rossetti, G., Martens, K., Baltanás, A. & Butlin, R.K. 1997. Congruence of morphological and molecular variability in non-marine Ostracoda (Crustacea). *in: Molecules and morphology in systematics* (Abstract of the meeting, Paris), 65.

Schön, I., Butlin, R.K., Griffiths, H.I. & Martens, K. 1998a. Slow evolution in an ancient asexual ostracod. *Proc. R. Soc. Lond. B,* **265:** 235-242.

Schön, I. & Martens, K. 1998. Opinion: DNA-repair in ancient asexuals: a new solution to an old problem? *J. nat. Hist.*, **32:** 943-948.

Schön, I., Verheyen, E. & Martens, K. 1998b. Speciation in ancient lake ostracods: comparative analysis of Baikalian *Cytherissa* and Tanganyikan *Cyprideis*. In: *Abstracts of the XXVII congress of the International Society of Limnology*.

Schulze, E.-D. & Mooney, H.A. (eds) 1994, *Biodiversity and ecosystem function,* Springer Verlag: 525 pp.

Sohn, I.G., 1987. The ubiquitous ostracode *Darwinula stevensoni* (Brady & Robertson, 1870), redescription of the species and lectotype designation. *Micropaleontology*, **33:** 150-163.

Sohn, I.G., 1988. Darwinulocopina (Crustacea: Podocopa), a new suborder proposed for non-marine Paleozoic to Holocene Ostracoda. *Proc. biol. Soc. Wash.,* **101 (4):** 817-824.

Solbrig, O. T. 1994. Plant traits and adaptive strategies: their role in ecosystem function. *in*: Schulze, E.-D. & Mooney, H.A. (eds), *Biodiversity and ecosystem function,* Springer Verlag: 97-116.

Straub, E.B. 1952. Mikropaläontologische Untersuchungen im Tertiär zwischen Ehingen und Ulm a.d. Donau. *Geol. Jb.*, **66:** 433-523.

Sylvester-Bradley, P.C. 1976. Speciation patterns in the Ostracoda. *Abh. Verh. naturwiss. Ver. Hamburg (N/F), Suppl.,* **18/19:** 29-37.

Tétart, J. 1978. Les garnitures chromosomiques des Ostracodes d'eau douce. *Trav. Lab. Hydrobiol. Univ. Grenoble,* **69-70:** 113-140.

Tilman, D., Knops, J., Wedin, D., Reich, P., Ritchie, M. & Siemann, E. 1997. The influence of functional diversity and composition on ecosystem processes. *Science,* **277:** 1300-1302.

Turner, G. F. 1994. Speciation mechanisms in Lake Malawi cichlids: a critical review. *in:* Martens K., Goddeeris B. & Coulter G. (eds), *Speciation in ancient lakes, Adv. Limnol.*, **44:** 139-160.

Vandel, A. 1928. La parthénogénèse géographique. *Bull. Biol. Franc. Belg,* **62:** 164-281.

Wakefield, M.I., 1994. Middle Jurassic (Bathonian) Ostracoda from the Inner Hebrides, Scotland. *Monogr. Paleontogr. Soc., Lond.*, **148 (596):** 1-89, 11pls.

Weider, L.J. 1993. A test of the 'general-purpose' genotype hypothesis: Differential tolerance to thermal and salinity stress among *Daphnia* clones. *Evolution*, **47:** 965-969.

Whatley, R.C. 1990a. The relationship between extrinsic and intrinsic events in the evolution of Mesozoic non-marine Ostracoda. *in*: Kaufmann, E.G. & Walliser, O.H. (eds), Extinction events in Earth history. *Lecture Notes in Earth Sciences,* **30:** 253-263.

Whatley, R.C. 1990b. Ostracoda and global events. *in*: Whatley, R. & Maybury, C. (eds), *Ostracoda and global events,* Chapman & Hall: 3-24.

Whatley, R.C. 1992. The reproductive and dispersal strategies of Cretaceous non-marine Ostracoda: the key to pandemism. *in*: Mateer, N.J. & Chen, P.J. (eds), *Aspects of non-marine Cretaceous geology,* China Ocean Press: 177-192.

Whatley, R.C. & Moguilevsky, A. (in press). The origins and early evolution of the Limnocytheridae. *in*: Crasquin, S., Braccini, E. & Lethiers, F. (eds), Proceedings of the 3rd European Ostracodologists Meeting. *Bull. Centr. Rech. Explo.-Prod. Elf-Aquit.*

Yin, Y. 1997. Contribution to the morphology and ecology of ostracoda Limnocytheridae and Candonidae: Comparative studies between Austrian and Chinese Crustacea. *PhD Thesis. Univ. Vienna*, 163 pp.

Yin, Y., Geiger, W. & Martens, K (in press). Effects of genotype and environment on phenotypic variability in *Limnocythere inopinata* (Crustacea, Ostracoda). *Hydrobiologia.*

Zheng Shu-Ying 1995. Non-marine Triassic and Jurassic ostracods from Tarim basin. *Acta Palaeontol. Sin.*, **34(6):** 722-730 + 4 Plts.

SUBJECT INDEX

TAXONOMIC INDEX

(Higher groups of non-ostracod taxa are indicated)